Living with
TECHNOLOGY

Living with TECHNOLOGY

MICHAEL HACKER

ROBERT A. BARDEN

DELMAR TECHNOLOGY SERIES

NOTICE TO THE READER

Delmar Staff

Editor-In-Chief: Mark W. Huth
Developmental Editor: Cynthia Haller
Managing Editor: Barbara A. Christie
Production Editors: Christine E. Worden
Ellen Guerci

Publications Coordinator: Karen Seebald
Design Coordinator: Susan C. Mathews
Photo Procurement: Catherine Smith

For information address Delmar Publishers Inc.,
2 Computer Drive West, Box 15-015,
Albany, New York 12212-9985

Printed in the United States of America
Published simultaneously in Canada
by Nelson Canada,
A division of International Thomson Limited

10 9 8 7 6 5 4 3 2 1

Library of Congress Cataloging-in-Publication Data

Hacker, Michael.
  Living with technology / Michael Hacker, Robert A. Barden.
    p. cm.
  Includes index.
  Summary: A textbook examining the history of technology and its
impact in such areas as communication, industry, transportation, and
medicine. Also discusses the effects of further technological
developments on the future.
  ISBN 0-8273-3248-3. ISBN 0-8273-3249-1 (instructor's guide)
  1. Technology. [1. Technology.] I. Barden, Robert A.
II. Title.
T47.H253 1988
600—dc19   87-27576
                                                      CIP
                                                       AC

CONTENTS

SECTION TWO COMMUNICATION 88

SECTION THREE   PRODUCTION   212

SECTION FIVE   CONCLUSION: LOOKING INTO THE FUTURE   482

Chapter 16 Impacts for Today and Tomorrow   483

PREFACE

Technology has been part of human life for over a million years, but in recent decades technology has caused more remarkable change to occur than ever before in history. Today, we all are *Living With Technology*. We now are completely surrounded by technological systems. We have become dependent upon the products and services modern technology provides. All industries and all people will continue to be greatly affected by new technological developments. Ours is a highly technological society.

No matter where we live, technology affects us in our everyday lives.
(Courtesy of Pat Morrow)

TECHNOLOGY AND INDUSTRY

The impact of technology on industry is visible everywhere. Automobile manufacturers are increasingly making use of robots to paint and weld auto bodies. Computer-aided drafting (CAD) has increased the productivity of drafters in architectural and engineering firms. In the printing and publishing business, typesetting has become almost entirely computerized. Within the airline industry, technology continues to spur progress. A comprehensive airline network connects our nation. Most cities are now served by commercial jet services. Jet planes like the supersonic transport (SST) travel from Paris to New York in three and one-half hours.

TECHNOLOGY EDUCATION

Technology Education as distinct from "technical education" or "vocational education" is the means through which education about technology is provided as a part of all students' general education. The goals of other technically or occupationally oriented programs are intended to provide students with job-specific skills. A significant distinction must be drawn between programs specifically designed to provide vocational training, and those designed to provide technological literacy as a part of a fundamental, liberal education for all people.

School is the ideal place to begin to foster an understanding of technology, and the role it plays in shaping our culture. Students should learn, at an early age, that technology is a human endeavor. Whether technology is used to benefit people or to destroy our society is entirely a human decision. While learning about technological systems and the resources they use, students can also learn about human responsibility in the control and application of technology.

Today, success in industry hinges upon effective use of technology.
(Courtesy of Chrysler Corporation)

ORGANIZATION OF THIS BOOK

Living With Technology is designed to help students become technologically literate. The text is divided into five sections comprising sixteen chapters. The first section (Chapters 1–3) introduces the student to the impact of technology on our lives, provides an overview of generic technological resources, and focuses on problem-solving methods as applied to technological systems.

The second section is devoted to a study of communication technology. Chapters 4-7 cover electronics, computers, graphic communication, and electronic communication.

Section Three (Chapters 8-12) is devoted to production technology. Material processing, manufacturing, construction, and managing production are addressed in detail.

In Section Four (Chapters 13-15), students examine how technology is used to provide energy, power, and transportation.

The final section (Chapter 16) presents a look at how the future will be affected by developments in communications, manufacturing, construction, energy, and biotechnology.

SPECIAL FEATURES

Living With Technology includes many special features to assist the reader in becoming technologically literate.

- The use of four-color printing in the 500 photographs and illustrations throughout the book helps to clarify the explanation of technological concepts.
- Each chapter begins with a set of major concepts. These prepare the reader to be aware of key ideas while reading.
- The major concepts are repeated in the margin adjacent to the description in the text.
- A series of carefully chosen activities provides motivational connections between technological concepts and practical applications.
- Mathematics and science concepts are explained throughout the text.
- Key words are highlighted.
- Complete summaries are included at the end of each chapter. These review each of the major concepts.
- Information about careers within each of the major technological clusters is provided, as well as an overview of the future of the workforce in the United States.
- The text includes boxed inserts containing features of special interest.
- A glossary of terms with accurate definitions of words is provided.

ACKNOWLEDGMENTS

There are some colleagues, friends, and family who deserve our deepest and most sincere thanks. Our wives, Alice Barden and Susan Trump Hacker, provided constant encouragement during the writing process. Their support never wavered, and their constructive ideas and editorial assistance have helped refine the product. Rob and Lisa Barden deserve thanks for their help in testing for readability. Grateful thanks are given to reading specialist **Myrna Stolls**, who reviewed the text's readability, and to **M. James Bensen, Steve Brady, Jerry Butler, Daryl Floit, Dwayne Hobbs, Daniel Householder,**

Franzie Loepp, L. Grant Luton, Donald Maley, Kenneth Phillips, Ernest Savage, Jule D. Scarborough, Barry Smith, William Wargo, and **Jerry Weddle** who reviewed the text outline. Thanks also to **Robert Biddick, Jerry Butler, Daryl Floit, Lloyd Gober, Steve Houck, Carl Iverson, L. Grant Luton, Robert Rosengarn, Margaret Rutherford,** and **Jerry Weddle,** who reviewed the manuscript and whose expertise and excellent advice were key ingredients in refining the text.

We wish to express our deep appreciation to the Delmar staff, particularly to our editor and friend Cynthia Haller who worked untiringly to ensure that her own standards of excellence were met. Thanks are also gratefully given to production editors Ellen Guerci and Christine Worden, to Catherine Smith for her help in acquiring photographs, to Karen Seebald for coordinating the production of this text, and to Mark Huth for his administrative support and photographic assistance.

There are two individuals to whom the authors are particularly grateful. Thomas Barrowman and Alan Horowitz are master technology teachers, and good friends to whom we owe special recognition. We appreciate sincerely their creative contributions to the activity sections of this text. In writing these engaging activities, Alan and Tom have broken new ground for our discipline.

Finally, we would like to offer a special note of thanks to all of our colleagues in Technology Education, in New York and across the nation, from whom we have learned so much, and who consistently inspire us with their dedication and professionalism.

ABOUT THE AUTHORS

Michael Hacker is a twenty-year veteran of secondary school teaching in Long Island, New York. As early as 1969, his technology-based junior high school program had received national attention. He has authored a dozen articles in national journals, consulted nationally, and has presented at numerous state and national conferences. He is past president of the New York State Technology Education Association and has actively served on various International Technology Education Association (ITEA) committees. In 1985, he was named an Outstanding Young Technology Educator by the ITEA. In his present capacity as Associate State Supervisor for Technology Education, New York State Education Department, his responsibilities include the development and implementation of Technology Education curricula.

Robert A. Barden is an electronics engineer specializing in the design and development of high capacity data, voice, and video communication systems. His work includes the integration of fiber-optic, microwave, and satellite technologies. Mr. Barden has served as a member of the National Advisory Council of the ITEA and has published articles and presented seminars on systems and technology for teachers and teacher trainers. He is a senior member of the Institute of Electrical and Electronics Engineers, Inc. (IEEE) and a member of its Committee on Precollege Mathematics, Science and Technology Education.

ACTIVITY ACKNOWLEDGMENTS

Because technology implies application of knowledge, a study of technology without an applications phase does not present a sufficiently clear interpretation of its dimensions. In order to make the study of technology exciting and relevant to students, a laboratory-based, activity-oriented approach should provide the vehicle through which content is presented. Living With Technology has been written to provides the conceptual base necessary to support such a hands-on instructional program. The text includes an outstanding series of student activities that reinforce major technological concepts and enhance problem-solving skills. These activities additionally provide an opportunity to apply mathematics and science concepts within a laboratory setting.

The activity contributions of the following individuals have made this fine series of activities possible.

Thomas Barrowman
Queensbury Middle School
Glens Falls, New York

Now and Then; Selecting and Using Resources; Tug O' War; Computers in Industry; International Language; Beam that Signal; Investment Casting; Checkerboard; Prefab Playhouse; Laser Construction; Troubleshooting; Scrambler; Satel-loon; Pole Position; Glassmaking; Cycloid Curve; Hovercraft

Alan Horowitz
Felix V. Festa Junior High School,
West Nyack, New York

The Transparent Cover-Up; It's About Time; Low-Power Radio Transmitter; Team Picture; Morse Code Communications; The Missing Link; T-Square Assembly Line; Dome Construction; Bridge Construction; One, if by Land; Hot Dog!; A Penny for Your Thoughts!; My Hero; Repulsion Coil; You're on the Right Track; Dust or Bust; Hurricane Alley; The Medusa Syndrome; Space Community

Ethan Lipton
California State University at Los Angeles
Los Angeles, California

Communicating a Message

Grant Luton
Schrop Middle School
Akron, Ohio

Robotic Retriever

Doug Polette
Montana State University
Bozeman, Montana

Production Company

Fred Posthuma
Westfield High School
Westfield, Wisconsin

Photograms; Muscle Power

A.R. Putnam
Indiana State University
Terre Haute, Indiana

Model Rocket-Powered Spacecraft; The New Computer; An Entrepreneurial Company; Installing Electrical Systems

Margaret Rutherford
Howell Intermediate School
Victoria, Texas

Space Station; Air Flight

Kenneth Welty
Illinois Plan for Industrial Education
Illinois State University
Normal, Illinois

Mapping the Future

SECTION 1

INTRODUCTION: WHAT IS TECHNOLOGY?

CHAPTER 1

TECHNOLOGY IN A CHANGING WORLD

MAJOR CONCEPTS

After reading this chapter, you will know that:

- Technology influences our routines.
- Science is the study of why natural things happen the way they do.
- Technology is the use of knowledge to turn resources into goods and services that society needs.
- Science and technology affect all people.
- People create technological devices and systems to satisfy basic needs and wants.
- Technology is responsible for a great deal of the progress of the human race.
- Technology can create both positive and negative social outcomes.
- Complex technological systems develop from more simple technologies.
- Technology has existed since the beginning of the human race, but it is growing at a faster rate today than ever before.

1

INTRODUCTION

For most of us, life consists of a variety of **routines.** At a certain time every morning, we wake up, wash up, and eat breakfast. Then we go to school or work. In the evening, we come back home for dinner, have an evening activity (such as TV, homework, or visiting friends), and go to sleep, usually at about the same time every night. The entire cycle of routines is completed in one day, and it begins again and repeats on the following day.

WAKE UP AT 7:00 A.M.

GO TO SLEEP AT 11:00 P.M.

WASH UP

HAVE AN AFTER-DINNER ACTIVITY

EAT BREAKFAST

RETURN HOME FOR DINNER

GO TO SCHOOL OR WORK

Diagram of a twenty-four hour cycle: a common daily routine

People in some parts of the world base their routines upon the solar cycle.
(Photo by Michael Hacker)

Why do we follow the routines that we do? Why do other cultures have routines that are completely different? In part, our routines are determined by our technology. In some parts of the world, for example, there is no electricity. Routines are therefore based upon the solar cycle. People awaken at sunrise and go to sleep very early at night.

We follow our routines almost automatically. Changing them too drastically makes us uncomfortable. For example, we would probably not like waking up at 4:00 a.m. and going to sleep at 9:00 p.m. Such a change would force us to make changes in other routines as well, such as the time we eat breakfast and the time we take a bath.

This book is about technology and how it affects people's routines and life-styles. Our routines and the way we live have been greatly influenced by the devices, products, and services that technology provides.

Technology influences our routines.

WHAT IS THE DIFFERENCE BETWEEN SCIENCE AND TECHNOLOGY?

Science and technology differ but are also related. Let us look at some examples of how science and technology interact.

Science is the study of why natural things happen the way they do.
Technology is the use of knowledge to turn resources into the goods and services that society needs.

- Scientists study how the earth was formed and what it is made from. Technologists use the materials found in the earth to make useful objects.
- Scientists study the relationship between the earth, the planets, and the stars. Technologists build and launch space shuttles and satellites.
- Scientists study materials under a microscope to learn why they have the characteristics that they do. Technologists create new materials with improved characteristics.
- Scientists discover the way the human body works. Technologists make artificial hearts and limbs.
- Scientists discover genetic codes that determine the nature of life. Technologists use those codes to develop new medicines and vaccines.

Scientists and technologists work as a team. Scientific discoveries are made useful by technologists, who apply new scientific knowledge to the solution of practical problems. Biologists, chemists, and physicists are scientists. Architects, engineers, product designers, and technicians are technologists.

Science and technology affect all people.

Scientists and Technologists

WHAT IS TECHNOLOGY?

Because technology affects so many areas of our lives, it is hard to define in one simple way. To try to understand its nature, we should look at it from several points of view.

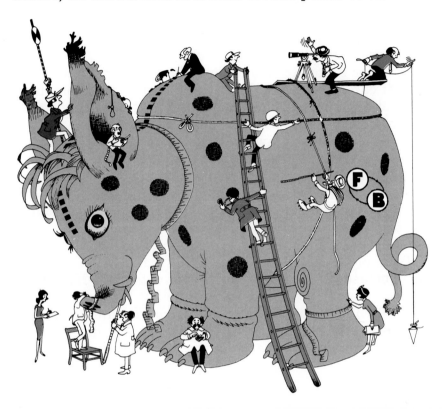

Technology is interpreted differently by different people.
(Courtesy of Feedback, Inc.)

Technology Is . . .

- The sum of all human knowledge, used to transform resources for the purpose of meeting human needs.
- Making things work better.
- A strategy for the survival of our species.
- The means by which people control or modify their natural environment.
- The practical application of a theoretical subject (like science).
- The application of knowledge and the knowledge of application (Melvin Kranzberg).
- That great growling engine of change (Alvin Toffler).
- A disciplined process that uses scientific, material, and human resources to achieve human purposes.
- Our major means of adjusting to our environment.

WHY STUDY TECHNOLOGY?

Technology has had an impact on all of our lives. The results of advances in technology—the telephone, television, automobiles, plastics, refrigeration, new medicines, polyester clothing, nylon, and so on—are all around us. However, they are often taken for granted. We have come to depend upon technology to fulfill our basic needs and to make our lives more comfortable, healthy, and productive.

People create technological devices and systems to satisfy basic needs and wants.

In order to use technology wisely, we must understand what it can do for us and what it is unable to accomplish. Technology provides many helpful products and services. It can, however, create some negative effects, such as pollution and crowded highways. Whether technology benefits us or not will be determined by the way we use it.

Technology has had a great influence on our society. Without modern technology our lives would be entirely different. Technological knowledge has made us able to shape our own environment. With technology, we can create a world different from the world we inherited from nature.

Technology is responsible for a great deal of the progress of the human race.

Technology has extended our natural capabilities. We can increase the power of our hands with tools like pliers and hammers. We can increase our height with ladders. We can swim under water with scuba tanks. We can fly, and we can survive in extreme cold. We can conquer disease and can make replacement parts for the human body.

HOW DOES TECHNOLOGY AFFECT OUR LIVES?

When we wake up in the morning and go into the bathroom to wash up, we hardly realize how much technology influences this ordinary routine. Washing up in the morning with warm water is something we simply expect to do. We adjust the faucet until the amount of water and the water temperature are just what we desire. We wash our hair,

A subroutine is part of a longer routine.

using shampoo with all kinds of chemical conditioners, and quickly blow it dry.

Our morning wash-up routine, then, is composed of a number of subroutines (showering, shampooing our hair, blowing our hair dry, and so on). These subroutines have developed only because technology makes them possible.

In George Washington's time, people bathed only a few times a year. It was difficult to heat enough hot water to fill a bathtub. As late as the 1930s, Americans living in isolated rural communities bathed in nearby creeks after working in the cornfields.

Morning Routines

MORNING ROUTINES NOW

(Courtesy of Kohler Co.)

Hot water for a bath or shower is provided by a modern system. The system circulates heated water through pipes to various places in the home.

(Courtesy of Kohler Co.)

Electrical appliances and new materials like formica make cooking and clean-up much easier.

MORNING ROUTINES THEN

In the middle and late 1800s, a charcoal-burning heater was immersed in a tub of cold water. The heater was removed when the water was warm enough for a bath.

(Courtesy of Science Museum Library, London)

Meals were cooked on wood or coal stoves. Eating breakfast required starting a fire and was a routine that took hours.

Technology Satisfies Our Needs

TECHNOLOGY SATISFIES OUR NEED TO PRODUCE FOOD

Agricultural Technology Now

(Courtesy of Sperry Corp.)

Modern machinery can harvest enough wheat in nine seconds to make seventy loaves of bread.

Agricultural Technology Then

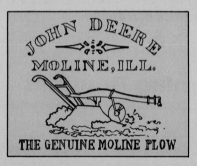

Years ago, farmers had to use human or animal muscle power to plow their fields.

TECHNOLOGY SATISFIES OUR MEDICAL NEEDS

Medical Technology Now

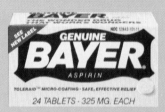

(Courtesy of Glenbrook Laboratories of Sterling Drug Inc.)

Various medications are available for headache relief.

Medical Technology Then

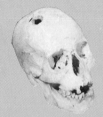

(Courtesy of Science Museum Library, London)

In primitive times, headaches were thought to be caused by spirits that invaded the head. In order to release the spirits, holes were drilled into the skull with drills made from shark's tooth or flint. The process was called trepanation. Surprisingly, patients occasionally survived more than one such procedure.

TECHNOLOGY SATISFIES OUR NEED FOR MANUFACTURED ITEMS

| Manufacturing Now | Manufacturing Then |

(Courtesy of Renault, Inc.)

(Courtesy of Ford Motor Co.)

In a modern automobile plant, cars are produced at the rate of one every minute.

In 1885 in Germany, Karl Benz built the first gasoline-driven motorcar. Henry Ford built his first car in 1896 and founded the Ford Motor Company in 1903.

TECHNOLOGY SATISFIES OUR NEED FOR ENERGY SOURCES

| Energy Sources Now | Energy Sources Then |

(Courtesy of Bob Klein)

Modern technology has brought us a variety of efficient methods for producing energy.

In the Middle Ages (about A.D. 1100), water wheels provided most of the power for production. The water wheel was used to grind corn or wheat in order to produce flour.

TECHNOLOGY SATISFIES OUR NEED TO COMMUNICATE IDEAS

| **Communications Now** | **Communications Then** |

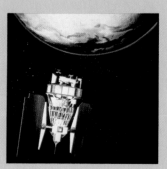

(Courtesy of NASA)

Today, space-based communications satellites bring television signals from distant countries to our homes. These satellites are like relay stations. They receive signals from one location on earth and rebroadcast them to other locations.

In 1774 in Geneva, Switzerland, George Lesage set up a telegraph using one wire for each letter of the alphabet. Static electricity generated by a friction machine was used to power the system. The telegraph sent a message along the wires into an adjacent room.

TECHNOLOGY SATISFIES OUR TRANSPORTATION NEEDS

| **Transportation Now** | **Transportation Then** |

(Courtesy of NASA)

(Courtesy of Smithsonian Institution—National Air and Space Museum)

Today we can ferry materials and devices like communications satellites back and forth between earth and outer space. In the future, people will be working in outer space. They, too, will be shuttled back and forth like space-age commuters.

In 1903 at Kitty Hawk, North Carolina, Wilbur and Orville Wright's biplane flew a distance of 852 feet and stayed aloft for 59 seconds.

WHEN DID TECHNOLOGY BEGIN?

There is a tendency to think that our times are technological times and that previous eras were not. In truth, people have been using and creating technology since prehistoric times, as you can see from the technological time line at the back of the book.

Many thousands of years ago, entire civilizations were using tools and developing techniques for using them to work on various materials. In fact, the technology of early peoples has provided us with a means of classifying historical periods.

The **Stone Age** refers to the period from about 1,000,000 B.C. to about 3,000 B.C. During this period stone was the major material used for tools.

Of course, not all the tools during this million-year period were made of stone. Many were made from animal bones and from wood. The earliest known dugout canoe is thought to have been made during this period (about 6,500 B.C.).

The **Bronze Age** (which began about 3,000 B.C.) was the period when people first made tools and weapons from bronze. Bronze is a mixture of copper and tin. Prior to the Bronze Age, copper had been used only for decorative purposes. During the Bronze Age, people discovered that copper ore could be heated along with charcoal to yield pure metal. Then, by melting other ores along with the copper, stronger metals like bronze could be produced.

The **Iron Age** was the period when iron came into common use. It began around 1200 B.C. in the Middle East and about 450 B.C. in Britain. The process of making iron from iron ore is called **smelting.**

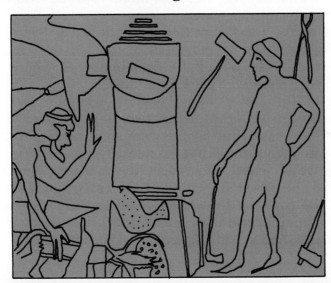

People have been using technology for over a million years.

The prehistoric iron-smelting furnace was a clay-lined hole in the ground. Iron ore and charcoal were placed in the hole. Air was then pumped in with a bellows. The air enabled the charcoal to burn at a high enough temperature to reduce the ore to a spongy mass of iron. This spongy mass was hammered into shape while it was red hot.

Prehistoric civilizations used tools made from bone, wood, stone, bronze, and iron. They used fire for cooking meat and for protection from wild animals. They developed the bow to extend their hunting capabilities. They used needles made from splinters of bone to transform animal skins into clothing.

During the Stone Age, there were almost no villages, and most people were living nomadic lives. They didn't live in villages. Instead, they wandered from place to place in search of food. When they used up whatever food nature provided, they would move to a new location. Techniques of growing food had not yet developed. People had to rely upon hunting animals and finding edible plants, fruits, and vegetables.

The plow, developed in Egypt, enabled people to grow their own food. People began to settle into towns and villages. Civilizations became agricultural rather than nomadic.

Ancient agricultural civilizations grew food crops, wove flax (a plant) into linen, and used oil lamps for artificial light. They had invented the wheel and wheeled vehicles, and they built roads and stone houses. They had systems of irrigation for agriculture and sewerage systems that improved health conditions.

The Romans made sure that people didn't bathe or drink water from the Tiber River, into which sewers emptied. Later civilizations were not as concerned with the pollution of the water supply. Disease often resulted from people drinking unsanitary water.

One of the most significant of the early technological advances was the development of the **water wheel.** The water wheel was one of the first instances where human muscle was replaced by artificial muscle.

Before the water wheel was in common use, slaves ground grain by hand, using two large stones. This was such a hard task that people were sometimes sent to work at the mill as a form of punishment.

The water wheel ushered in the machine age, since it provided the power for many of the early machines in the mills. The power of the water wheel was used for pumping water, grinding grain (to make flour), manufacturing textiles (cloth), and producing iron.

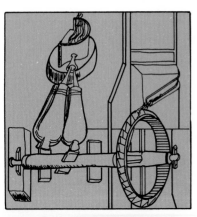

This water wheel operates a set of bellows for iron smelting.

The use of machinery instead of hand-operated tools was perhaps the most important technological development of the Middle Ages. In Britain, the water wheel was so important that there were 5,624 water wheels in operation. This is the equivalent of one water wheel for every fifty families.

THE INDUSTRIAL REVOLUTION

Beginning about 1750, a great flurry of inventive activity accelerated the events that took place during the Middle Ages. New ideas were spread into various industries throughout the Western world. This period marked the beginning of the Industrial Revolution and the factory system of work.

The Industrial Revolution was so named because of great changes that took place in the way products were produced. Before this time, craftspeople used their own tools and their own workshops to make things. With the availability of machines, however, wealthy businesspeople were able to set up factories. Using these factories, businesspeople could produce things faster and cheaper. Individual craftspeople, who could not afford to buy machinery, were unable to compete. Therefore, many went to work for an employer in a factory. They used the employer's tools and machines instead of their own and were paid a salary for their labor. The coming of the assembly line and mass production made manufacturing in factories an important part of the economy, both in Europe and in the United States.

Henry Ford is often called "the father of mass production" because he made the idea popular. In fact, however, the concept of the assembly line had been born more than a hundred years earlier. In England during the latter part of the eighteenth century, Richard Arkwright developed a manufacturing system by linking together several machines of his own design. His system converted raw cotton into thread in a sequence of steps. Each machine performed one task and prepared the material to be further processed by the next machine.

This machine helped convert raw cotton into thread.
(Photo by Michael Hacker)

Arkwright is credited with being the single individual most responsible for the development of the **factory system.** He built many mills, and his machines mechanized what had previously been manual work. His inventive genius and skill at managing people made a tremendous impact on the textile industry.

The factory system increased production to a very great degree. However, it brought about some undesirable social changes as well. Often, men, women, and very young children were employed for fourteen hours a day under crowded and unsafe conditions. This evil of the early factory system cannot be overlooked.

Technology can create both positive and negative social outcomes.

Because of poor factory conditions and cruelty to workers, laws establishing minimum working conditions and wages were passed. Child labor laws were also enacted. Such laws forbade employers to hire children below a certain age.

The Industrial Revolution brought about social change as well as technological change. Labor laws, trade unions, and schools grew out of the Industrial Revolution. Many parents left the home to work in factories during the day. Therefore, constructive activity and supervision had to be provided for the children. Public schools grew in number and increased in importance as a direct result of the factory system of labor.

Before the Industrial Revolution, technological change had taken place very slowly. During the Industrial Revolution, however, change was visible. People began to expect change.

Patent regulations helped increase the rate of technological change. These regulations protected people from having their inventions pirated. People could therefore make a profit from any technological improvements they made. The profit motive began to drive people to improve existing devices and invent new devices. Change became deliberate as people set out to improve things and make them work more effectively.

TECHNOLOGICAL CHANGE BECOMES RAPID

Technological change occurred at a rapid rate during the Industrial Revolution. This was largely because inventors, engineers, and scientists were able to make use of ideas that had already been developed. Since the year 1500, printing had been common in most parts of Europe. People could build upon past knowledge by reading about previous technological achievements in books and manuals. Because land and sea transportation systems were developing, printed materials were distributed far and wide. The sharing of knowledge enabled people to combine ideas. Old ideas could be used as building blocks for newer and more powerful technologies.

Complex technological systems develop from more simple technologies.

Combining Technologies

Ship-building techniques + the steam engine = the steamship

Optics + chemistry = photography

The telegraph + printing technology = newspaper

Internal combustion engine + wagon and carriage construction technology = automobile

Synthetic materials + medical technology = artificial heart

Kite and glider technology + lightweight gasoline engine = airplane

New materials (titanium) + jet engine technology = spy aircraft

Photographic technology + satellite technology = exploration of the earth.

Robert Fulton's steamboat, Clermont, sailed from New York City to Albany in the early 1800s. (Courtesy of The New York Public Library—Astor. Lenox and Tilden Foundations)

Cameras carried by satellites can determine the best locations to prospect for oil or natural gas. (Courtesy of NASA)

EXPONENTIAL CHANGE

During the prehistoric period, technological change came very, very slowly. Remember that it took many thousands of years for people to change from stone tools to metal tools. As time passed, people had new ideas and invented new tools and devices. The greater the number of ideas and devices people had, the greater was the possibility that ideas could be combined and lead to new ideas.

The rate of change became faster and faster throughout history. Today, it is faster than it has ever been. Some people say that our knowledge is doubling every four years. We say there is an **exponential rate of change** because change is happening at an ever-increasing rate.

Let's use an example to explain what is meant by exponential rate of change. Maria found a store that was selling cassette tapes at a very good price. She convinced her mother that she should be allowed to buy ten cassettes every month. Her brother, Steven, was jealous. He worked out an arrangement that permitted him to buy only two cassettes the first month, but to double the number he bought each month thereafter. Who do you think had more cassettes after six months?

First, let's look at Maria's purchases. At the end of the first month, Maria owned ten tapes. Each month, Maria bought the same number of tapes as she did the month before. So, after six months, Maria owned sixty cassette tapes. Since her tape supply increased by the same rate (ten tapes) every month, we can say that the change in the number of tapes was **linear** (like a straight line).

Let's see how Steven's tape purchases compare. Remember, he started out with a purchase of two tapes. He doubled his purchases each month thereafter. Steven's tape supply

MONTH NUMBER	NUMBER OF TAPES PURCHASED EACH MONTH	TOTAL NUMBER OF TAPES OWNED
1	10	10
2	10	20
3	10	30
4	10	40
5	10	50
6	10	60

Maria's tape purchases:
a linear rate of change

MONTH NUMBER	NUMBER OF TAPES PURCHASED EACH MONTH	TOTAL NUMBER OF TAPES OWNED
1	2	2
2	4	6
3	8	14
4	16	30
5	32	62
6	64	126

Steven's tape purchases:
an exponential rate of change

increased faster and faster every month, unlike Maria's, which increased at the same rate every month. Steven's tape supply increased exponentially. It wasn't long before their mother realized she would be spending all of her money on Steven's cassette tapes and put an end to the arrangement.

When a quantity changes at an increasing rate, we say that it changes **exponentially.** Like Steven's cassette collection, our technical knowledge has grown exponentially since prehistoric times. Remember that the period between the Stone Age and the Bronze Age was about one million years. It took that long for human beings to gain enough technological knowledge to produce their tools from metal instead of stone, bone, and wood.

We can get an idea of how slow technological change used to be by looking at a technological time line. A **time line** is a type of graph. It shows how much change has occurred over a given period of time.

As people began to accumulate new knowledge, they were able to combine techniques and ideas and share information. The development of writing, for example, was responsible for a great increase in technological change. Writing permitted people to share their ideas more widely.

Writing caused an increase in technological change. Many new developments began to take place. The wheel was developed. Iron was manufactured. Sailboats that could travel with the wind were built.

Over the last hundred years, the rate of technological change has become extremely rapid. It is still getting faster and faster. Many important technological developments have occurred in recent years.

After one year, Steven would have purchased over 4,000 tapes.

Technology has existed since the beginning of the human race, but it is growing at a faster rate today than ever before in history.

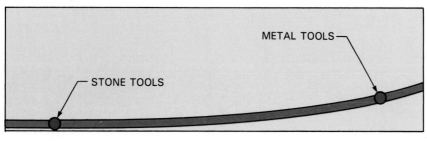

STONE TOOLS

METAL TOOLS

1,000,000 B.C.
THE STONE AGE

3000 B.C.
THE BRONZE AGE

Time line: From Stone Age to Bronze Age. Technology was changing exponentially during the prehistoric period, but the rate of change was slow. It was almost linear. Many years passed with only a few major technological achievements.

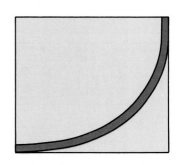

Exponential change occurs when the rate of change continues to increase.

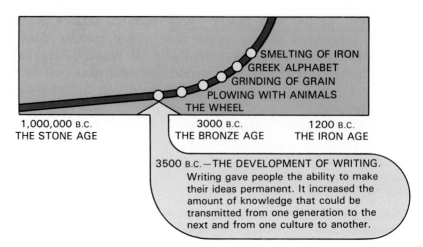

Time line: From Stone Age to Iron Age

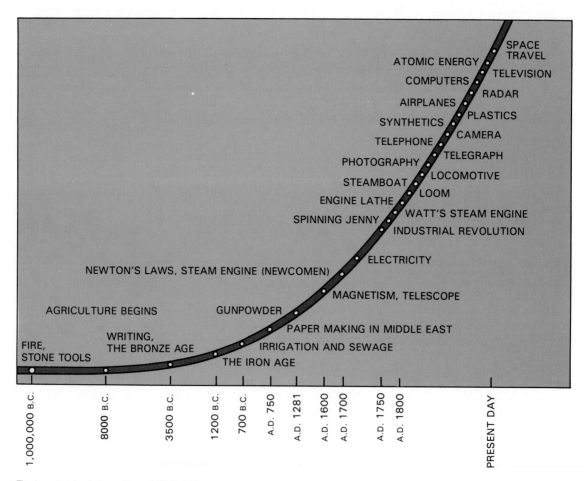

Technological time line: 1,000,000 B.C. to present day

SHIFT FROM AGRICULTURAL TO INDUSTRIAL TO INFORMATION SOCIETY

Technology has created three distinct eras. In the **agricultural era,** most people lived off the land. Many of the tools and discoveries were related to improving methods of tilling the soil. The **industrial era** began with the Industrial Revolution in the late 1700s. During the industrial era, a great many mechanical devices were invented. Today, we are in an **information age.** Many of today's inventions are based upon electronics and the computer.

During the agricultural era, many people were farmers. They produced their own food. They used their own muscle power or animal power to do chores like pumping water and plowing the fields.

During the industrial era, many people were employed in factories. Machines replaced human muscle and animal power. Steam and electricity were used to run motors, which in turn ran machinery. The factories depended upon human labor to run the machines.

Most modern societies have now entered the information age. In this period, many good jobs require educated workers who have lots of knowledge.

Because the pace of technological change has increased so rapidly during the last few decades, technology has become a major driving force in our society. In order to

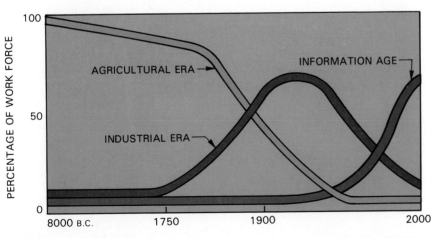

Shift from agricultural era to industrial era to information age

make sense of our world, we need to know about technology and understand how it affects the way we live. We need to understand technology so that we can control it. We need to become **technologically literate.**

WHAT IS TECHNOLOGICAL LITERACY?

A person who is literate in English can read and write. A person who is literate in technology understands technology. Technologically literate people know that technology is not magic. Rather, it is created by people to solve problems and satisfy human needs. We study technology so that we can make more informed technological decisions.

What kinds of technological decisions do we have to make? One type relates to our role as purchasers, or consumers of technological goods and services.

As a citizen living and working in a democracy, you have the opportunity to influence the way technology is applied. By writing letters, voting, and speaking publicly, each of us can encourage the development of helpful technologies. We can also oppose technological uses that we feel would be harmful to people or to the environment.

A technologically literate person is a good consumer of technology.
(Courtesy of Tandy Corp.)

SUMMARY

Technology is as old as the human race itself. Since prehistoric times, technology has satisfied people's basic needs and wants. Our biological needs for food and medical assistance, our physical needs for clothing, shelter, and manufactured products, and our need to communicate information are all satisfied through technological means.

Technology has been so important to the development of the human race that we sometimes classify historical periods in terms of the technology that was available. During the Stone Age, people used tools and devices made primarily from stone. During the Bronze Age and the Iron Age, tools and devices were made from metal. The Industrial Age began with the development of industrial machines.

Before the Industrial Revolution, technological change took place very slowly. During the industrial age, the profit motive caused people to make many improvements and develop new devices. Technological change became very rapid.

Technological change is increasing at an exponential rate. Each new technological development can be used as a seed from which new ideas can spring. Ideas can also be combined to form new ideas. New technological devices and systems therefore develop.

Civilization has gone through three important stages. In the agricultural era, most people worked as farmers. In the industrial era, most people worked in factories. In the present era, called the information age, technology affects almost everything we do. The purpose of this book is to provide you with the kind of technological literacy you will need to live successfully in our technological world.

As you study more and more about technology, you will see that there are similarities among all technological developments. Whether you are considering primitive technologies like plowing with animal power or ultramodern technologies like communication satellites, certain resources are essential in each and every case. The next chapter will introduce you to resources for technology.

THE TRANSPARENT COVER-UP

Setting the Stage

The time is the year 3000. You've been hired as head archaeologist for a dig on planet Earth. Your most interesting find so far has been the ancient recording device known as a *book*. If you use an opaque coating to protect the book, you can't see its cover. How can you protect the book?

Your Challenge

Design a durable, well-fitting, and *transparent* cover to protect this book. You must treat the book as you would a valuable archaeological find. No tape may be used directly on the book, but the cover may be taped to itself.

Procedure

1. Using the scissors, carefully cut the *sides* of the plastic bag. (Do not cut along the bottom).
2. Unfold the bag on a flat surface.
3. The unfolded bag should measure about 10 ⁹/₁₆" × 22".
4. If necessary, trim the plastic to size.
5. Open the front cover of the book and fold about 3" of the shorter side of the plastic bag over the edge of the cover.
6. Close the book, turn it face down, and repeat step 5 for the back cover.
7. Close the book and pull all the wrinkles out of the plastic bag. (Be careful not to overstretch the bag.)
8. Use 2" lengths of transparent tape to hold the plastic on each corner of the book. Tape the plastic to *itself.* DO NOT USE TAPE DIRECTLY ON THE BOOK.
9. For identification, tape to the cover a small label with your name neatly lettered (or typed) on it.
10. Your teacher may wish to have the class develop a computer *data base* of students enrolled in the course. The labels can then be run off on a printer.

Suggested Resources

Resealable food storage bag—gallon size (10 ⁹/₁₆" × 11")
Transparent tape
Scissors

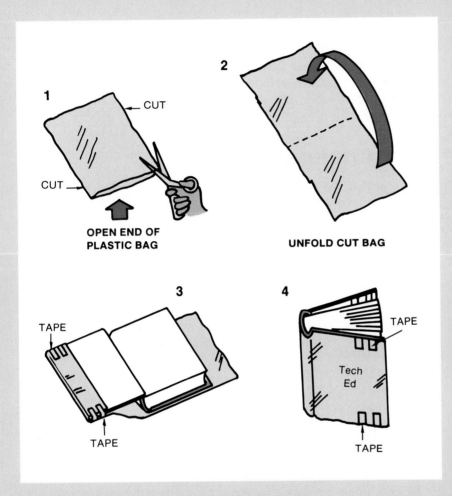

1 CUT

CUT

OPEN END OF PLASTIC BAG

2 UNFOLD CUT BAG

3 TAPE

TAPE

4 TAPE

Tech Ed

TAPE

Technology Connections

1. Technology has been around since the human race began. It is growing at a faster rate today than ever before. How have books helped with this growth?
2. Technology is the use of knowledge to turn resources into goods and services that society needs. From what natural raw material was the plastic bag probably made?
3. How many transparent materials can you name? Tell whether they are natural or synthetic materials.
4. What does *opaque* mean? Can you name some opaque materials?
5. What is a *computer data base?* What data would you include in a data base designed to print book labels for students in your class?

Science and Math Concepts

▶ *Archaeology* can be defined as the study of the material remains of past human life.
▶ *Transparent* can be defined as clear or easily seen through.
▶ *Opaque* can be defined as not able to be seen through.
▶ *Natural* can be defined as produced by nature.
▶ *Synthetic* can be defined as produced artificially or made by human beings.

23

NOW AND THEN

Setting the Stage

An open house is being planned for the architectural firm where you are employed. You have been asked to set up a display showing how your firm designed a building in the past and how it is done now. Can you show how technology has changed designing buildings from drawing with pencils to using computers? How has the computer changed other fields such as business, medical, and transportation?

Your Challenge

In teams of two, research a historic machine, tool, or process. One student will demonstrate and explain the historic machine, tool, or process. The other student will demonstrate the modern machine, tool, or process. Be sure to use visual aids when giving your report.

Procedure

1. Form teams of two.
2. Pick a machine, tool, or process you and your partner would like to research.
3. Find out more about the topic you've picked by collecting the following data:
   - pictures of past and present forms
   - dates used
   - operating principles, past and present
   - how widely the machine, tool, or process was/is used
   - how the machine or process was/is made
   - how the machine, tool, or process was/is operated safely
   - how efficiently work was/is accomplished
   - how safe the tool or process was/is
   - what environmental impacts the tool or process did/does have
4. Create visual aids that show two different periods in the development of the tool or process.
5. Check your data and organize a presentation with your partner.
6. Give your presentation to your classmates. Demonstrate the safe use of the tool or process.

Suggested Resources

Lab library
School library
Local industries
Older people
Museums
Local craft shops
Wood
Plastic
Metal
Glues
Cardboard
Photos

Technology Connections

1. Complex technological systems develop from more simple technologies.
2. Was the original machine, tool, or process developed because of a want or a need? Did this want or need change with the modern version?
3. Describe the technological developments that caused the tool or process you studied to be changed.
4. How might the tool or process you've chosen change in the future?

Science and Math Concepts

▶ Many machines and processes make use of the principles of the six simple machines: lever, wheel and axle, pulley, screw, wedge, and inclined plane.

REVIEW QUESTIONS

1. Explain the difference between science and technology.
2. Describe how technology influenced your routine this morning.
3. List five physical needs that people have and explain how technology helps satisfy those needs.
4. List five biological needs that people have and explain how technology helps satisfy those needs.
5. Explain how technology satisfies our need to communicate ideas and process information.
6. Give two examples of technologies that have developed from more simple technologies.
7. Draw a technological time line that illustrates how technology is growing at an exponential rate.
8. Identify a technological system and discuss how it has made life easier for people.
9. Define agricultural era, industrial era, and information age.

KEY WORDS

Agricultural	Exponential	Resources	Technology
Bronze Age	Industrial	Routines	Technological
Change	Revolution	Science	change
Communications	Iron Age	Stone Age	
Construction	Manufacturing	Subroutine	
Energy	Mass production	System	

SEE YOUR TEACHER FOR THE CROSSTECH PUZZLE

CHAPTER 2

RESOURCES FOR TECHNOLOGY

MAJOR CONCEPTS

After reading this chapter, you will know that:

- Every technological system makes use of seven types of resources.
- The seven resources used in technological systems are people, information, materials, tools and machines, energy, capital, and time.
- Since there is a limited amount of certain resources on the earth, we must use resources wisely.
- Solving technological problems requires skill in using all seven resources.

TECHNOLOGICAL RESOURCES

Resources are things we need to get a job done. Think, for example, about hamburgers. McDonald's restaurants have sold over fifty billion of them. Whether you are cooking fifty billion or just one small hamburger, you need resources to get the job done.

What do we need? First, we need **people.** We need people to raise and butcher cattle, to grind up the meat, and to cook the hamburgers.

We also need **information.** We must know what food cows eat, how to take care of the cattle, how to keep the meat fresh, and how to cook it. We need to know cooking temperatures, types of seasoning, and the length of time needed to cook the meat.

In addition, we need hamburgers. The hamburgers are the **materials** that we process. We start out with materials (hamburgers, spices, and seasoning) and process these materials until they are good to eat.

What about **tools?** Do we need tools to cook hamburgers? Certainly! The tools are a stove or grill and cooking utensils.

We have to have **energy** to cook the meat. Perhaps the energy comes from the charcoal. Perhaps it is gas or electrical energy in a kitchen stove.

We also need to buy the meat, the spices, the charcoal, and the tools. We need money. Money is a form of **capital.** There are other forms of capital that we will discuss in this chapter.

Every technological development makes use of seven types of resources.

People eat the hamburgers and create a demand for the product. (Courtesy of McDonald's Corporation)

We need information about how to keep cows healthy.
(Courtesy of Cetus Corporation)

Tools are an important resource for preparing food.
(Courtesy of McDonald's Corporation)

Finally, **time** is necessary. If you want to eat cooked meat, it takes time. In France, many people eat raw hamburger (called steak tartare). Even if the meat is uncooked, however, it takes some time to prepare.

Cooking hamburgers requires using resources. We named seven types: people, information, materials, tools and machines, energy, capital, and time. For one hamburger or a billion hamburgers, the same seven resources are needed. In fact, **every technological process involves the use of these seven resources.** Whether we are cooking hamburgers, plowing fields, or talking over the telephone, we need the same seven resources.

The seven resources used in technological systems are people, information, materials, tools and machines, energy, capital, and time.

PEOPLE

Let's look at each of these seven technological resources a little more closely. First, people are needed for technology. In Chapter 1, we saw how technology can satisfy our needs for transportation, manufactured goods, improved communication, and medical assistance. In fact, people's needs can stimulate the development of technology.

It is people who design and create technology, using their knowledge and intelligence. New technologies are being created constantly by people who combine previous knowledge with new ideas. Human intelligence and creativity are the keys to new technological development. For example, in 1876, Alexander Graham Bell invented the telephone. Seventy-one years later, people at Bell Laboratories invented the transistor and the Microelectronic Age began.

People can make policies that promote technological growth. For example, the Soviet Union launched the first orbiting space satellite, Sputnik, in 1957. The United States government then decided that it wanted to win the space race by putting a man on the moon before the Soviet Union. In 1958, NASA (National Aeronautics and Space Administration) was created. In 1969, U.S. astronauts, as part of the Apollo program, planted a United States flag on the moon.

Of course, people provide the labor upon which our industries depend. It is because of the efforts of so many workers that we can provide as many products and services as we do.

People's needs for exercise have caused the development of clothing that makes running easier. (Courtesy of NASA)

(Courtesy of NASA)

People are also the consumers of technology. People consume billions of hamburgers and buy millions of television sets, automobiles, and stereos. People spend billions of dollars on entertainment and travel millions of miles each year on public highways, railroads, and airlines. As creators, promoters, and consumers, people are a very important resource for technology.

INFORMATION

Every technological activity requires a base of information. Not only do we need to know what to do, but also how to do it. Technology has developed rapidly during the last several decades because of an explosion of information. Information is now doubling every five years. All this new information is being shared worldwide because of our communications technology. Shared information provides people with new ideas, and new ideas create new knowledge.

We use the information resource in many ways. When surgeons perform operations, their knowledge of procedures and techniques is extremely important. They need to know exactly what to do first, how to make the incision, what tools to use, and so on. In previous times, capital and machines were the most important technological resources. In our era, the resource of information has become extremely important.

Information can be thought of as data that is processed. Data is raw facts and figures. In order to convert data into information, the data must be collected, recorded, classified, calculated, stored, and retrieved. **Data processing** is the act of turning data into information.

Computer files, books, and museums are sources of information. Information by itself is not valuable, however, until we make use of it. We need to process information by

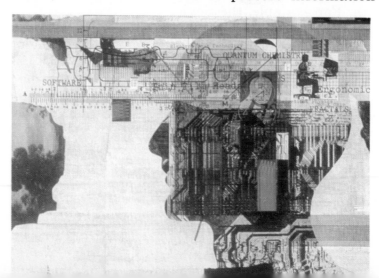

Only people have knowledge. People create knowledge from information. (Copyright IBM Corporation, 1983)

collecting it, considering what it means, and applying it to satisfy our needs or wants. Human intelligence turns information into knowledge.

MATERIALS

Most people think of materials when they hear the word "resources." Materials are indeed a very important resource for technology. **Natural resources** are materials that are found in nature. These include air, water, land, timber, minerals, plants, and animals. Natural resources that will be used to make finished products are called **raw materials.**

Countries that are rich in natural resources have an abundant supply of raw materials. The United States is rich in some material resources but has used up much of its supply of other materials. We have a great deal of timber, oil, coal, iron, and natural gas. However, many important resources must be imported from other countries. We import most of our chromium, platinum, and industrial diamonds from South Africa. Our nickel comes from Canada and Algeria. Our cobalt is imported from Africa and Europe. Our aluminum comes from South America and Australia.

Raw Materials

There are two kinds of raw materials: **renewable raw materials** and **nonrenewable raw materials.** Renewable raw materials are those that can be grown and therefore replaced. Wood is a renewable raw material because we can grow more trees. Since natural rubber (latex) comes from trees (Hevea trees, primarily found in the Far East and South America), it too can be considered a renewable raw material. Animals and plants are other examples of renewable resources.

Nonrenewable resources cannot be grown or replaced. Therefore we must take special care in consuming them. Oil, gas, coal, and minerals are nonrenewable resources. Once we use up our supplies of these resources, there are no more.

Limited and Unlimited Resources

Some resources are available in great supply, like sand, iron ore, and clay. Other resources are in short supply. When we have a choice, we try to use plentiful materials instead of materials that are scarce. Fresh water is a resource that is very scarce in some parts of the world. Some people believe that the shortage of water will become a very serious problem in the future.

Since there is a limited amount of certain resources on the earth, we must use resources wisely.

Renewable and Nonrenewable Raw Materials

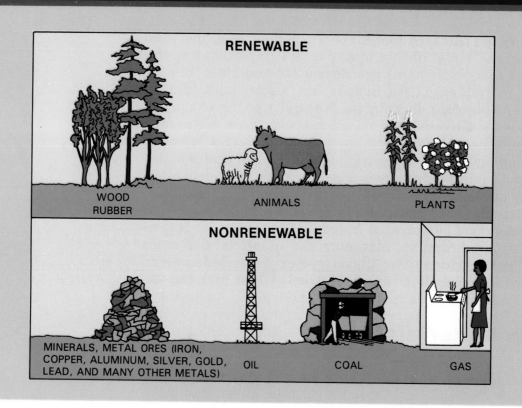

Synthetic Materials

The world is rich in natural resources, but they cannot last forever. People have therefore used technology to make substitutes for some nonrenewable resources. Materials made in the laboratory are called **synthetic materials.** Many materials that we use every day are synthetics. Plastic materials like acrylic, nylon, and Teflon are not found in nature and have to be produced chemically. Other synthetic materials include industrial diamonds, dacron, rayon, gasoline, and fiberglass.

Synthetic materials are sometimes better than raw materials because they can be made more cheaply. They can also be made stronger, lighter, and more long-lasting than the natural materials that they replace. In addition, synthetic materials can be produced with exactly the kinds of properties or characteristics that we desire. We can produce glass that conducts electricity or clothing that

Sails made from synthetics (Mylar™ and Kevlar™) helped speed the yacht "Freedom" to victory in the America's Cup races. (Courtesy of DuPont Company)

New synthetic fibers provide bright colors and exciting styles for clothing. (Courtesy of DuPont Company)

The largest fabric structure in the world is the Haj Airline Terminal in Saudi Arabia. (Courtesy of OC Birdair)

repels water. Perhaps most importantly, synthetics can replace scarce raw materials. They can therefore help us conserve our natural resources.

TOOLS AND MACHINES

Tools include hand tools and machines. People have been using tools for over a million years. Some tools are very simple, like hammers and screwdrivers. Some are more complex, like microscopes. Others, like some machines, include many different moving parts. Tools extend our natural capabilities.

Although some forms of animal life can use tools, only human beings can use tools to make other tools.

We use tools to fix things around the house. (Photo by Michael Hacker)

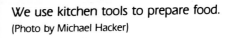

We use kitchen tools to prepare food. (Photo by Michael Hacker)

(Courtesy of Stanley Tools, Division of The Stanley Works, New Britain, CT 06050)

An electric sewing machine is an example of an electromechanical device.
(Photo by Michael Hacker)

All this early juke box machine needs is for the human to put two cents in the slot. Quite a bargain! (Courtesy of Smithsonian Institution)

Tools are a resource that may be used to process or maintain other resources.

Hand Tools

Hand tools are the least complicated tools. They rely on human muscle power to make them work. Hand tools extend the power of human muscle.

Machines

Machines change the amount, speed, or direction of a force. Early machines were mechanical devices. They made use of the principles of the six **simple machines** (lever, wheel and axle, pulley, screw, wedge, and inclined plane). Early machines used energy from human, animal, or water power.

Many modern machines have moving mechanical parts. Some machines, such as televisions and stereos, involve the movement of particles of electricity, called electrons. Some machines use electrical energy to move mechanical parts (like machines that use electric motors). These machines are called **electromechanical** devices.

Modern automatic machines can sometimes operate without much human control. Automatic machines need people only to start them and watch over them to make sure that they are working properly.

Electronic Tools and Machines

Electronic tools are used for testing and troubleshooting electrical circuits. One modern electronic tool is the computer. Computers are used to process and communicate

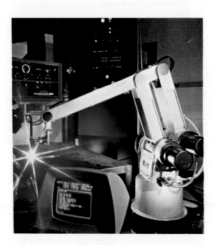

This welding robot can follow a seam between two pieces of metal. (Courtesy of General Electric Research and Development Center)

Six Simple Machines

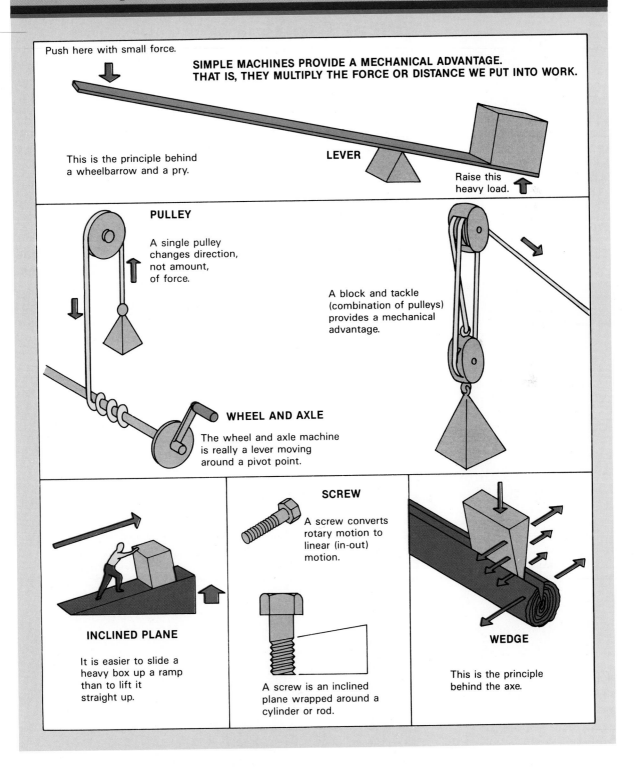

Push here with small force.

SIMPLE MACHINES PROVIDE A MECHANICAL ADVANTAGE.
THAT IS, THEY MULTIPLY THE FORCE OR DISTANCE WE PUT INTO WORK.

This is the principle behind
a wheelbarrow and a pry.

LEVER

Raise this
heavy load.

PULLEY

A single pulley
changes direction,
not amount,
of force.

A block and tackle
(combination of pulleys)
provides a mechanical
advantage.

WHEEL AND AXLE

The wheel and axle machine
is really a lever moving
around a pivot point.

SCREW

A screw converts
rotary motion to
linear (in-out)
motion.

INCLINED PLANE

It is easier to slide a
heavy box up a ramp
than to lift it
straight up.

A screw is an inclined
plane wrapped around a
cylinder or rod.

WEDGE

This is the principle
behind the axe.

This computerized system controls the flow of over one-half million gallons of fuel oil through pipelines in Texas. (Courtesy of DuPont Company)

information. They are also used to control industrial machinery. Computerized devices have been responsible for saving tremendous amounts of energy and labor.

Optical Tools

Optical tools extend the power of the human eye. These tools use lenses to magnify objects. Microscopes and telescopes are examples of optical tools. A great deal of scientific research has been accomplished through the use of optical tools. Today, powerful optical tools help us learn more about the genes that are our body's basic building blocks.

One modern optical tool is the **laser.** Lasers transmit very strong bursts of light energy. The light energy can be used for accurate measuring, cutting, welding, and even for communicating messages over long distances. Even though the laser was invented only recently, it now has important applications in many fields, including medicine, material processing, and defense.

Lasers can be used to cut through pieces of metal. However, they can be controlled so accurately that they can also be used to weld broken blood vessels in the human eye without causing damage.

Hot laser pulses cause a metal surface to soften and vaporize. (Courtesy of General Electric Research and Development Center)

This powerful microscope can magnify the surface of a piece of metal as much as 140,000 times. (Courtesy of General Electric Research and Development Center)

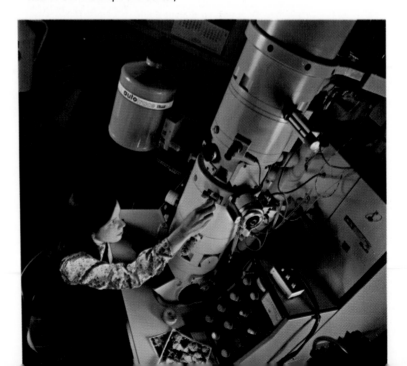

ENERGY

Countries like the United States use a tremendous amount of energy. The energy is used to manufacture products, to transport goods and people, and to supply our household heating, cooling, and lighting needs. Some energy resources are in great supply and are practically unlimited. Others are in limited supply and can get used up.

Renewable energy sources are those that can be replaced. Human and animal muscle power and wood are examples of renewable energy sources. **Limited energy sources** are those that cannot be replaced once we use them up. These sources include coal, oil, natural gas, and nuclear fission (atomic energy). Most of the energy we use comes from limited energy sources. **Unlimited energy sources** are those that we have more of than we can ever use at least in a practical way. These sources include solar, wind, gravitational, tidal, geothermal, and nuclear fusion.

If we go back far enough in time, most energy can be traced to the sun. People and animals obtain their energy by eating meat and vegetables. Meat and vegetables are sources of stored solar energy. Coal, oil, and gas are byproducts of plant and animal matter that has decayed over a very long period of time. The sun keeps the earth in orbit and therefore creates gravitational force. Gravitational force causes water to fall for hydroelectric power. The sun also causes wind.

Rather than thinking of most of our energy sources as solar energy, it is convenient to classify them into categories. The six energy categories are human and animal muscle power, solar energy, gravitational energy, geothermal energy, chemical energy, and nuclear energy.

Human and animal muscle power is still in use today, especially in developing nations. **Solar energy** provides wind energy, heat, and light energy. **Chemical energy** comes from sources such as wood, and fossil fuels like coal, oil, and gas. Chemical energy can also be stored in batteries. **Gravitational energy** comes from tides and falling water. Heat deep inside the earth provides **geothermal energy.** **Nuclear energy** comes from the conversion of radioactive matter into energy. We can either use these sources of energy directly, or we can convert them into other forms of energy, such as mechanical, electrical, and light energy.

The sun is the original source of much of the energy on earth. We cannot create more energy. We can only change matter into energy or change one form of energy into another form.

Wind energy is actually a form of solar energy.
(Courtesy of U.S. Department of Energy)

Capital resources are necessary for any technological act. (Courtesy of PPG Industries)

CAPITAL

Capital is one of the seven technological resources. To produce a product, set up a communication system, grow food, or transport goods or people, we must use capital resources. Any form of wealth is a capital resource. Cash, shares of stock, buildings, pieces of equipment, and land are all capital resources.

Capital is needed to finance companies. To raise capital, a company may sell stock to private investors. Each share of stock has a certain value. When investors buy stock from the company, they provide the company with operating capital (money with which to do business). These investors in turn become part owners (shareholders) in the company.

The shareholders expect that the company will make a profit. If it does, the shares of stock they own may be worth more. The company may even distribute some of the profit back to the shareholders as **dividends.**

Companies can also borrow money from banks. When money is borrowed, the banks charge **interest.** That means that the amount of money paid back is more than the amount of money borrowed. When a company borrows money and pays interest, it expects that its profits will be high enough to justify the cost of the interest.

TIME

In prehistoric eras, time was measured in periods based upon the rising and setting of the sun. There was no artificial lighting available. When people hunted for food or farmed their own land, work was done only during periods of daylight. It didn't matter much when a task got done, as long as it got done during the period of sunlight.

In the Middle Ages, monks, feeling the need to pray at regular times, invented water clocks. Later, clocks became common sights on city halls. Soon the world was ruled by clocks and time schedules.

In the agricultural era, time was measured in days. In the industrial era, time was measured in hours, minutes, and seconds. In the information age, time has become a very precious resource. We can accomplish a great deal in a very short time using modern machines and devices. Electronic technology, for example, permits a huge amount of data to be processed in a very short time. Electronic impulses flow through computer components at speeds near the speed of light. Data is processed in nanoseconds (billionths of a second). The increasing importance of time as a resource is one of the striking trademarks of our information age.

Time Measurement Throughout the Ages

In the agricultural era, people needed to know what part of the growing season it was. Since they did not need precise time measurement, a sundial or a burning rope clock was adequate. A huge stone structure at Stonehenge, England is believed to have been built in ancient times to be used as a calendar. Certain stars and planets seen through spaces between the stones indicated various important days.

During the Middle Ages, monks devised water clocks for telling time. In the industrial age, daily schedules became important. More accurate measurements of hours and minutes were necessary. People kept time using mechanical clocks and watches. These timepieces were based on a mainspring, pendulum, and balance wheel.

In today's information age, it is necessary to make exact measurements of small fractions of a second. Some of the tools used for these precise measurements include the quartz crystal clock, the atomic clock, and the oscilloscope.

Stonehenge (Photo by Susan Warren)

(continued on page 40)

(Time Measurement, continued)

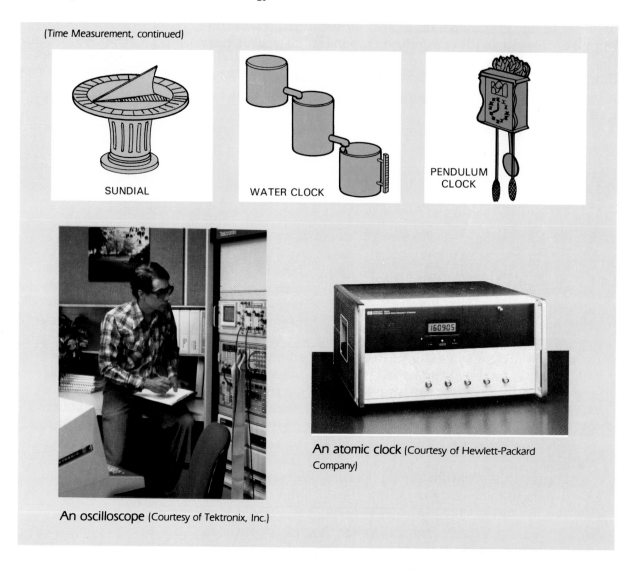

SUNDIAL

WATER CLOCK

PENDULUM CLOCK

An atomic clock (Courtesy of Hewlett-Packard Company)

An oscilloscope (Courtesy of Tektronix, Inc.)

SUMMARY

Every technological activity involves the use of seven types of resources. Whether the technology is very simple or extremely complex, the same seven resources are necessary. These resources are people, information, materials, tools and machines, energy, capital, and time.

People's needs drive technology. Human beings design and create technology using their knowledge and intelligence. People can make policies that promote or limit technological growth, and people are the consumers of technology.

We use information to solve problems, to create new knowledge, and to develop procedures and techniques. Information can be turned into knowledge by people who give it meaning. Information comes from raw data. The data is processed by collecting, recording, classifying, calculating, storing, and retrieving it.

Materials found in nature are called raw materials. Raw materials can be processed into useful products. Renewable raw materials are those that can be grown and therefore replaced. Nonrenewable raw materials get used up and cannot be replaced. People must use nonrenewable resources with care. Synthetic materials are those made by human beings. These materials can be produced with characteristics that we desire.

Tools extend the capabilities of people. Hand tools extend the power of human muscle. Optical tools extend the power of the eye. Computers extend the power of the brain. Machines are tools that change the amount, speed, or direction of a force. Most modern machines involve moving parts. Machines that use electrical energy to move mechanical parts are called electromechanical devices. Automated machines can operate without much human control. Electronic tools, particularly the computer, have been responsible for saving tremendous amounts of time, energy, and labor.

Energy sources are either renewable, limited, or unlimited. Renewable energy sources include human and animal muscle power and wood. Limited energy sources include oil, gas, coal, and nuclear fission. Unlimited energy sources include solar, wind, gravitational, tidal, geothermal, and nuclear fusion.

Capital is any form of wealth, such as money or property. Capital can be cash, shares of stock, buildings, pieces of equipment, or land.

Time is an especially important resource in the information age. Electronic circuits can carry huge amounts of data in billionths of a second. Time is therefore very precious.

The age we live in is a complex one. If we understand how technological resources are used and develop the ability to use them wisely, we will be better able to function as creators and consumers of technology.

Solving technological problems requires skill in using all seven resources.

SELECTING AND USING RESOURCES

Setting the Stage

Your expedition ship is forced to land at an unknown, unexplored jut of land. The landscape is bleak, offering only snow and ice. You and your fellow travelers must make shelter for winter protection. Fortunately, your experience in the technology laboratory has clued you into the best materials to use and how to process the materials.

Your Challenge

Build a model of a futuristic shelter using as many of the following materials, fastening devices, glues, and solders as possible. As you work, develop a material information chart to use for future technology projects.

Procedure

1. Using the chart supplied by your teacher, record the results of each of your investigations as you build your shelter. You will use this data in designing and building future technology projects.
2. Find and record the density of each material.

$$\text{Density} = \frac{\text{Mass (or Weight)}}{\text{Volume}}$$

3. After your teacher has given you instructions on the safe use of machines and tools, separate the materials into various shapes.
4. Choose screws, bolts, and rivets for assembling the shelter. Determine the correct drills to use for the fastening devices by trying them. Wax the wood screws for easier assembly. Record the drill sizes used for each fastening device.
5. Test different solders using soldering guns, irons, and torches. Determine which materials can be soldered and try soldering them. (Use these processes only after your teacher has given you instructions on the safe use of these materials and tools.)
6. Fasten assorted materials with glues and record how strong each joint is.
7. Classify each material you used as renewable, nonrenewable, or synthetic.

Suggested Resources

Steel
Sheet metal
Copper wire
Maple
Pine
Acrylic
Polyethylene
Aluminum
RH wood screw
FH wood screw
Sheet metal screw
Machine screw
Stove bolt
Pop rivet
Tinners rivet
60/40 solder
Copper tubing
50/50 solder
95/5 solder
Epoxy
White glue
Hot glue
Duco glue

42

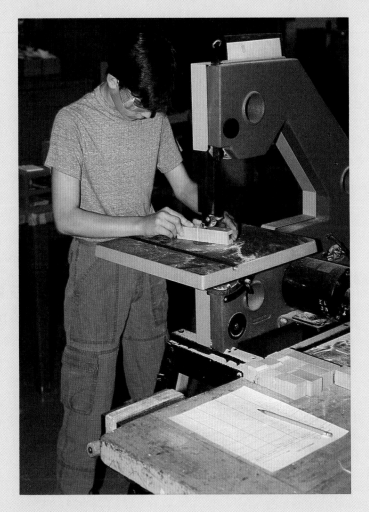

Technology Connections

1. Finishes protect materials and add color and texture. How might you finish your shelter?
2. Research several additional materials, fastening devices, and glues. Add the results of this research to your recorded data.
3. List the seven resources of technology. Explain how you used each resource in the building of your shelter.

Science and Math Concepts

▶ For industrial purposes, mass and weight are ordinarily considered equal. To find the density of a material, use the formula

$$\text{Density} = \frac{\text{Mass (or Weight)}}{\text{Volume}}$$

▶ To find the weight of a given volume of material use the formula:

$$\text{Weight} = \text{Volume} \times \text{Density}$$

▶ In the English system, density is expressed in pounds per cubic foot or lb./ft.^{3}

▶ In the metric system, density is expressed in grams per cubic centimeter or gm/cm.^{3}

43

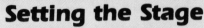

IT'S ABOUT TIME!

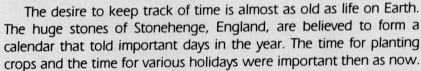

Setting the Stage

The desire to keep track of time is almost as old as life on Earth. The huge stones of Stonehenge, England, are believed to form a calendar that told important days in the year. The time for planting crops and the time for various holidays were important then as now.

Throughout history, people have viewed time as a valuable resource. Today we must often measure amounts of time as small as one-billionth of a second. As computers process more and more data at higher speeds, time as a resource becomes more and more important.

Your Challenge

Design and build a clock using a battery-operated quartz movement.

Procedure

NOTE—Do not use any tools or machinery until you have been told how to use them safely. Always wear safety glasses when working in the technology lab.

1. Using proper drawing techniques, design a clock body to be cut out of a piece of wood. Cut a full-scale *template* from heavy paper or cardboard.
2. If computer aided drawing (CAD) is available, transfer your hand-drawn sketch onto the system. Use the printed output as a template.
3. Make a detailed list of ALL the supplies and resources you need to make the clock. Estimate the amount and cost of: supplies, labor (minimum wage), and machinery and tools. Correct this list as the activity continues.
4. Make a detailed step-by-step list of procedures you will follow to make the clock. Again, correct the list as the activity continues.
5. Use the template you made to transfer the shape of your clock onto the wood.
6. Cut the shape out.
7. Drill a 3" diameter hole halfway through the clock body using a multispur drill bit.
8. Use a ⁵⁄₁₆" twist drill to make a hole the rest of the way through the clock body *concentric* to the 3" hole.
9. Use a file, sandpaper, sanders, etc. to shape and smooth the clock body. A router can also be used to shape the edges on the clock body. *A thorough sanding is very important.*

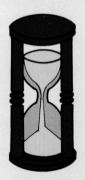

Suggested Resources

Safety glasses and lab aprons
¾" hardwood or softwood at least 6" wide
Multispur drill bit—3" diameter
Twist drill—⁵⁄₁₆"
Router
Jigsaw or band saw
Drill press
Stains and finishing supplies
Clock parts—quartz action, numbers, hands, washers, nuts, sawtooth hanger
'AA' alkaline battery

10. Colored pencils and acrylic paints can be used to decorate the clock. Stains may also be applied.
11. Apply at least two coats of polyurethane transparent finish to the clock body. Rub down the finish with #280 grit sandpaper between coats.
12. Adhere the self-stick clock numbers, install the quartz movement, clock hands, and saw-tooth hanger.

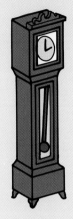

Technology Connections

1. The seven resources used in technological systems are people, information, materials, tools and machines, energy, capital, and time. Describe how you used each of the seven technological resources to design and build your clock.
2. Time is an important resource in the information age. The growing use of computers has made time even more important. Why?

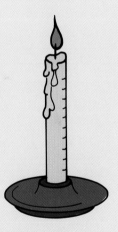

Science and Math Concepts

▶ Some computers can process data in *nanoseconds*. A nanosecond is one-billionth of a second (1/1,000,000,000 or 10^{-9} seconds).
▶ *Concentric* circles have the same center.

45

REVIEW QUESTIONS

1. What are the seven resources common to all technologies?
2. Define a machine in your own words. Then use your definition to determine whether the following items are machines:
   a. a baseball bat
   b. software for a video game
   c. a radio
   d. a ramp for wheelchairs
   e. a hand-operated drill
   f. a wrench
3. Wood is a renewable resource. Does that mean that we can cut down all the trees we need? Explain your answer.
4. What are two advantages of synthetic materials over natural materials?
5. Name three of the limited and three of the unlimited sources of energy.
6. If you wanted to start a company, how could you arrange to get capital?
7. How do knowledge and information differ?

KEY WORDS

Capital	Geothermal	Machines	Resources
Coal	Hydroelectricity	Material	Solar
Energy	Inclined plane	Nuclear energy	Synthetic
Finite	Information	Oil	Time
Gas	Laser	People	Tools

SEE YOUR TEACHER FOR THE CROSSTECH PUZZLE

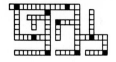

CHAPTER 3

SYSTEMS AND PROBLEM SOLVING

MAJOR CONCEPTS

After reading this chapter, you will know that:

- People design technological systems to satisfy human needs and wants.
- All systems have inputs, a process, and outputs.
- In a technological system, a technological process combines resources to provide an output in response to a command input.
- We can use the basic system model to analyze all kinds of systems.
- We use feedback to make the actual result of a system come as close as possible to the desired result.
- Systems often have several outputs, some of which may be undesirable.
- We can combine subsystems to produce more powerful systems.
- Formal methods (systems) are used to solve technological problems.
- Problem specifications include design criteria and limitations.
- A good solution often requires making compromises (trade-offs).
- Our values and society's values affect technological decisions.
- We try to reach technological solutions that take human and environmental needs into consideration.
- Modeling techniques are useful problem-solving tools.

47

SYSTEMS

We are surrounded by marvelous examples of technological systems. These systems range from pocket calculators to supercomputers; from personal systems of transportation to complex systems of highways; and from telephones to satellite communications systems. Our world is made up of thousands of these technological systems. Some are large, like metropolitan subway systems and the national telephone system. Some are small, like home stereo systems and personal computing systems.

Each of these technological systems satisfies a human need or want. We design and build technological systems to accomplish **desired results.** A **system** is simply a means of achieving a desired result. There are similarities among all technological systems. For example, every technological system has an **input,** a **process,** and an **output.**

People design technological systems to satisfy human needs and wants.

All systems have inputs, a process, and output.

Modern technological systems are all around us, touching our lives in many ways.

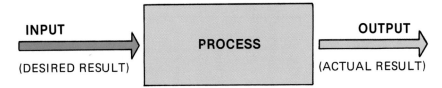

Diagram of a basic technological system

THE INPUT

The **input** is the command we give a system. The input command is the **desired result,** or the result we want the system to achieve. For example, when we turn on the television set, we are giving it an instruction. The input command is: "Provide picture and sound." Our desired result is to have the television give us a good picture and clear sound.

Let's look at another example. A car moves when we "tell" it to by stepping on the gas. In this case, the input command, or desired result, may be: "Go 30 miles per hour."

THE PROCESS

The **process** is the action part of a technological system. The process is how the system will achieve the desired result.

In our automobile example, the process includes both the car and the driver. The process combines the seven technological resources to produce the desired result. Energy (stored in the gasoline), machines (the car itself), people, information (supplied by the driver), and other resources are combined to make the car go 30 miles per hour.

THE OUTPUT

The **actual results** of a system are those that the process produces. We call these actual results the **output** of the system. We hope that the actual results are what we desired in the first place. For example, we hope that our car will, in fact, go 30 miles per hour.

In a technological system, a technological process combines resources to provide an output in response to a command input.

The process combines the seven technological resources to produce the desired result The output of this system is the speed that the car actually goes.

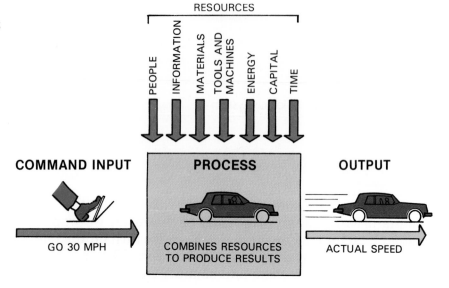

THE BASIC SYSTEM MODEL

We can use the basic system model to analyze all kinds of systems.

All systems include an input, a process, and an output. Feedback is added to provide a better way of controlling the system. The basic system model can be used to describe any technological system.

FEEDBACK

How does the driver know when the car is going the desired speed? The driver looks at the speedometer. The speedometer tells how fast the car is going. When the car's speed reaches 30 miles per hour, the driver eases up on the gas pedal. The driver then drives so as to just maintain 30 miles per hour.

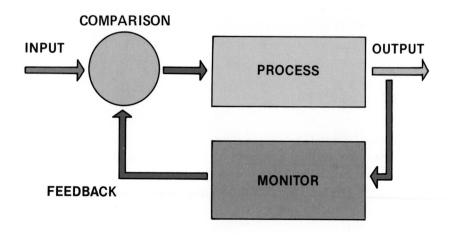

The basic system model includes input, process, output, and a feedback loop.

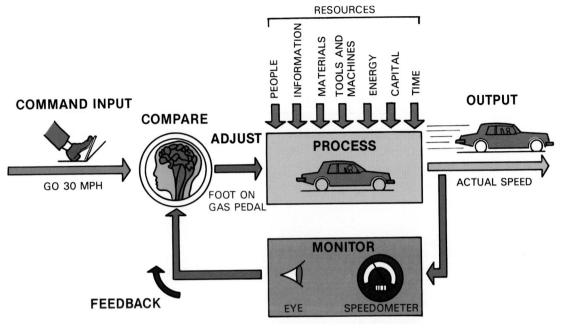

The combination of the speedometer, the driver's eye, and the driver's brain forms the system feedback loop.

We can use a monitor to watch a system's performance. A **monitor** provides feedback about the system's actual performance. This feedback permits us to compare actual performance to desired performance. Then, we can adjust our system to give us better results. Feedback allows us control over a system's output. For this reason, systems with feedback are sometimes called **control systems** or **feedback control systems.**

An example of a feedback control system is a door that opens automatically in a supermarket. Automatic doors often use an electric eye (a photocell) which is set to see a

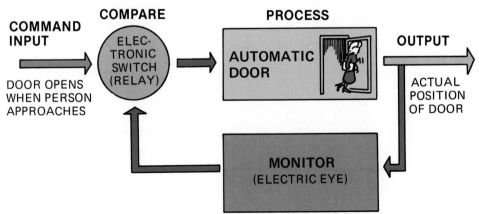

System diagram of an automatic door in a supermarket. The feedback is provided by a photoelectric cell (an electric eye).

certain light level. When a person walks by, the light level drops. The falling light level is a kind of feedback. This feedback is used to control an electronic switch called a relay, which causes the door to open.

Another example of a technological control system is a sump pump. Some people live in areas where the water from the ground can seep into the basements of their homes. In houses like this, they use a sump pump to remove the water. The sump pump uses feedback. The water level in the basement is monitored by a float. When the water level rises, the

Instant Feedback, Medieval Style

In the Middle Ages, teaching machines were used in the training of knights. The knight on horseback charges a wooden figure that is mounted on a pivot. If the knight strikes the shield exactly in the center, the wooden figure will fall over. If the knight strikes the shield off-center, the figure will swing around and hit the knight with a club. Instant feedback—instant learning.

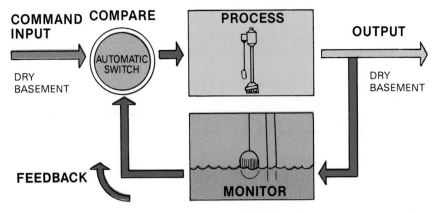

System diagram of a sump pump. The feedback and control are supplied automatically by a float and switch.

IN HOT WEATHER, EVAPORATING SWEAT COOLS THE BODY DOWN. IN COLD WEATHER, SHIVERING WARMS THE BODY UP.

float rises and turns a switch on. The switch controls the pump. When the water level falls, the float falls and turns the pump switch off.

Even our own bodies contain systems. For example, our bodies maintain a temperature of about 98.6° Fahrenheit. The input command to the body's temperature regulation system is the desired temperature, 98.6° Fahrenheit. Maintaining that temperature involves the action of the muscles, skin, blood, and the body "core." The output is the actual body temperature.

Sweating and shivering are two of the body's many feedback control systems.

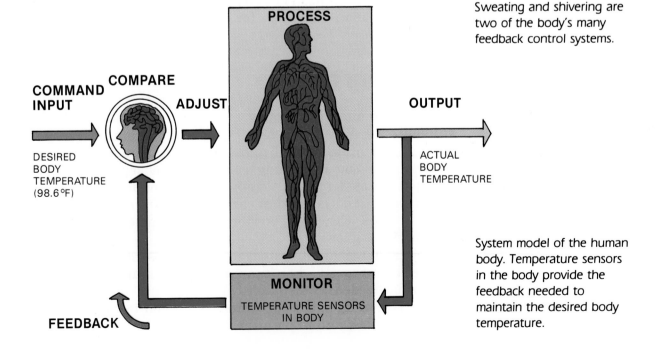

System model of the human body. Temperature sensors in the body provide the feedback needed to maintain the desired body temperature.

The body has many other control systems. These are systems that regulate sugar level, heartbeat, oxygen collection, and other important activities. These control systems keep our body conditions just about the same, even though outside conditions may be changing.

MULTIPLE OUTPUTS

Many times, a system produces several outputs. This may happen even if we were only interested in producing one output when we designed the system. A coal-burning power plant is designed to produce electricity. However, it also produces heat, smoke, ash, noise, and other outputs.

Some of the extra outputs are useful. For example, the heat produced can be used to heat nearby buildings. Other of the outputs may be undesirable, however. We may have to take steps to reduce or eliminate them, even if we get less electricity in the process.

The outputs that we get from a system can be either expected or unexpected. Which they are depends on whether we thought ahead carefully enough to anticipate them. They can also be either desirable or undesirable. We can have these different kinds of output in all combinations.

When designing systems, engineers must take into account all expected outputs. They must also try to foresee unexpected outputs. To eliminate undesirable outputs while retaining desirable outputs, engineers sometimes have to change their designs. They may have to accept less of the desired output in exchange for reducing some undesirable outputs.

Systems often have several outputs, some of which may be undesirable.

SUBSYSTEMS

Systems are often made up of many smaller systems called **subsystems**. When you are trying to understand a large system, it is often helpful to break it down into its smaller subsystems. You can study each of them separately and then put the system back together. For example, let's look at a transportation system that carries goods from Los Angeles to New York City. The system uses vehicles (trucks) to transport the goods.

The transportation system relies on subsystems in order to operate properly. For example, a communication subsystem is needed to dispatch the trucks and report breakdowns. A management subsystem is needed to hire drivers and mechanics, and to plan the best routes.

Four Kinds of Output

The outputs from a power plant illustrate the four possible types of output.

1. The **expected, desirable output** from the power plant is electricity. The additional output of heat may also have been expected. If we can do something useful with it, it is also desirable.
2. **Expected, undesirable outputs** from the power plant are noise and smoke.
3. An **unexpected, desirable output** was found at one power plant. The plant discharged some of its heat into a river, making the water warmer. Tropical fish flourished in the river near the plant, creating an attraction.
4. An **unexpected, undesirable output** of plants built some time ago and still operating is their contribution to acid rain. Acid rain actually falls hundreds of miles or more away from the plants.

This power plant produces all four kinds of output.

Subsystem tree diagram for a large transportation system. Each subsystem can be broken down into smaller subsystems.

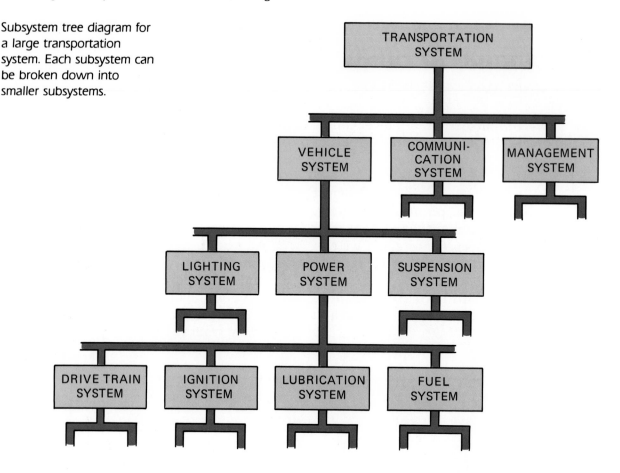

Each of the subsystems can be further broken down into smaller subsystems, creating a subsystem tree. The vehicle system includes lighting, power, and suspension subsystems. The power system includes the drive train, ignition, lubrication, and fuel subsystems.

We can combine subsystems to produce more powerful systems.

PROBLEM SOLVING

Human beings have always been faced with problems and have always tried to solve them. Physical and biological needs, like the needs for shelter, food, clothing, and health care, cause us to invent technological means to make life easier. The solutions we employ are based upon the most up-to-date resources available.

Headaches, for example, have been a problem for human beings for a long time. For thousands of years people have tried to eliminate headaches. Do you remember the trepanning remedy discussed in Chapter 1? People drilled holes in the patient's head to let the evil spirits escape.

Trepanning was based upon the theory that evil spirits in the head were the cause of pain. If people had had more knowledge about the actual cause of headache pain, trepanning might never have existed.

Our ancestors had limited knowledge and techniques with which to solve problems. We are now able to solve complex problems using modern tools and techniques. We can also apply the knowledge that people have gained over thousands of years.

Today's problems also relate to human needs and desires. Although many more resources are available, the system we use to solve problems is just about the same system people used many years ago. The basic system model can also be used to help us solve different kinds of problems.

Formal methods (systems) are used to solve technological problems.

THE PROBLEM-SOLVING SYSTEM

The problem-solving system consists of six steps:

1. Define the problem clearly.
2. Set goals (desired results).
3. Develop alternative solutions.
4. Select the best solution.
5. Implement the solution.
6. Evaluate the actual results and make necessary changes.

Step 1 is defining the problem. To solve the problem, we must first clearly understand it. We must be sure that we know exactly what the problem is. Then we can figure out what to do about it.

Step 2 is setting goals. Setting goals helps us determine exactly what we want the solution to accomplish. Our goals are our desired results. Our goals must include the exact requirements (specifications) of the problem. Specifications tell us "how fast," "how far," "how big," "how much can it cost," and so on.

Steps 3, 4, and 5 are the process part of the system. Remember that the process is the action part of our problem-solving system. The process involves developing, selecting, and trying out different alternatives. These three steps tell us how we will achieve our desired results.

Step 3 is developing alternatives. We try to come up with many different ideas that may provide alternative solutions. **Step 4 is selecting the best solution.** Of all the alternatives, we try to pick the best one. **Step 5 is**

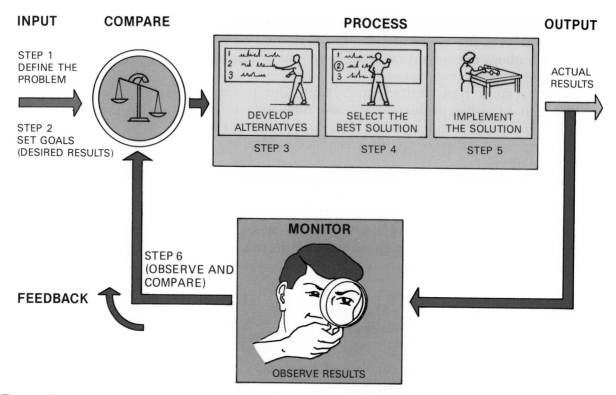

The problem-solving system has the same system diagram as technological systems.

implementing the solution. In this step we try our solution and obtain some actual results.

Step 6 is evaluating the actual results and making necessary changes. The last step in our problem-solving system is to evaluate our actual results. Then we can make any modifications that are necessary. We evaluate by observing (monitoring) the actual results. The actual results are then compared to the desired results. This comparison shows us whether or not the actual results are what we wanted. If our actual results are what we desired, our solution is acceptable. If our actual results are not what we desired, we have to make some changes.

USING THE PROBLEM-SOLVING SYSTEM

Let's consider a problem that might be presented in a technology class. Suppose we have to design and construct a rubber-band-powered vehicle. This vehicle must be able to carry a raw egg safely over a distance of 50 feet on a smooth, level surface. We want our vehicle to travel faster than other competing vehicles.

How would we go about solving this problem? One method is to use our six-step problem-solving system. First, we must define the problem and set goals to determine how to solve it. We must clearly identify our desired results.

Step 1: Define the Problem

In this case, the problem is to win a school competition by constructing an egg-carrying device. To more clearly define the problem, however, we must consider it very carefully. We must recognize that winning the competition means:

1. We must design and construct a device powered by rubber bands.
2. The device must carry an egg without breaking it.
3. The device must be faster than any other student's device.

Once we really understand the problem, we are in a much better position to solve it.

Let's assume we have now clearly defined the egg transport problem. We understand that the problem is to win the competition by designing the fastest method of transporting a raw egg over the race course without breaking the egg.

We can now decide what we'd like to do about the problem. Our feelings and our attitudes will affect this decision. We might feel that the problem is dull, uninteresting, or unimportant. If we do, we might choose not to try to solve it at all.

Most of the time, however, problems that are presented to us need to be solved. Sometimes we want to solve them because they are important to us. Sometimes, we need to solve them because parents, teachers, or friends ask us to. Sometimes, we want to solve problems because they are challenging and are fun to solve.

Step 2: Set Goals and Consider the Specifications

Once we decide to try to solve a problem, we have to decide exactly what it is we want to accomplish. We have to set goals. We may have several goals in mind for the egg transport problem. Perhaps we want

- to have fun.
- to improve our technical skills.

The goal is to provide a solution that works better than anyone else's solution.

- to get a good grade.
- to impress our friends.

Our major goal, of course, is to build a device that will win the competition.

By setting clear goals, we determine what results we want to occur. These goals are our desired results. Setting goals is an important part of our problem-solving system. It helps us understand exactly what we want to accomplish and why.

Our goals should take into account any special requirements imposed by the problem. The list of specifications must include all the requirements of the problem. These requirements are called **design criteria.** If we were building a house, our design criteria would include the kinds of rooms, number of stories, and whether the house will have a basement and an attic.

Some of the design criteria for our egg transport problem might include the following:

1. The vehicle must carry a medium-size raw egg without breaking it.
2. The vehicle must be painted attractively.
3. The vehicle must have an identification number.
4. The vehicle must travel faster than any other competing vehicle.

Can we try to meet our goals any way we please? Can we decide, for example, to build a device that uses a jet engine? It would certainly be the fastest vehicle, but would the solution be acceptable to the judges?

Problems generally have certain limitations. The specifications should include these limitations. The following are some of the limitations imposed by the egg transport problem.

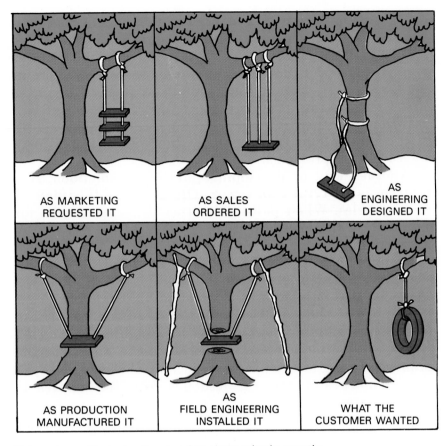

This cartoon illustrates the need to state criteria exactly.

1. The vehicle must cost no more than two dollars.
2. The vehicle must weigh no more than 1 pound.
3. The vehicle must be able to fit into a 12″ × 6″ × 4″ container.
4. The vehicle must be powered by no more than four #6 rubber bands.

Problem specifications include design criteria and limitations.

The specifications tell us what conditions the solution must meet. If we understand the problem specifications and have set our goals, we are in a good position to begin the problem-solving process.

Step 3: Develop Alternatives

We can now begin to suggest some possible solutions, keeping in mind the problem specifications. There is generally more than one solution to a problem. We can usually propose several good ideas, each of which might work quite well. These ideas are called **alternatives.**

Edison used feedback from his failures to come up with a successful solution.

(From the collections of Henry Ford Museum and Greenfield Village)

Developing alternative solutions is one of the most important parts of the problem-solving process. How do we come up with alternatives? One way is by using our **past experience.** We think of things we know worked before in similar situations. These things are likely to work again.

We can also use the past experience of others. We can ask people for their ideas. Or, we can do research by going to the library and reading about how other people have solved similar problems.

Another way of developing alternative solutions to a problem is by **trial and error.** Often we use this method to do puzzles. When putting together a jigsaw puzzle, for example, we try putting various pieces together. Eventually we wind up with the right pieces together.

Thomas Edison developed many electrical devices, including a practical electrical lighting system. Part of his work involved developing a light bulb that could burn continuously for hours. He tried hundreds of different materials for the light bulb filament without success. Finally, he tried carbonized (burned) thread. Edison used the trial-and-error problem-solving method. He was successful because he took careful notes and used the feedback from his failures to modify his approach.

A third way of developing alternative solutions to a problem is to use what psychologists call **insight.** Have you ever had an idea just pop into your head? These inspirations are usually accompanied by the "Aha!" response. "Aha! I've got it!" The solution comes just that suddenly. When you use insight to solve a problem, try to think about the problem from all angles. Try to be as creative and outrageous as possible in proposing solutions.

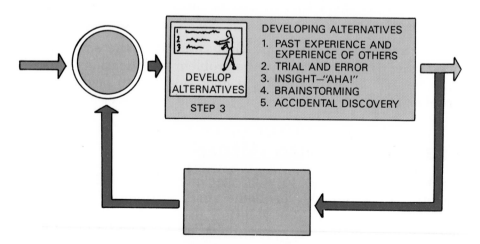

Five ways of developing alternatives

A fourth way of coming up with alternative solutions to a problem is **brainstorming.** A group of people trying to solve a problem together might use brainstorming. During the brainstorming, each person in the group proposes an idea. One group member writes the ideas down. No one is allowed to criticize anyone else's idea during the brainstorming process, no matter how foolish the idea seems.

The brainstorming process is intended to stimulate creative thinking. People should feel free to share any wild ideas they come up with. Sometimes one person's outlandish idea will open up someone else's mind to a totally new approach. After many ideas have been proposed, the group reviews them all. The best ideas are then developed further.

Still another way alternative solutions are discovered is by **accident.** Some of the most important discoveries have come when the inventor or scientist had gone as far as possible and still not solved the problem. A chance happening then provided the answer. In other instances, the solution to one problem was discovered by someone who was looking for the solution to another problem.

Step 4: Select the Best Solution

Suppose that several alternative solutions have been found through one or more of the methods just described. We now need to find out which alternatives might be good solutions. We do this by studying each alternative to find out what disadvantages it might have. We then try to make each alternative work as well as it can. This helps us determine if it will really be a good solution.

Normally, each alternative solution requires some careful thought before it can be accepted or rejected. We must find out all we can about each alternative. We may do research, using books, experts, and museums. We may also do original experimentation ourselves.

Once we have researched our alternatives, we are able to make some educated decisions. We now know which ideas may work and which are not worth considering further. We can identify those alternatives that will best meet the problem's specifications.

The problem specifications may cause us to reject some alternatives. We must develop a solution that takes any limitations into account. For example, in our egg transport problem, perhaps the vehicle must be made only of certain materials (materials limitation). Maybe we may purchase materials for no more than two dollars (capital limitation).

Perhaps construction must be completed within three weeks (time limitation).

Some limitations are imposed by the problem specifications. Other limitations are imposed by the resources themselves. Certain resources may be readily available, but not appropriate for our needs. Other resources may be very appropriate, but not as readily available or more expensive. All seven technological resources have limitations. Before we choose resources to use in a solution to a problem, we must consider each resource's limitations.

Once we have narrowed down our alternatives, we need to consider the best possible form of each alternative. Suppose, for example, we have decided that we could make our egg transport vehicle out of wood, metal, or styrofoam. Styrofoam is a lightweight material that can be carved and shaped easily, but it is not very strong. Wood is heavier, stronger, and a little more difficult to work with. Metal is heavier still, very strong, but harder to shape. To explore each of the alternatives more fully, let's build one test vehicle out of each material.

When we make the wooden vehicle, we find that it weighs too much and therefore doesn't go very fast. To reduce the weight, we cut away portions that are not really needed. Each time we make the vehicle a little lighter, it goes a little faster. We try to make it as lightweight as possible so that it can go at its maximum speed.

Making an alternative work as well as it can is called **optimization.** When we optimize an alternative, we improve it as much as we can before we compare it to other alternatives.

After we optimize the performance of the wooden vehicle, we optimize the performance of the styrofoam and metal vehicles. In each case we try to achieve a good balance between weight, speed, and strength.

Many technological limitations had to be overcome for a man to be put on the moon.
(Courtesy of NASA)

Optimizing an alternative

Finally, to select the best of our optimized alternatives, we compare them. We ask ourselves, "Which alternative is better? Which one best meets the problem specifications?" We rate each alternative on how well it meets the problem specifications and how well it achieves our goal. We compare the actual results to the desired results for each alternative.

Very often we are unable to come up with a perfect solution and have to make compromises. We wanted to

Scoring the Alternatives

One way of deciding which alternative is the best solution is to score the alternatives. A sample of scoring is shown below. In this case, a rating of 0 means a poor score, 1 is fair, 2 is good, 3 is very good, and 4 is excellent.

When we compare the actual results of the three alternatives with the desired results (goals), we find that the best solution is to build the vehicle out of styrofoam. Of the three alternatives, the styrofoam vehicle best meets the specifications. It is therefore the best solution.

SPECIFICATION	WOOD	STYROFOAM	METAL
Carries egg safely	3	4	2
Painted attractively	4	3	4
Weighs no more than 1 lb.	4	4	0
No larger than 12″ × 6″ × 4″	4	4	4
Has identification number	4	4	4
Faster than other vehicles	2	4	1
Uses four #6 rubber bands	4	4	4
Total score	25	27	19

Cost versus performance is a trade-off we often experience.

A good solution often requires making compromises (trade-offs).

build our vehicle lighter because our research showed that lighter vehicles go faster. When we made the vehicle lighter by removing material, however, the vehicle lost strength. Our final solution was a compromise. In a compromise, we give up one thing (in this case, strength) to gain something else (in this case, speed). We arrive at what we feel is the best combination of the two. In problem solving, this kind of compromise is called a **trade-off**.

Trade-offs always involve at least two factors. Complicated technological problems may require trade-offs of many factors. One very common trade-off is cost versus performance.

Step 5: Implementing the Solution

Once we have chosen the best solution, we can begin to put it in place. We are now at the problem-solving phase called **implementation**. Implementation means trying out the solution under actual conditions. In our egg-carrying problem, we are ready to race.

Implementation may involve the construction of a device (as in the egg-carrying problem), the production of a new wonder drug, or the construction of a new power plant. Implementation may be a lengthy and expensive step. It can take years to complete at a cost of millions or even billions of dollars. Before we implement such large and expensive solutions, we need to have studied the problem, the alternative solutions, and all possible impacts thoroughly.

Step 6: Evaluating the Actual Results

After we have implemented our solution, we must evaluate it. Observing (monitoring) the results helps us improve the design or construction of the solution. The

feedback we get when we observe our results allows us to compare our actual results to our desired results. Then we can know what adjustments or changes have to be made to our solution. Evaluation is very often a continuous process. We receive feedback and evaluate performance throughout the life of a solution.

There are two important things to remember about the problem-solving system. The first is that we can't begin to solve a problem until we understand it clearly. The second is that we don't have to be satisfied with the results we get the first time. We can use the feedback we get from observing our results to make changes. We can continue to improve performance until we come close to the desired results.

DIFFERENCES BETWEEN SCHOOL PROBLEMS AND PROBLEMS IN THE OUTSIDE WORLD

The technological problems presented in school require the same kind of analysis as the problems of the outside world. The experience we gain by solving problems in school helps prepare us to solve other problems. However, there are differences between the carefully designed problems we are given by our teachers and the problems posed by complex real-world conditions.

Social and Environmental Concerns

When engineers try to solve technological problems, they must consider what effects their solutions will have on society and the environment. They should not, for example, plan to construct an airport close to a residential area. The noise and pollution would interfere with the lives of the residents.

In the real world, the needs of a society or community often influence technological solutions. A good technological solution meets the needs of people and the environment.

Politics

Often, solutions to problems in the outside world are affected by politics. Various groups have special interests. These interests cause them to favor or oppose a particular solution. For example, some groups of people have opposed the development of nuclear power plants. These people claim that radioactive material could poison the environment in the case of an accident. They also point out that there is no effective way to dispose of radioactive waste.

Other groups favor nuclear power development. They believe that nuclear energy can provide a partial solution to our need for increased energy. They feel it will make our country self-sufficient so that we don't have to rely on nations with unstable governments for our oil supply. These people are convinced that the risks are at a minimum. They are willing to trade off possible risks for the benefits of an increased energy supply.

Risk/Benefit Trade-Offs

A common trade-off made in solving large, complex problems is a risk/benefit trade-off. In order to obtain certain benefits, we accept a certain level of risk. We usually try to keep the risk as low as possible. In fact, we may not implement the solution if the risk is too high. When you travel by car, you are accepting a very low risk of being injured in an accident. In exchange, you receive the benefit of being transported rapidly and in comfort.

Need for Continued Monitoring

Solutions to complex, real-world problems often require constant, long-term monitoring for unexpected outcomes.

For example, between 1958 and 1961, many pregnant women took a drug called thalidomide to help them relax. Only after several years did doctors realize that the drug had harmful side effects. Some of these women had babies born with birth defects. The birth defects were later traced to the drug.

When thalidomide was first given, it seemed to be a fine solution to the problem of nervousness. Only after continued monitoring and evaluation did people realize that it was not the optimal solution at all. The effects of a technological solution are occasionally not known until some time after the solution is implemented. It is therefore necessary to keep on monitoring and evaluating the results of technological systems.

Values

Our values influence the way we make decisions. The way we feel about something makes us decide in favor of or against it. If we feel that automobiles are simply devices for transporting our families and ourselves as inexpensively as possible, we might decide to buy a very basic car that gets good gasoline mileage. If we feel that automobiles are really neat gadgets and that driving is lots of fun, however, we might decide to buy a sports car or a luxury car.

Our values and society's values affect technological decisions.

Culture

Our culture influences our values and therefore affects our technology. In South Africa, for example, two official languages are spoken: English and Afrikaans (which is similar to Dutch). South African law states that all official

Our values affect our choices.

(Courtesy of Bob Thomas and Associates, Inc.)

(Courtesy of American Motors Corp.)

communications must be delivered in both languages. Television was not introduced in South Africa until the late 1970s because of the difficulty of developing Afrikaans programming equal to the amount of English programming available. This is an example of how cultural tradition and values influence the development of technology.

Normally, problems that can be solved in a school technology laboratory do not have political, environmental, or cultural implications. Problems at this level are not as complex as the technological problems found in society. Societal problems impose additional limitations related to social, environmental, and political factors.

We try to reach technological solutions that take human and environmental needs into consideration.

Problem Solving At Work

Problem solving in the real world is often a group effort. In corporate industry, think tanks, task forces, committees, project teams, and product groups are common. Bringing people together in small groups helps to make problem solving more effective. Hearing the ideas of the individual group members stimulates the others in the group to think. Additionally, groups bring together people with different skills and varied experience. Each person in the group has a chance to make a contribution. The combined effort helps the project succeed.

Students can accomplish a great deal when they work as a team. (Courtesy of Michael Hacker)

Team work is an important part of the Hewlett Packard Company's philosophy and style of doing business. While recognition is based on each person's contribution, the company emphasizes working together and sharing rewards. (Courtesy of Hewlett Packard Company)

At Turner Corporation, meetings are held among engineers, managers, and support staff before a construction project is begun. (Courtesy of The Turner Corporation, photograph by Michael Sporzarsky)

MODELING

When solving problems, we must develop and test alternative solutions. When we were designing the egg-carrying vehicle, we built three models out of three different materials. What happens, though, if it is very expensive or dangerous to actually test several alternative solutions? To understand options without actually trying them, engineers often construct **models** of solutions.

Modeling techniques are useful problem-solving tools.

Modeling Techniques

Modeling covers a wide range of activities. It can be thought of generally as any activity that aids in studying a solution without actually having to implement it. There are five types of models.

1. **Charts** and **graphs** describe in detail the performance of a proposed solution to a problem.

2. **Mathematical models** describe the operation of a system by mathematical equations that predict system performance.

3. **Sketches, illustrations,** and **technical drawings** visually represent the designer's ideas so that they may be understood by others. The thought required to

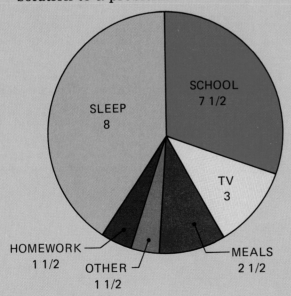

This pie-chart describes the number of hours a student spends on various activities.

$$V = C \log_e \frac{M_o}{M_t}$$

V	= ROCKET VELOCITY
C	= EXHAUST VELOCITY
M_o	= INITIAL MASS
M_t	= MASS AT TIME "t"

This equation predicts the velocity of a rocket during flight.

(continued on page 72)

draw an idea and the discussion of drawings with others are good ways to refine proposed solutions.

4. **Working models** simulate the performance of a proposed solution. Working models can be partially functional (only part of the system is modeled) or fully functional. They can be made of the actual material to be used or of a different, more easily formed material. They can also be either full size or made to scale size (larger or smaller than the proposed solution).

5. In **computer simulation,** a computer performs the mathematical modeling. Often a graphics program that displays results pictorially is used. **Computer-aided design (CAD)** is a modeling tool used by engineers, designers, and drafters (see Chapter 6).

Men creating a model of an automobile (Courtesy of Ford Motor Company)

This model tests wind effects on the city of Boston, Massachusetts. Engineers used data from the model to design buildings that would not create harmful wind effects in the city. (Reprinted from TECHNIQUE magazine, Summer 1985, with permission. Copyright Data General Corporation)

This flight simulator allows pilots to experience the sensations of flight without leaving the ground. (Courtesy of Boeing)

Suppose, for example, that several alternative solutions are proposed for the design of a large power plant. It might take several years and much money to build such a project. It is simply not feasible to build and test each alternative solution. Instead, the planners construct models. Models are used to test ideas without risking a great deal of time, capital, or public safety.

SUMMARY

People design technological systems to satisfy human needs and wants. A system is simply a method of achieving the results that we desire.

A system may be discussed in terms of its input (desired results), process (way of achieving the desired results), and output (actual results). For a system to work properly, we need to provide a means of monitoring the actual result and comparing it with the desired result. We then use this feedback to adjust the operation of the system.

Systems often have multiple outputs. When we design systems, we must think about any possible undesirable outputs. We may have to modify a system design to reduce or eliminate the undesirable outputs.

Large systems are often made up of many smaller subsystems. Examining each subsystem by itself can be useful in understanding a large, complex system.

The system concept is useful for solving problems. The problem-solving system can be thought of as having six steps:

1. Define the problem clearly.
2. Set goals (desired results).
3. Develop alternative solutions.
4. Select the best solution.
5. Implement the solution.
6. Evaluate the results and make necessary changes.

Defining a problem means understanding the problem's exact nature and its specifications. To develop alternative solutions, we can use past experience, trial and error, insight, brainstorming, and accidental discovery.

Real-world problem solving includes a need for continued monitoring and a need to be concerned with the environment, politics, values, culture, and risk/benefit trade-offs. The limits of each solution need to be recognized and each alternative optimized. In the real world, this is often done using modeling techniques.

MODEL ROCKET-POWERED SPACECRAFT

Setting the Stage

When an unmanned spacecraft is launched on a mission, feedback control systems are operating on board the spacecraft. These systems cause the spacecraft to go where we want it to, and return to be recovered and flown again.

Several subsystems must work together to make the mission successful. First, a low-speed guidance system must control the craft until it gains speed. Then, a high-speed guidance system must take over. When the craft reaches its destination, a system must turn it around for the return trip. Finally, a recovery system must bring it back to earth safely.

Your Challenge

Build a model rocket-powered spacecraft with on-board feedback control systems. These systems must allow the spacecraft to fly successfully and be recovered to fly again.

Suggested Resources

Paper mailing tape—2" wide and 20" long

One wooden dowel rod—¾" in diameter and 12" long

Two pieces of wooden dowel rod—¾" in diameter and ½" long

One paper soda straw—2" long

One strip of tagboard—1½" × 10"

Three pieces of heavy thread (crochet thread is good)—each 10" long

One sheet of light plastic (bags from drycleaners are a good source)—12" × 12"

Six pieces of masking tape—½" × ½"

Two cork stoppers—#4 and #6 sizes

Soft elastic cord—⅛" wide and 10" long

One model rocket engine (Estes A8-3 or B6-4)

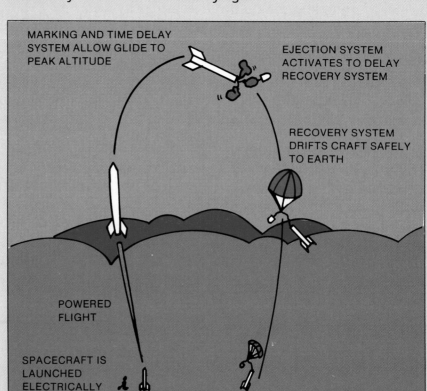

MARKING AND TIME DELAY SYSTEM ALLOW GLIDE TO PEAK ALTITUDE

EJECTION SYSTEM ACTIVATES TO DELAY RECOVERY SYSTEM

RECOVERY SYSTEM DRIFTS CRAFT SAFELY TO EARTH

POWERED FLIGHT

SPACECRAFT IS LAUNCHED ELECTRICALLY

RECOVERY

Procedure

1. A few safety rules must be followed:
   a. Engines must only be ignited electrically and by remote control. Each package of engines contains instructions.
   b. Rockets must never be fired indoors or in a congested area. A launch rod must be used, and no one should stand closer than ten feet to the launch area.
   c. Rockets should never be recovered from power lines or dangerous places.
   d. All vehicles should be tested for flight stability before being flown for the first time.
2. To make the body tube, tear or cut a sheet of notebook paper in half so that you have two sheets, each 4¼" × 11".
3. Carefully roll one piece of paper around the 12" dowel rod and glue it together with white glue. You have a paper tube 11" long. Do not glue the tube to the rod, and do not remove the rod.
4. Cut one end of the mailing tape at an angle of 45 degrees.
5. Wet the tape (do not soak it) and carefully spiral the tape around the paper tube on the dowel. The angled edge should be started along the top edge of the paper tube. Be careful not to glue the tape to the rod. As you reach the lower end of the tube, slide the rod up inside the paper tube and trim the tape along the edge of the paper. Remove the finished body tube from the rod to dry.
6. To make the nose cone, glue the two corks together and to one of the ½" dowel pieces as shown. When dry, shape the nose cone with sandpaper.
7. Cut the plastic into an 8" hexagon and attach the shroud lines (crochet thread) at the corners with the masking tape squares. This is the parachute.
8. Drill a ¼" hole in the remaining dowel piece, and glue it firmly inside the body tube, 2½" from the bottom of the tube.
9. Using the fin pattern, cut three fins of tagboard. Glue them near the end of the body tube. When dry, reinforce the joints with more white glue.
10. Glue the straw to the body tube between two fins.
11. Attach the elastic shock cord to the top of the body tube. Staple the free end of the shock cord and the parachute shroud lines to the nose cone.

12. Before you launch your spacecraft, test it for flight stability. Place an engine in the craft and balance it at mid-point on the end of a string. If it flies straight when you swing it around in a circle, it will fly straight when you fire it. If it does not fly straight, add weight to the nose and try again.

SAFETY NOTE: Prepare and fly your rocket only as directed by the instructions that came with the engine. Use cotton balls or tissue paper for flame-proof wadding.

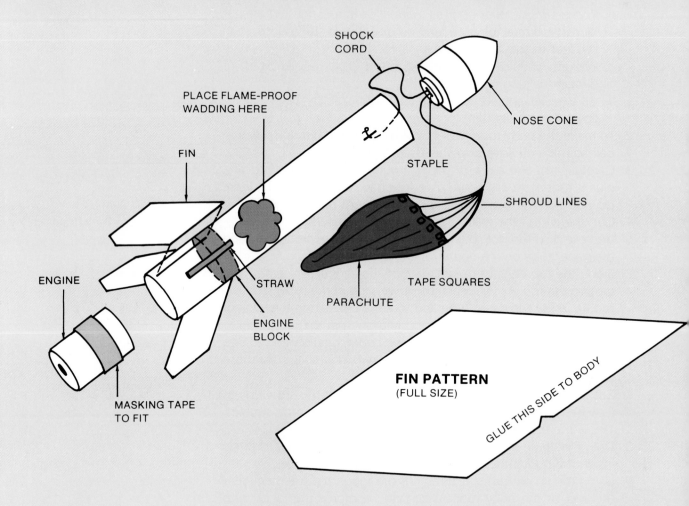

SHOCK CORD

PLACE FLAME-PROOF WADDING HERE

NOSE CONE

FIN

STAPLE

SHROUD LINES

ENGINE

STRAW

TAPE SQUARES

ENGINE BLOCK

PARACHUTE

MASKING TAPE TO FIT

FIN PATTERN
(FULL SIZE)

GLUE THIS SIDE TO BODY

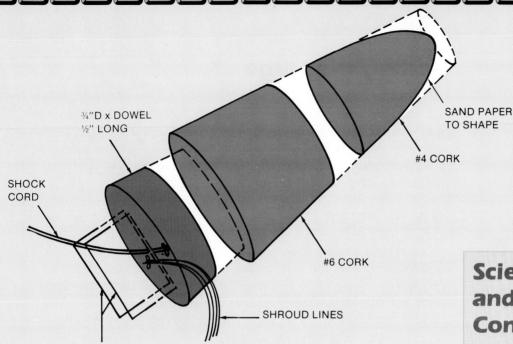

³⁄₄"D x DOWEL
½" LONG

SHOCK
CORD

SAND PAPER
TO SHAPE

#4 CORK

#6 CORK

SHROUD LINES

2 STAPLES

Technology Connections

1. Which part of your craft acts as the low-speed guidance system? The high-speed guidance system? The recovery system? How is feedback provided to the recovery system when it is needed?
2. Systems with feedback control are used by most industries today. What kind of feedback does the entertainment industry use? What about the advertising industry?
3. Some outputs of some systems are undesirable. What are some undesirable outputs from the air transportation industry?

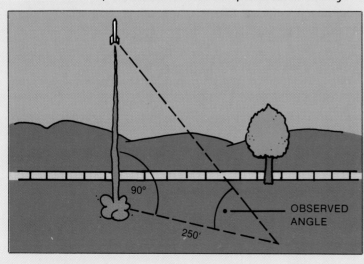

90°

250'

OBSERVED
ANGLE

Science and Math Concepts

▶ Systems are sometimes complex combinations of many other systems.
▶ To find the maximum altitude your spacecraft obtains in flight, a principle of trigonometry can be used which states that all of the angles and sides of any triangle can be found if any three of the parts, including one side, are known. Measure a distance of 250 feet from the launch site and track the flight with a protractor. Observe the angle of the maximum altitude of the flight from that spot. Look in a table of trigonometric functions and find the tangent of the angle. Multiply the tangent of the angle times 250 (the distance from the launch site) to calculate the altitude.

TUG O' WAR

Setting the Stage

The county fair is coming up this weekend. One of the organized activities you have entered is the tractor pull. With only 20 seconds to win, you must pull the opposing tractor over a center line, using a tow line fastened to the rear of the tractors. Because you are the defending champion, you find yourself out in the hot sun, making sure your tractor is ready to compete.

Your Challenge

Using the problem-solving system outlined in Chapter 3, design and construct a vehicle that can pull another student's vehicle across a center line, using an attached line. The vehicles must begin the competition near the center line. Tow lines are attached to the rear of each vehicle. At the end of 20 seconds, the marked center of the string will be checked to determine the winner.

Procedure

1. Make a full-size drawing of the top view of your vehicle.
2. Draw the motor placement and pulley sizes. (Hint: You can reduce the rpm of the final drive pulley by increasing its size. (This will also increase the torque or pulling power at the wheels.)
3. Make the chassis for your vehicle.
4. Turn the wood wheels and pulleys on the metal lathe, using the jig supplied by the teacher.
5. Mount the motors on the chassis by bending sheet metal strips over the motor and fastening them to the chassis. Be sure that the strips are long enough to be attached to the chassis.
6. Consider bearing surfaces. Mount the pulleys and wheels. Soft copper tubing can be used for bearings. Welding rod makes excellent axles.
7. Run motor wiring neatly to the rear of your vehicle. Color code your wires so that when the control is hooked up, switches will start the motors in the same direction.
8. Design a towing subsystem that will make your vehicle better than the other vehicles.
9. Hook your vehicle to the challenger's vehicle and test your design skill.

Suggested Resources

2 toy dc motors
2 battery packs
8 'C' cells
Rubber bands (belts)
Sheet metal
Acrylic
Wood
Assorted fasteners
¼" OD soft copper tubing

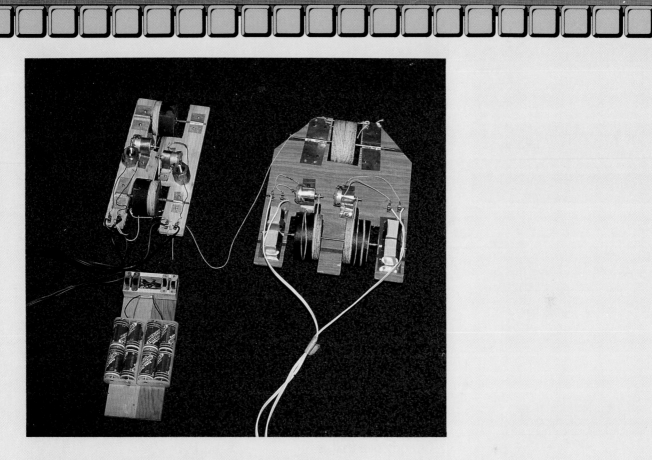

Technology Connections

1. Problem solving begins by identifying the problem. In this activity, what was the problem?
2. What goals did you set? What specifications did you have?
3. What alternative solutions did you come up with?
4. Why did you choose your solution?
5. What feedback did you receive? How would you change your solution for the next time?
6. Lubricants help reduce friction. Are there areas on your vehicle where you could apply lubricating compounds? What type of lubricant would you use?
7. Increasing the weight of the vehicle may give you an advantage. How will this help? What vehicles are designed with heavy chassis?

Science and Math Concepts

▶ The rate of doing work is called **power.** The basic unit of electrical power is the watt. 746 watts = one horsepower.
▶ The speed ratio of one pulley to another is found by dividing the diameter of the small pulley into the diameter of the large pulley. If the drive pulley is the smaller pulley, the result is a speed reduction. If the drive pulley is the larger pulley, there will be a speed multiplication.

REVIEW QUESTIONS

1. What are the six steps in the problem-solving system?
2. What are five ways of coming up with alternative solutions?
3. You are in the market for a new AM-FM stereo. You would like to buy one that is rated at a power output of 50 watts per channel. You have $200 to spend, but you find that the stereo you want costs $300. What are some trade-offs you could make?
4. When we solve major technological problems in society, political and environmental factors must be considered. Identify three examples of major technological problems. Describe how political or environmental concerns would influence their solutions.
5. Your egg-transporting device provides you with feedback about its design: during tests, the egg falls off and breaks. What are five other kinds of feedback you could get from preliminary testing?
6. Feedback helps us optimize solutions to problems. Give an example of
   a. feedback you've received recently at school.
   b. feedback you've received from a parent or a friend.
   c. feedback that can help you solve a technical problem.
7. The automobile has become a very important system of transportation. Identify an output resulting from the development of the automobile that is
   a. expected and desirable.
   b. expected and undesirable.
   c. unexpected and desirable.
   d. unexpected and undesirable.
8. Using the basic system diagram, model the operation of a nuclear power plant. Indicate the input (desired result), the process, the output (actual result), monitoring, and comparison.
9. Why would engineers choose to model the operation of a space-based defense system before actually constructing it? What kinds of modeling techniques might they use?
10. Identify several subsystems of a radio communication system.

KEY WORDS

Actual results	**Feedback**	**Modeling**	**Specifications**
Alternatives	**Goals**	**Monitoring**	**Subsystem**
Control	**Implementation**	**Output**	**Systems**
Desired results	**Input**	**Problem solving**	**Trade-off**
Evaluation	**Insight**	**Process**	**Trial and error**

**SEE YOUR TEACHER FOR
THE CROSSTECH PUZZLE**

TOMORROW'S JOBS

In Chapter 1, the shift from an agriculturally based to an industrially based to an information-based society was described.

During the industrial age, many jobs required people to do physical labor, such as working on factory assembly lines operating large, noisy machines. In the information age, many good jobs require people with a great deal of knowledge. The jobs that pay the most are those that require people to use their heads, not their muscles. For example, engineers and computer programmers are paid more than clerks and fast food workers.

WHAT THE FUTURE WORK FORCE WILL LOOK LIKE

Workers can be divided into four categories. These categories are production workers, management and administrative workers, service workers, and professional and technical workers. During the industrial age, more people worked in production than in any other category. Today, there are fewer production workers, but more management and administrative workers, and more professional and technical workers. These workers have jobs that require knowledge and a good education.

WHAT EMPLOYERS LOOK FOR

Besides looking for workers with knowledge and skills, employers want to hire people who have good work attitudes. A good work attitude includes the ability to get along with other people, the desire to do a good job, and the ability to get things done on time. Sometimes, employers feel that a good work attitude is even more important than good technical skills. Employers can often teach their workers how to do technical things, but an employee with a good attitude toward work will always be an asset to a company.

EXAMPLES OF WORKERS WITHIN OCCUPATIONAL CLASSIFICATIONS

Production Workers

Laborers
Machine operators
Assemblers
Welders
Machinists

Management, Administrative, and Clerical Workers

Managers
Proprietors
Clerical workers
Computer operators
Retail trade workers
Finance workers

Service Workers

Fast food workers
Hospital workers
Security guards
Personal service workers (hair stylists, tour guides, etc.)

Technical and Professional Workers

Programmers
Engineers
Technicians
Teachers
Lawyers
Health care workers

The graph shows the continuing decrease in workers engaged in production, an increase in management and administrative workers, average growth for service workers, and a steadily increasing growth in technical and professional jobs. (Data from Dr. Dennis Swyt, National Bureau of Standards.)

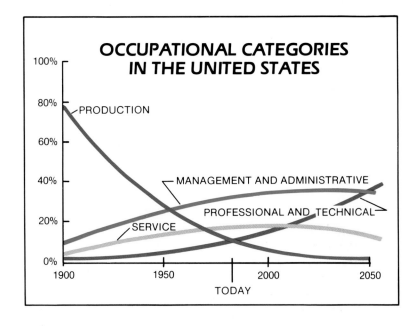

OCCUPATIONAL CATEGORIES IN THE UNITED STATES

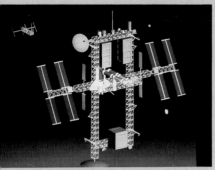

(Courtesy of NASA)

Equipment and Supplies

Model lumber
Toothpicks
Empty plastic pop bottles
Straws
Lego® Systems
Aluminum foil
Aluminum foil trays
Laminated foil or mylar
Glue sticks
White glue
PVC glue
Latex paint
Spray paint
L'eggs® pantyhose containers
Construction paper
Cardboard
PVC pipe
Transfer letters
Cylindrical containers
Sandpaper
Background music tapes
Blank VCR tape
VCR camera
Tape recorder
Glue gun
Coping saw
Scissors
Computer

SPACE STATION

Objectives

When you have finished this activity, you should be able to:
- Identify the modules and main components of the space station.
- Describe the function of modules and components of the space station.
- List justifications (reasons) for the space station.
- Design and construct a model of the space station.
- Discuss international cooperative efforts in the planning and construction of the space station.

Concepts and Information

Rockets, space shuttles, space stations, and the national aerospace plane all are a part of the Space Age. During the Space Age, technologies have increased at a very high rate. Prior to this time in history, technologies existed from the very simple ones of the Stone Age to the more complex during the Industrial Age. Just as in the past, technological systems in this Space Age require resources. These resources are people, information, materials, tools and machines, energy, capital, and time. The space station, a transplant of science and technology in outer space, is one of our most advanced technological systems.

Skylab was the United States' first space station. It was a converted third stage of a moon rocket. Three crews of astronauts visited Skylab during its existence.

The Soviet Union has a space station in orbit that they intend to maintain permanently. The Soviet space station is known as Mir, a Soviet word for peace. It is to become a complex of space-based factories, construction and repair facilities, and laboratories.

On January 25, 1984, President Ronald Reagan announced a ten-year plan to develop a permanent space station. Along with the United States, Japan, Canada, and the European Space Agency would contribute in this international effort to launch a space station in early 1990s.

Maintaining United States leadership in space is the main reason for our commitment to this endeavor. NASA (National Aeronautics and Space Administration) sees the space station as a facility in space with many purposes. The space station will be made up of laboratories for scientific research, observatories to study the earth and the sky, and a garage-like facility to repair and service satellites and spacecrafts.

The space station will have as one of its modules a manufacturing plant that will produce metallic super alloys for construction. This plant will also produce pure glass for laser and optical uses, super-pure pharmaceutical chemicals, and superior crystals for electronic systems. Another important reason for the space station is to further develop space-based communication systems. It also will serve as a construction site to assemble structures too large to be carried by a space shuttle. In the future, it can serve as a base for vehicles that will send and retrieve payloads to and from a higher orbit.

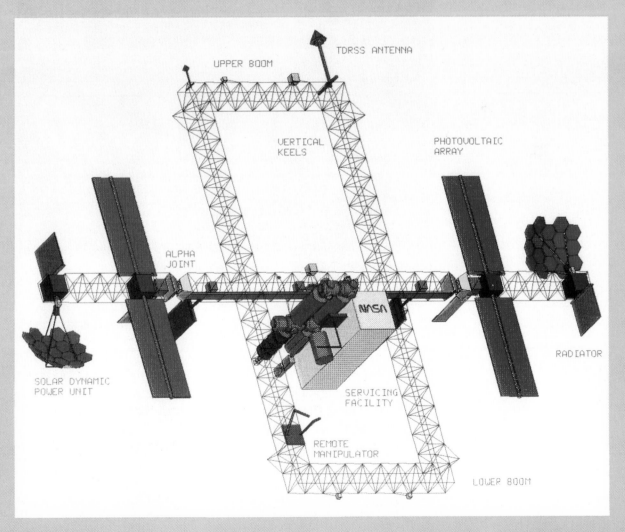

(Courtesy of NASA)

As shown, the space station features a dual keel or latticework beam structure. The two vertical beams are joined at the top and bottom by horizontal beams. Near the middle is a long horizontal beam that supports the arrays of solar cells, which convert sunlight into electricity. It also supports a solar dynamic power unit that is located at the end of the horizontal beam. This system is made up of hexagonal mirrors that collect solar heat. Solar heat drives an electricity-generated turbine. Attached to the upper boom is the TDRSS (Tracking and Data Relay Satellite System) antenna. This is part of a satellite communication system that makes voice exchange and record data flow possible. Radiators for dissipating heat are attached to the horizontal beam that supports the array of solar cells (photovoltaic array).

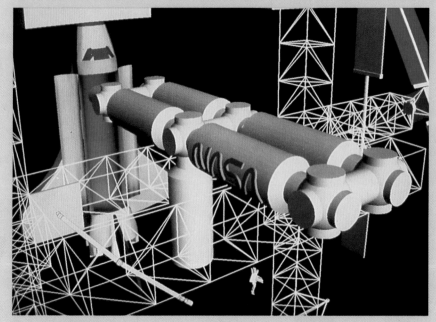

(Courtesy of NASA)

Near the center of the dual keels are the pressurized modules. These modules are linked together with external airlocks and tunnels. The modules, which are identical in external shape, are about 13.6 meters (45 feet) long and 5 meters (15 feet) in diameter. With atmospheric conditions near that of the earth, these cylindrical modules will serve as a habitat (living quarters) and laboratories. Astronauts will be able to work in shirt sleeves. Near the modules is the servicing facility where repair and assembly will occur.

Japan, Canada, and the European Space Agency are cooperating with the United States in the development of the space station. Japan is designing a research and development laboratory. The European Space Agency is also designing a laboratory and a platform. Canada is designing a mobile servicing center equipped with a manipulator arm.

This remote manipulator arm will be used to assemble and maintain the space station.

The space station is a venture into the future where vehicles will roam through space collecting and transmitting information back to earth, lunar, or space stations. As an international effort, it will be the beginning of a mission to Mars. The space station will be a future factory where pharmaceuticals, pure glass, super metallic alloys, and superior crystals are produced. As an outpost in space, it will continue the exploration of the newest frontier.

Activity

Space stations are similar to a service station. You can go there to refuel your vehicle, have it serviced or repaired, or get supplies. Parts of the service station are operated by people. Other parts of the service station are maintained automatically. Like a service station, space stations have communication and energy systems. A service station is usually contained on a pad of concrete and asphalt. Gravity holds all the parts onto the concrete and asphalt. Space stations exist over 300 miles above the earth where gravity is very weak. Modules and components must be attached to some form of framework. All parts of the station in the sky must be shuttled up in a fixed position in space. Using this information, design and build a model of a space station that could be used in a proposal you present for a real space station.

(Photo by Dennis Moller)

Procedure

Space Station Design

After you have read the concepts and information, you will become a member of a team. As a team you are to:

1. Study the concepts, information, and literature provided.
2. Select a name for your space station team. Each member of the team is to submit one design. As a team, choose one of the designs. One member of the team is to sketch the final form of the design the team has selected. Use the computer to generate a drawing of the space station.
3. Decide upon materials, tools, and equipment needed. List the materials needed and get them from your instructor or bring them from home.
4. Whether it be a manufacturing plant on earth or a space station in outer space, when a company wishes to construct the structure or some component, a proposal or bid is submitted. Your team will go through a similar process. As you formulate your ideas, begin a proposal for your team's concept of a space station. Include the following:
    a. Front cover for the proposal (be sure you have chosen a name for your space station team).
    b. List of subcontractors (family members or a member of the community who helps you in any manner).
    c. Drawing of the space station.
    d. List of all materials, tools, and equipment you will need to use.
    e. Procedure you will follow in constructing the space station.
    f. List of all modules and components, and functions of each.
    g. List the desirable and/or unique features of your space station.
5. After receiving proper instruction on safety and approval from your instructor, construct the framework to which all modules and components will be attached. Wear safety goggles.
6. Construct modules and components for the space station.
7. Paint all structures, modules, and components of the space station. (Follow manufacturer's requirements for ventilation.)
8. Attach transfer labels to identify your space station and its parts.

Space Station Proposal

1. Using the materials collected under Item 4, have a member of the team use a word processor and graphic program to write up your proposal.
2. Write up a script you can follow in presenting your concept of the space station on videotape.
3. Design and construct a setting for the videotape presentations. (One setting can be used by all teams.)
4. Rehearse the presentation. Have your instructor or a member of the team videotape the presentation.

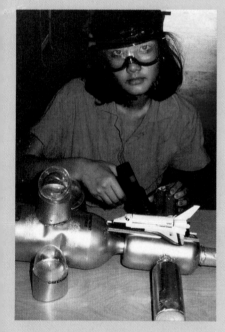

(Photo by Dennis Moller)

(Photo by Dennis Moller)

Review Questions

1. What are the functions of the following modules and components of the space station: habitat; laboratories; servicing facility; solar dynamic power unit; vertical keels; photovoltaic arrays; remote manipulator; radiator; and TDRSS antenna.

2. List eight possible uses of the space station that would justify its existence.

3. Which foreign countries are involved with the United States in the construction of the space station? What are the contributions of each?

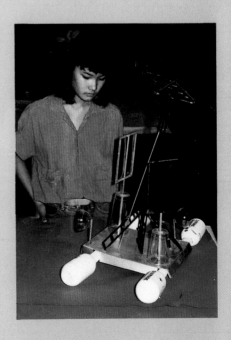

(Photo by Dennis Moller)

SECTION 2

COMMUNICATION

CHAPTER 4

THE ELECTRONIC COMPUTER AGE

MAJOR CONCEPTS

After reading this chapter, you will know that:

- The use of electronics has revolutionized all aspects of technology.
- Electric current is the flow of electrons through a conductor.
- Electronic circuits are made up of components. Each component has a specific function in the circuit.
- Integrated circuits are complete electronic circuits made at one time on a piece of semiconductor material.
- Computers are general-purpose tools of technology.
- Computers use 1s and 0s to represent information.
- Computers have inputs, processors, outputs, and memories.
- Computers operate under a set of instructions, called a program. The program can be changed to make the computer do another job.
- Computers can be used as systems or can be small parts of larger systems.

ELECTRONICS

The use of electronics has revolutionized all aspects of technology during the last hundred years. This has happened because people have learned how to use electricity to work with information. Information represented by electrical signals can be quickly and easily sent over wires or through the air by radio. It can be shared by many people at the same time, or privately between only two people. Electronics also allows people to communicate with machines. Such communication makes machines even more useful.

The use of electronics has revolutionized all aspects of technology.

Electronics in Our World

The electric light has extended our day. (Courtesy of General Signal Corporation, Stamford, CT)

Electric appliances in the kitchen allow us to keep our food longer and prepare it more quickly. (Courtesy of Maytag)

Other electric appliances have reduced the amount of time we spend on work, giving us more free time. (Courtesy of Sears Roebuck & Company)

Electronic entertainment has changed the way we spend our free time. (Courtesy of RCA)

THE SMALLEST PIECES OF OUR WORLD

All materials are made up of tiny particles called **atoms.**
Materials that are made up of atoms of only one type are
called **elements.** Iron and carbon are examples of elements.
Only 103 elements are known to exist. An atom is the smallest piece of an element that can exist and still have all the
properties of that element. Atoms are so small that you cannot see them, even with the most powerful optical microscope.

Often, atoms of different types are combined to form
compounds. When a compound is formed from two or more
elements, it has properties all its own. It may be nothing
like any of the elements that make it up. For example,
sodium (a metal that is poisonous if eaten and that reacts
violently when it comes in contact with water) combines
with chlorine (a poisonous gas). The compound formed is
sodium chloride, which is better known as common table
salt. Because of the many ways elements can be combined,
millions of compounds can be made from the 103 elements.

Atoms are made up of even smaller particles. These particles, however, do not have the properties of the element.
They have their own general properties. Atoms have a
heavy center part called a **nucleus.** The nucleus is made up
of **protons** and **neutrons.** Small particles called **electrons**
circle the nucleus very rapidly.

The number of protons in an atom determines the kind of
element it will be. For example, atoms with 13 protons in the
nucleus are aluminum. Atoms with 29 protons in the nucleus
are copper. Other elements you may know are oxygen and
nitrogen, which make up the air we breathe, and metals
such as gold, silver, nickel, and lead.

Both protons and electrons have electric charges. Protons are positive, and electrons are negative. Neutrons have
no charge (that is, they are neutral). In most natural atoms,
the number of protons equals the number of electrons. The
positive charges cancel out the negative charges, leaving
the atom with no charge.

Particles with opposite charges are attracted to each
other. Particles with the same charges repel each other
(push each other away). Negative electrons are thus attracted to the nucleus because of the protons' positive charge.
However, the rapid motion of the electrons around the nucleus keeps them from crashing into the nucleus.

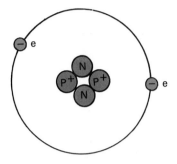

A helium atom has two
protons and two neutrons
in its nucleus. The atom
also has two electrons in a
shell (orbit) around the
nucleus.

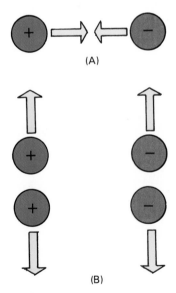

Objects with different
charges attract each other
(A). Objects with similar
charges repel each other (B).

ELECTRIC CURRENT

Electric current is the flow of electrons through a conductor.

In some atoms, the electrons are held tightly to the atom. In other atoms, some of the electrons are easily pulled away from the atom and may move from one atom to another. Materials whose atoms give up some electrons easily are called **conductors.** Materials that hold tightly to their electrons are called **insulators.** In a conductor, electrons can move from one atom to another. In an insulator, each atom's electrons are tightly bound. An insulator's electrons do not move freely from one atom to another.

Electrons may flow through a thin wire, a solid piece of material, or even through the air (a lightning bolt is a good example of this). The measure of electric current flow is the **ampere, or amp.** Smaller units of measurement are the milliamp (one-thousandth of an ampere) and the microamp (one-millionth of an ampere). One ampere is equal to about six billion billion (6×10^{18}) electrons flowing past one point in one second.

When you want to move something along the surface of a table, you have to push on it, or exert a force on it. In a similar way, to get a current to flow (to get electrons to move from one atom to another within a conductor), a force has to be applied. This force, which is called an **electromotive force,** is measured in **volts.** Without voltage, no current will flow.

While many materials are conductors, some are better conductors than others. For a given voltage (force), more current will flow through a good conductor than will flow through a poor conductor. **Resistance** is the opposition to a flow of current. It is the measure of how good a conductor is. A material with a high resistance is a poor conductor. A material with a very low resistance is a good conductor. The unit of resistance is the **ohm.**

In a metal wire, electrons are free to move from one atom to the next. Electric current is the flow of electrons through the wire.

A battery is a source of electromotive force. You probably know that a battery has a positive terminal and a negative terminal. When a battery is connected to a conductor, such as a light bulb, negative electrons in the wires are attracted to the positive terminal (opposites attract each other) and repelled from the battery's negative terminal. This is how a current, or flow of electrons, is set up.

Electrical Current Flow

The flow of electrical current through a wire is similar in concept to the flow of water through a thin pipe. In an electric circuit, the electromotive force might be supplied by a battery (A). In the water pipe, the force might be applied by a person pushing a piston (B).

In both cases, a greater force will make a larger current flow. With the same force, if the water pipe is made bigger in diameter, it will offer less resistance to the water, and more water will flow (C). In a similar way, if a larger wire is used, it will offer less resistance to electron flow, and more electrical current will flow.

The equation that describes current flow in an electrical circuit is called Ohm's Law. It states that

$$\text{Current (amps)} = \frac{\text{Voltage (volts)}}{\text{Resistance (ohms)}}$$

If voltage (force) gets larger, current gets larger. If resistance gets larger, current gets smaller.

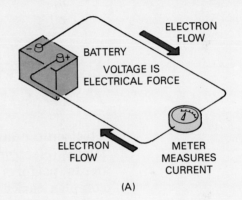

(A)

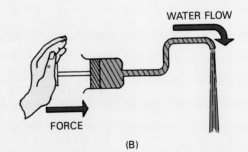

(B)

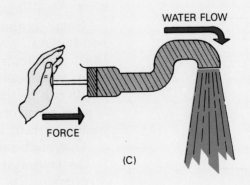

(C)

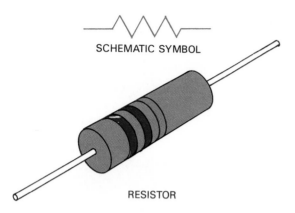

SCHEMATIC SYMBOL

RESISTOR

On many resistors, color-coded bands indicate the resistance in ohms.

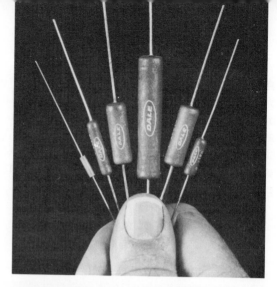

Resistors come in many shapes and sizes. They are used in many different kinds of circuits. (Courtesy of Dale Electronics, Inc., Columbus, NE)

ELECTRONIC COMPONENTS AND CIRCUITS

Electronic **components,** or parts, control the flow of electricity (electrons) and perform useful tasks. Some components perform very simple tasks. Others perform very complex tasks. By connecting the components together in different ways, we can make **circuits.** Circuits perform a wide variety of useful functions. Designers show circuits by using special circuit drawings called **schematics.** In a schematic, each component is shown by a **symbol.**

Electronic circuits are made up of components. Each component has a specific function in the circuit.

One of the simplest electronic components is the **resistor.** A resistor has a very well-defined resistance and can be used to control current flow in a precise way. Using a simple equation called Ohm's Law, a circuit designer can always calculate the value of the resistor needed to limit the current in a circuit to any desired value. Resistors are available in a wide range of values. They vary from less than one ohm to tens of millions of ohms.

Some materials are neither good insulators nor good conductors. These materials are called **semiconductors** (half conductors). The most widely used semiconductor element is **silicon.** Many different types of electronic components can be made using semiconductors. One very common type of semiconductor component is called a **diode.** A diode allows current to flow in one direction but not in the other.

One of the most important electronic components today is the **transistor,** which was invented in 1947. The word "transistor" is short for "**trans**fer re**sistor.**" A transistor

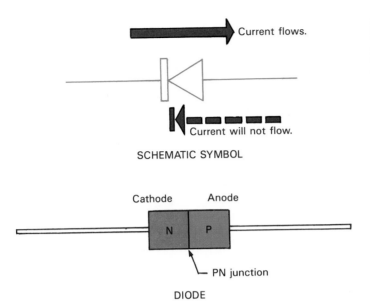

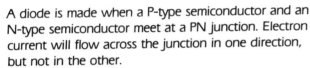

SCHEMATIC SYMBOL

Current flows.

Current will not flow.

Cathode   Anode

N   P

PN junction

DIODE

A diode is made when a P-type semiconductor and an N-type semiconductor meet at a PN junction. Electron current will flow across the junction in one direction, but not in the other.

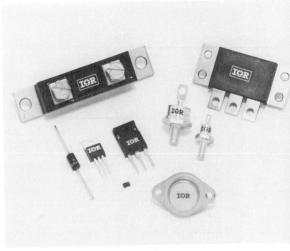

Small diodes are used in digital circuits and in radio detection circuits. Large diodes are used in circuits that supply power or control large currents. (Courtesy of International Rectifier)

allows a small amount of current to control the flow of a much larger amount of current. This control capability can be used to control large amounts of current, as in the control of an electric motor. It can also be used to control the storage of a small amount of electric charge used to represent information, as in a computer memory circuit. The transistor itself is very small (a small square wafer, several thousandths of an inch long by several thousandths of an inch wide). It is usually packaged in a larger metal or plastic container for ease of handling.

There are many components that are similar to resistors and transistors, but are made to perform very special jobs. For example, a **thermistor** is a resistor whose resistance changes with temperature. Thermistors can be used to make electronic thermometers, or as sensing elements in feedback control circuits for ovens or refrigerators. A **photoresistor** is a component whose resistance changes with the amount of light hitting it. Photoresistors can be used to automatically turn on lights when it gets dark. They also measure light in other applications.

Many other components are available for use in building electronic circuits. These include capacitors, inductors, LEDs (light-emitting diodes), which glow red, yellow, or green when current flows through them, batteries, and electromechanical devices, such as switches, motors, and generators.

Transistors

A transistor has two PN junctions. The transistor pictured is an NPN type because it has a thin P-type semiconductor sandwiched between two N-type semiconductors. PNP transistors are made in the same way, but with an N-type semiconductor in the middle. A small amount of base current in the transistor will control a much larger collector current. Collector current can be made larger, smaller, or even turned off by a small amount of current change at the base.

Even though transistors are very small (.01 inch × .01 inch is not uncommon), they are put into larger containers for protection, ease of handling, and removal of heat.

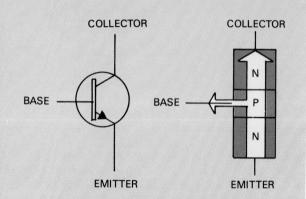

(Courtesy of Hewlett-Packard Company)

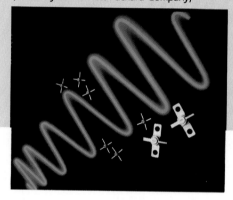

A soldering iron melts the solder, joining the wires to the component terminals. Soldering provides a good mechanical and electrical connection. (Courtesy of the Cooper Group)

Printed Circuits

Circuits are groups of components connected together to perform a specific function. Components may be connected together with wires. The wires are often soldered to the components. **Soldering** is a method of joining two wires together by melting a metal called solder on them. Solder has a very low resistance, so it makes a good electrical connection between the wires.

When wires are used to connect different components, care must be taken to insure that the wires do not accidentally touch each other. This would cause an unintended flow of current from one part of the circuit to another (a short circuit). To prevent this, wires often have a cover of insulation on them. The insulation keeps current from accidentally flowing from one wire to another.

Early electronic circuits used large components that were connected to other components by several wires. Each wire

Components are inserted into holes in the printed circuit board and then soldered in place. (Courtesy of Universal Instruments)

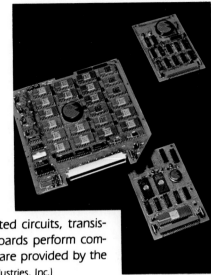

These completed printed circuit boards contain integrated circuits, transistors, resistors, and other electronic components. The boards perform complete circuit functions. Connections to the components are provided by the paths plated on the boards. (Courtesy of Tadiran Electronic Industries, Inc.)

was soldered by hand at both ends. Electronic components, however, have become smaller over the years. It has become more difficult to hand solder wires to the components in the small spaces available. In addition, manufacturers desired a method for rapidly making the same circuit over and over again, without errors, quickly, and at low cost. To meet these needs, the printed circuit board was developed.

A **printed circuit board** is a thin board made of an insulating material, such as fiberglass. On one or both sides, a thin layer of a good conductor, usually copper, is plated right on the board. Patterns etched in the copper form conducting paths. Holes for mounting components are drilled in the board. The components are then soldered to the conducting paths on the board.

The conducting paths are photographically placed on the board. Therefore, many boards can be made with the same circuit appearing every time, with no mistakes or variations. After all the components are mounted on the board, they can all be soldered at one time by an automatic soldering machine. This machine produces rapid, uniform soldered connections. Almost any type of component can be mounted on a printed circuit board.

Integrated Circuits

One of the most important inventions of the twentieth century was the integrated circuit. An **integrated circuit** provides a complete circuit function on a single piece of semiconductor material. Integrated circuits are often less than one-tenth of an inch long by one-tenth of an inch wide.

Size and Complexity of Integrated Circuits

Integrated circuits have become increasingly complex since they were invented in 1958. Single circuits are able to perform more tasks and more complex tasks. One measure of a chip's complexity is a count of the number of transistors it uses. People often talk about four types of integrated circuit size and complexity:

1. SSI—Small Scale Integration: several dozen transistors
2. MSI—Medium Scale Integration: up to several hundred transistors
3. LSI—Large Scale Integration: up to several thousand transistors
4. VLSI—Very Large Scale Integration: up to one hundred thousand or more transistors.

Each of the squares on these round wafers is a complete LSI integrated circuit containing thousands of transistors. Wafers are usually two to four inches in diameter and contain dozens or hundreds of integrated circuits.

(Courtesy of Matsushita/Panasonic)

Transistors, diodes, resistors, conducting paths, and other circuit components are made on the integrated circuit, or **chip,** at the same time.

A chip is laid out by an engineer. The engineer makes a drawing of the chip, several hundred times larger than the actual circuit. When the engineer is satisfied that the layout is correct, it is photographically reduced. This photographic reduction is called a **mask.** The mask is used to put patterns on a thin wafer of semiconductor material by means of a process similar to making a photograph. Many tiny identical circuits are made at one time on a round wafer, which is several inches in diameter.

On the very largest integrated circuits, entire computers can be built, as can other very complex, special-purpose circuits. Computers that occupied entire rooms only twenty years ago now fit onto the top of one desk, thanks to integrated circuits. In addition, whole assemblies that were once difficult to design, build, and test are now routinely used as small, inexpensive components in larger systems.

At the same time as the integrated circuit was providing more and more functions in a smaller and smaller space, the cost per function was also coming down very quickly. If similar performance advances and price decreases had occurred in the automobile industry over the last ten years, a Rolls Royce would today cost $500 and would get 1,500 miles per gallon.

Analog and Digital Circuits

Information can be represented by electricity in several ways. One often-used method is to have a voltage vary directly according to the information it is representing. For example, suppose a voltage represents a person's speech. The voltage would get larger as the person talked louder. This voltage is said to be an analog of (similar to) the speech it is representing. It is called an **analog signal.**

An **analog** is one thing that is similar to another thing. For example, a soccer ball is an analog of the Earth. The two are similar in shape, but they are obviously quite different in other ways. An electronic circuit that works with analog signals is called an **analog circuit.** Voltages in analog circuits usually change very smoothly, as do the things in nature that they represent (such as the loudness of a person's voice).

When information must be very accurate or must be sent over long distances, analog circuits are sometimes not good enough. Because they sometimes have to deal with small voltages (representing quiet music or speech, for example), they are subject to noise. If the information must be represented very accurately, a different technology, called **digital technology,** is used. Digital circuits are not as subject to noise as analog circuits.

In digital circuits, information is first coded into a series of 0s and 1s. In the simplest code, a voltage above a certain value is coded as a 1. A voltage below that value is coded as a 0. Each 1 or 0 is called a **bit,** which is short for **binary digit.** Binary refers to the number system that has only two numbers, 0 and 1.

COMPUTERS

One field that electronics and microelectronics have changed radically is the field of computers. For thousands of years, people have used machines to help with arithmetic calculations. Some early calculators were the Chinese abacus,

Integrated circuits, like transistors, are packaged in larger plastic or metal containers for protection and ease of handling. Often, the more complex the integrated circuit, the more input and output connections it needs. This requires the package to be larger. (Courtesy of Hitachi America, Ltd., Semiconductor & IC Division)

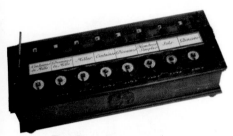

The Pascaline was a mechanical adding machine invented by Blaise Pascal in 1645.
(Courtesy of The Computer Museum, Boston, MA)

still used today, and Napier's bones, invented in 1617 to perform multiplication. The first mechanical adding machine was invented by Blaise Pascal in 1645. It used sets of wheels, activated by a needle, to add numbers together and indicate a result.

The first general-purpose calculator was developed in the mid-1800s by Charles Babbage. His "analytical engine" could be instructed, or programmed, to perform different sets of mathematical operations. Ada Lovelace, Babbage's co-worker, is credited with being the world's first computer programmer.

Herman Hollerith devised a way of automating the U.S. Census of 1890. Information was coded in different positions and combinations of holes punched into cards. The information could then be tabulated electrically by a machine, rather than by hand. Hollerith's census tabulator was a complete system. It had a machine to punch the cards (input), a tabulator for sorting the cards (processing), and a counter to record the results (output). A sorting box rearranged the cards for reprocessing (feedback). In 1911, Hollerith's Tabulating Machine Company became a division of a company that later became International Business Machines Corporation (IBM). IBM is the largest computer company in the world today.

The ENIAC

The ENIAC (Electronic Numerical Integrator And Computer) used 18,000 vacuum tubes. It was ten feet tall, three feet deep, and 100 feet long. The vacuum tubes gave off light and heat that attracted moths. The moths became entangled in the wires and moving parts of the electromechanical relays. People had to regularly clean the moths out, a process known as "debugging." Modern computers don't have vacuum tubes or relays, but people still say they are "debugging" the computer when they are finding and fixing problems.

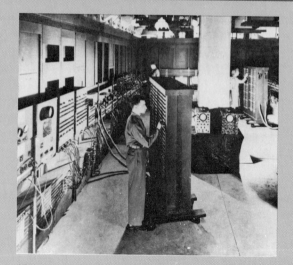

(Courtesy of Sperry Corporation)

The rapid development of integrated circuits over the last twenty years has made computers available to everyone. The electronic computer has been in use since the late 1940s. However, it was not until entire computers could be put onto one chip (the **microcomputer**) that they came into household use.

Today, computers and microcomputers surround us, even though we don't always recognize that they are there. Microcomputers are used in many electronic games, including the ones found at video arcades. They are found in machines used to conduct business, such as automated bank teller machines (ATMs) and cash registers. They are found in small appliances around the home, such as programmable ovens and VCRs. They are found in automobiles, where they control the amount of gas and air entering the engine. They are also, of course, found in home computers used for playing games and word processing.

Most of the computers that we are familiar with are very small computers, even though they may be very powerful. There are also very large computers, which are used by government, businesses, and researchers to do large, complex jobs. For instance, such computers might record the income tax forms sent in each year by all taxpayers or help engineers design new airplanes. These large computers have many things in common with the very small computers we find around us in our homes.

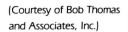
(Courtesy of Bob Thomas and Associates, Inc.)

(Courtesy of The Maytag Company)

(Courtesy of Whirlpool Corporation)

Computers, large and small, are used in all aspects of our daily lives.

WHAT IS A COMPUTER SYSTEM?

Computers can add, subtract, multiply, and divide very quickly. Calculators can do the same tasks with similar speed. Computers can control the operation of an automobile engine. Car engines, however, work fairly well with mechanical controllers. What makes a computer unique?

A computer performs its work according to a list of instructions, called a **program.** The program can be changed at any time. The computer is thus a general-purpose tool. A programmer makes it do a given job by providing it with the proper instructions. If the program (set of instructions) is changed, the computer can do a different job. The computer is thus under **program control.**

Computers are general-purpose tools of technology.

Computers use 1s and 0s to represent information.

Almost all computers in use today are digital electronic computers. In digital computers, information is represented in binary form as bits (1s and 0s). Any number of quantity, no matter how large or how small, can be represented by a binary number, a sequence of 1s and 0s. Because the information is stored and worked on digitally, it remains very accurate.

Bits are organized into groups of eight to make them easier to deal with. These groups of eight bits are called **bytes.** Each byte can represent one of 256 different characters (numbers, letters, punctuation, or other information). In talking about computers, we often talk about amount of data in terms of bytes, kilobytes (one KByte = one thousand bytes), and megabytes (one MByte = one million bytes).

DECIMAL	BINARY
0	0000 0000
1	0000 0001
2	0000 0010
3	0000 0011
4	0000 0100
5	0000 0101
6	0000 0110
7	0000 0111
248	1111 1000
249	1111 1001
250	1111 1010
251	1111 1011
252	1111 1100
253	1111 1101
254	1111 1110
255	1111 1111

This chart shows some conversions from decimal (base ten) numbers to binary (base two) numbers. Eight-bit binary bytes are shown.

The Computer Processor

All computers have certain common component parts. The first of these is the **processor.** The processor is the heart of the computer. It controls the flow of data, the storage of data, and the way the computer works on the data. The processor reads the program (set of instructions) and converts the instructions to actions. These actions might include adding two numbers, comparing two numbers, or storing a number or letter.

The **power** of a processor refers to how many instructions the processor can handle in a second. Personal computers can handle hundreds of thousands of instructions per second. Large computers used in business or research can handle millions of instructions per second (MIPS). Very fast computers handle instructions through different paths at the same time. This increases the number of instructions handled per second. These very fast computers can handle hundreds of millions of instructions per second.

Memory

The place where the program is stored is called the **memory.** The memory also stores the information being worked on at any one time. Almost all modern computers use integrated circuit memory circuits, sometimes referred to as semiconductor memory. Today, a single chip 3/8 inch × 5/16 inch can store up to 125,000 characters. New memory chips are being developed with even higher capacities.

The **main memory** stores the program and the information currently being worked on. When a computer is referred to as a 64 KByte computer, the size of the main memory, or **RAM**, is being described as 64 kilobytes. Modern personal computers have main memory sizes of 8 KBytes (small lap-top portables) through more than 1 MByte (larger desk-top personal computers). Large computers used by businesses and universities have main memories of many megabytes.

In addition to the main memory, a computer needs to have another, much larger memory. This other memory, called **storage** or **secondary storage,** stores information for use at a later time. Secondary storage is designed to be

(Courtesy of Radio Shack, a division of Tandy Corp.)

(Courtesy of Apple Computer, Inc.)

(Courtesy of Prime Computer, Inc.)

These three computers have very different main memory sizes. The small lap-top computer has 24 kilobytes of memory; the desk-top personal computer has 128 kilobytes of memory; the large superminicomputer has up to 16 megabytes of memory.

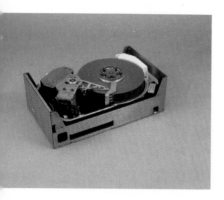

This fixed disk unit contains several disks stacked one on top of another. There is a separate head for each disk. (Courtesy of Pertec Peripherals Corporation)

very large so that many different types of information (or large amounts of the same type of information) can be stored.

Secondary storage includes floppy disks, hard disks, and magnetic tape. On each of these, data is stored magnetically. The surface of the disk or the tape (called the **medium**) is coated with a very thin layer of iron oxide, a magnetic material. A very small electromagnet called a **head** is placed near the tape or disk as the disk is moving. A voltage applied to the head will magnetize the tiny pieces of iron oxide. Information is thus put onto the medium. This is called "writing to disk" or "writing to tape." When information is needed from the disk or tape, the electromagnetic head senses the magnetic fields stored in the iron oxide coating of the medium. The head turns them into electrical impulses. This is called "reading from disk" or "reading from tape."

The amount of data that can be stored in secondary storage is almost limitless. If one disk or tape is full, it can be removed, and another one can be put in its place. A typical floppy disk, the secondary storage medium most often used with personal computers, can store up to 360 KBytes of information, or 360 thousand characters. This is roughly equivalent to 200 pages of typed text. A small hard disk, which is used in many personal computers, can hold up to 10 MBytes of information, or 10 million characters. This is over 5,000 pages of typed text. There are hard disk drives that contain several large disks, with a total storage capacity of several hundred MBytes. Tapes have similar, very large storage capabilities.

Another technology that holds a great deal of promise for the near future is **optical disk** storage technology. Optical disks can be used to store computer data, just as they are used in audio compact disks and in video disks (see Chapter 7). The emerging part of this technology is in how to write to the disk, in addition to reading from the disk. Optical disks

A single optical disk, 8 inches in diameter, can store as much information as 15,000 sheets of paper. The disk and its drive can retrieve the information within 0.5 second. (Courtesy of Matsushita/Panasonic)

have very large memory capability. Manufacturers who are developing them talk about billions of bytes, or gigabytes (GBytes).

Computer Input

In order for the information that the computer stores and processes to be useful, it has to be exchanged with people or other machines outside the computer. This exchange is called **input/output,** or **I/O.** There are many different ways we exchange information with a computer. The many I/O techniques have arisen because of the many ways computers are used.

Information provided to a computer is called **input.** The form of input that you are probably most familiar with is through a **keyboard.** A keyboard is very similar to a typewriter keyboard. People can use a keyboard to give instructions to (program) a computer. A keyboard can also be used to provide raw data to a computer. The data can then be processed by means of a program already stored in the computer. Using a keyboard, people can work with a computer on an **interactive** basis. Short instructions can be given, with immediate results coming from the computer.

Tapes, hard disks, and floppy disks can also provide input to a computer. Tapes and disks can contain a program or data. The computer can transfer, or **load,** the data into its main memory. Using tapes and disks, people can develop new programs for computers. The programs can then be sold and transferred to other people who have similar, or **compatible,** computers.

Still another form of input is through an optical character reader. This device can scan a page of printed text. It recognizes the letters and numbers on the page. It then converts them into a code of bytes that can be understood by the computer. This is one method used to put many pages of typewritten text into a computer's secondary storage for future use.

A more recently developed form of input, and one that is still being developed, is speech recognition. Some systems are equipped with special devices that can recognize some spoken words. The devices convert the words into a series of bytes. This technology is still in its developmental stage. The total number of words most devices of this kind can recognize is fairly small (up to several hundred). These devices work better with some people's voices than with others. It is expected that voice input will be a very important form of information exchange with computers in the near future.

Computers have inputs, processors, outputs, and memories.

Bar Code Readers

A specialized form of input device is the bar code reader. The bar code reader directly reads labels on boxes and various products. You may be familiar with bar code readers. They are used in supermarkets in many parts of the country today.

A special code, called the Universal Product Code (UPC), consists of a series of vertical lines of varying thickness. A UPC symbol is printed on each product's packaging. At the check-out counter, a light-sensitive device called an **optical scanner** picks up light reflected from the white spaces between the bars of the symbol. As the product is scanned, the black-and-white stripes are converted to on/off pulses of electricity.

Each product has its own code. The computer can determine which product it is, how much it costs, and whether tax should be charged for it. The computer shows the salesperson the price. In addition, the computer might also keep a record of how many of that product have been sold and whether more should be ordered from the supplier.

The Universal Product Code (UPC) uses thick bars, thin bars, and blank spaces to encode information about a product. (From The Way the New Technology Works by Ken Marsh)

Computer Output

Computer output can also be in many forms. Probably the most common is the **video monitor,** or **CRT screen.** CRT stands for cathode ray tube. A cathode ray tube converts electric signals to visual images. CRTs are used in televisions to produce the picture you watch. They are also used in electronic test equipment called oscilloscopes, as well as in video monitors. Video monitors can be used to display text (letters, numbers, and punctuation), graphics (pictures), or a combination of the two. They can be monochrome (black and white, green and white, etc.) or full color.

Often, a keyboard is combined with a monitor. This makes a device that can both provide input to a computer (from the keyboard) and display output from a computer (through the monitor). The combination is called a **terminal.** Sometimes, other input devices are combined with the keyboard to make special-purpose terminals. Examples of this type of input device are the **mouse,** the **light pen,** and the **touch sensitive screen.**

Another very commonly used computer output device is a **printer.** A printer records the computer's output on a piece of paper. The paper may then be mailed to someone else or saved in a filing cabinet. Such a paper record of a computer's output is called a **hard copy.** Printers used in small computer systems are usually **dot matrix printers** or **daisy wheel printers.** Dot matrix printers may be used to print letters, numbers, and punctuation. They may also be used to make drawings. Because they use dots, and not continuous lines, their letters and pictures are not of very high quality. However, the dot matrix printer is acceptable for many applications.

Special Computer Input Devices

Some computer input devices depend on an interaction among a person, a computer output device, and the input device. The mouse, for example, senses its position on a table and places an arrow or other indicator on the screen. The operator places the arrow at the desired location by rolling the mouse on the table. When the arrow is at the right spot, a push button can be pressed to issue a command.

Another example of input/output interaction is the light pen. The light pen senses the light given off by the screen. The touch screen, another input device, can sense the approach of a finger.

A mouse combines output with input.
(Courtesy of Apple Computer, Inc.)

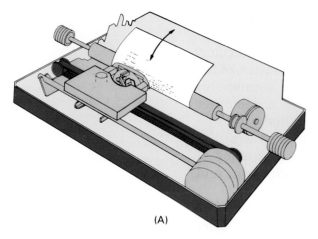

(A)

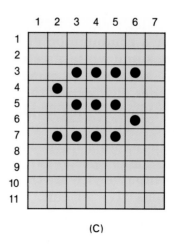

(C)

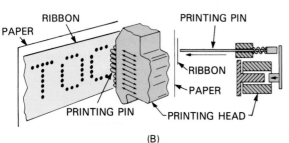

(B)

(A) A dot matrix printer has a print head that contains a group of pins. (B) The pins can be moved by electromagnets to strike a ribbon, making impressions on paper. (Parts A and B from Brightman & Dimsdale, *Using Computers in an Information Age,* © 1986 by Delmar Publishers Inc.) (C) The letter "s" is made by activating a number of pins in the right pattern. The matrix shown here is 11 × 7. Other matrix sizes are used.

Daisy wheel printers provide higher quality printing of letters, numbers, and punctuation. They cannot, however, create graphics. Daisy wheel printers are generally more expensive than dot matrix printers. They also usually operate somewhat more slowly.

Large computers use printers that can print much faster than the dot matrix or daisy wheel printers used with personal computers. These printers can print whole lines at a time (line printers) or whole pages at a time (page printers).

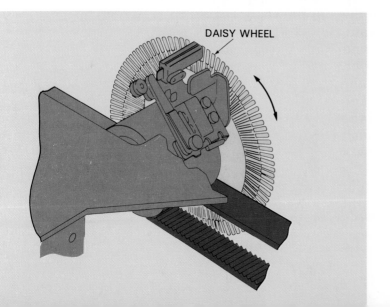

A daisy wheel contains many "petals." Each petal has a raised character on it. A daisy wheel printer positions the correct character, and strikes it. This pushes the character onto an inked ribbon, which leaves an impression on the paper. (From Brightman and Dimsdale, *Using Computers in an Information Age,* © 1986 by Delmar Publishers Inc.)

One of the most difficult problems with these printers is how to handle the paper they use, because it moves through the printer so fast. Another type of printer used with large computers is the **laser printer.** The laser printer can produce very high quality print and graphics at a very high speed.

Similar to a printer is another output device called a plotter. A plotter uses one or more pens. The position of the pens may be controlled by the computer to create drawings on a piece of paper. Plotters are often used in systems that create, modify, and store drawings (CAD, or computer-aided drafting systems).

Audio output takes the form of tones, beeps, music, and voice. Tones and beeps can be used to signal to the operator that the end of a page has been reached. They can also indicate that the operator has issued an improper command to the computer. Music can be generated inside the computer by a device called a **synthesizer.** A synthesizer can produce a wide range of tones, volume, music, and even percussion (sounds like drums). Speech can be produced by a voice synthesizer. Modern voice synthesizers can work with a large vocabulary. They actually use the rules of pronunciation to generate voice sounds. Telephone companies use voice synthesizers to give the time of day and to give directory assistance (information) phone numbers.

Computers can be connected to each other or to terminals located at a distant location. This kind of information exchange is called **data communications.** In order to use computers to send data communications, a special device called a **modem** is often used. Modems send computer data over standard telephone lines. Sequences of tones represent the bits of information to be sent. An inexpensive modem can send up to 1,200 bits per second over telephone lines. At 1,200 bits per second, one page of typed print would take about 12 seconds to send. With more expensive and complex modems, 9,600 bits per second can be sent over telephone lines. At 9,600 bits per second, it would take 1.5 seconds to send a page of typed print.

COMPUTERS, LARGE AND SMALL

Computers come in all sizes, from a single chip less than one square inch in size to a whole room full of large cabinets and equipment. Computers also come with a large variety of capabilities. Some are slow processors, while others are very fast processors. Some have a very small memory, while others have a very large memory. A user must select the right size computer for the job at hand. Sometimes it is useful to speak of a computer's size and capabilities by putting the computer into one of four categories. The categories are microcomputer, minicomputer, mainframe computer, and supercomputer.

Microcomputers are usually found in appliances, automobiles, and personal computers. They are sometimes as small as one chip. More often, however, they are a collection of integrated circuits including a microprocessor chip. Their memories are sometimes very small (one KByte), but memories may reach up to about one MByte in some "supermicrocomputers." Microcomputers handle data and instructions 8 or 16 bits at a time (1 or 2 bytes at a time). They are often referred to as "8-bit machines" or "16-bit machines," based on how many bits at a time the processor can handle. Some supermicrocomputers can handle 32 bits (4 bytes) at a time.

Minicomputers are somewhat larger than microcomputers. Minicomputers are typically used by several users in a small company, a department of a larger company, or a school. They handle 16, 24, 32, or more bits at a time. They often come with large disk or tape secondary storage devices.

Mainframe computers are the large computers used by large companies, government agencies, and universities for their administrative work. Such computers are used for making out payroll checks, keeping personnel records, keeping track of orders, and/or maintaining a list of all the items kept on hand in a warehouse. Mainframe computers handle data and instructions 32, 36, 48, and 64 bits at a time. Often they have very large secondary storage devices (hard disks and tapes) attached. Mainframes can execute millions of instructions per second.

A microcomputer
(Courtesy of NCR Corporation)

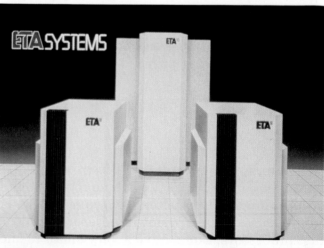

A supercomputer
(ETA Systems' ETA[10] Supercomputer)

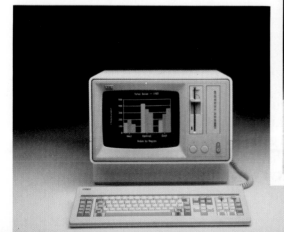

Supercomputers are the very fastest and largest computers. They are most often used for research, for analyzing satellite data, or for other very large problems. Supercomputer speed is most often measured by the number of complete multiplications or divisions (floating point operations) per second (FLOPS) the computer can perform. One example of a floating point operation is $1.23 \times 2.6 = 3.198$. Today's largest supercomputers can perform several billion floating point operations per second (GFLOPS). Supercomputers are very expensive to purchase and to operate.

The definition of each of these categories of computers is constantly changing as computers get more and more powerful. The power of a computer that was considered to be a supercomputer only twenty years ago is now available in a desk-top personal computer. People argue about whether this trend will continue through the year 2000. We can only guess what desk-top computers will be capable of then.

COMPUTER SOFTWARE

Computer systems are general-purpose tools of technology. How each one is used depends on its input/output devices and on how it is programmed. Computer software exists at different levels. Three important types of software are operating systems; applications programs that are purchased ready to use; and applications programs written by the user in a programming language.

The computer's **operating system** allows the user to control and access the computer's memories (disk, tape, semiconductor, and so on), printers, and other **peripheral** (attached) devices. The operating system also makes the computer's components available to other software, such as applications programs. Operating systems must be designed with the characteristics of the computer in mind, as well as the likely uses the computer will have. Sometimes, more than one operating system is available for a computer. The user must then choose which one is needed. The choice is based on the kind of tasks the computer will be expected to perform. Examples of operating systems used in personal computers are MS-DOS™ and TRSDOS™.

An applications program provides instructions to the computer to perform a specific, well-defined task. Applications programs include computer games, word processors, income tax preparation programs, car engine control programs, and rocket design programs. Applications programs are stored on tape or disk. They may be purchased for use on personal computers or for use on mainframe computers.

Computers operate under a set of instructions, called a program. The program can be changed to make the computer do another job.

When selecting an applications program, a user must make sure that it will work with (is **compatible** with) the computer's operating system. Many applications programs are available for use with a number of operating systems. These applications programs can then be used on different computers.

Sometimes an applications program is not available to do a job. In this case, a custom-written applications program may have to be written. This may be relatively easy to do, or it may be a very large effort. Some custom-written programs take many people several years to finish. Custom-written applications programs are written in one of many programming languages available today. Each programming language has unique features that make it useful for writing certain kinds of applications programs.

Today, the most commonly used programming language is **BASIC** (**B**eginner's **A**ll-purpose **S**ymbolic **I**nstruction **C**ode). It is used by students, businesspeople, and hobbyists. Its use is so widespread largely because it is relatively easy to learn and use. BASIC is available for use on most small personal computers as well as on large mainframe computers.

```
10 INPUT "What is your name ";N$
20 INPUT "Please tell me a number";A
30 INPUT "Please tell me another number";B
40 PRINT "Thank you,";N$;", the product of";A;"and";B;"is :";A*B;"."
50 PRINT "The Quotient of";A;"and";B;"is:";A/B;"."
60 PRINT "The sum of";A;"and";B;"is :";A+B;"."
70 PRINT "The difference between";A;"and";B;"is :";A-B;"."
80 END
```

This simple BASIC program instructs the computer to record the name of the operator and to ask for two numbers. The computer will then multiply, divide, add, and subtract the numbers.

(Program courtesy of R. Barden, Jr.)

```
RUN
What is your name ? Mary
Please tell me a number? 8
Please tell me another number? 2
Thank you,Mary, the product of 8 and 2 is : 16 .
The Quotient of 8 and 2 is: 4 .
The sum of 8 and 2 is : 10 .
The difference between 8 and 2 is : 6 .
```

The actual exchange between the operator and the computer is shown here. The entries in red are made by the operator. All others are output from the computer. The entry "RUN" starts the program.

Pascal is increasing in popularity as a general-purpose programming language. It is often taught to students as their second programming language, after BASIC. Pascal is named after Blaise Pascal, the mathematician who invented the first mechanical adding machine in the seventeenth century.

Other languages in common use today are **C, COBOL** (**CO**mmon **B**usiness-**O**riented **L**anguage), and **FORTRAN** (**FOR**mula **TRAN**slator). Another language, **ADA,** is named after Ada Lovelace, the first computer programmer. **LISP** (**LIS**t **P**rocessor) is a language of increasing importance. It is used in artificial intelligence programs.

Artificial intelligence is the imitation of human thought by computers. In most computer programs, instructions follow very strictly from one place to another, under the unchanging rules of the program. In artificial intelligence programs, answers may be arrived at with incomplete information. Sometimes, the answer arrived at is wrong. The computer, however, can learn from a mistake if it knows that it made one. Artificial intelligence is still in its infancy, but it is already being used to control manufacturing configurations and to make some business decisions. The Digital Equipment Corporation (DEC), a very large manufacturer of computers, uses an "expert system" (a form of artificial intelligence) to configure many of the computers that it sells.

COMPUTERS AND THE SYSTEM MODEL

A computer is a sophisticated technological system. Like other technological systems, it is helpful to think of its operation in terms of the system model. The very terms used for computer components are similar to system terms. Input devices provide both command inputs and resource inputs. For example, an operator at a terminal may provide a command input by keying in a short application program. He or she may provide a resource input by keying in data.

A computer's processor acts on the resources in response to the command. The processor is thus clearly the process in the system model. The output of the computer is also the output of the system. Feedback may be provided either by a person, or automatically through hardware or software, depending on how the computer is being used.

Computer Applications

Computers, as general-purpose tools, are often small parts of other, larger systems. In a manufacturing system, a computer may supply the command inputs to an automated production line. The same computer might also compare the output of the process (using appropriate sensors) with the desired input. The computer then indicates any necessary adjustments. In a bill processing or a mailing system that automatically addresses and mails thousands of letters per day, the computer actually forms the process portion of the system. Its output is envelopes, letters, and bills.

A computer is often used to help people form a feedback loop in a complex system. One example of this is a central control center in a transportation system, such as a train switching center, a freeway traffic control center, or an air traffic control system.

In some systems, a computer acts in several roles. An example is a microcomputer that controls a microwave oven. Based on information supplied by the cook, the computer sets a power level. The computer can also turn the oven on, sense when the food is done, and either shut off the oven or keep it on at lower power to keep the food warm.

A control computer for automated manufacturing and warehousing systems (Courtesy of Gould Inc.)

The control center for Buffalo, NY's light rail transit system. (Courtesy of General Signal Corporation, Stamford, CT)

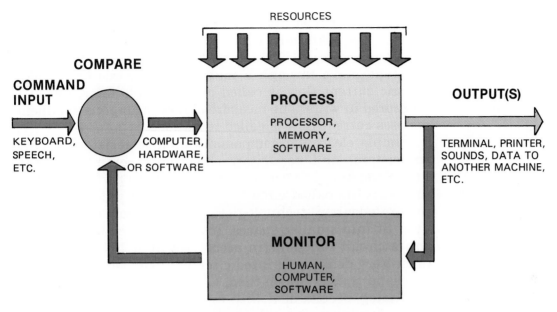

The general system diagram can be readily applied to computers.

Computers can be modeled as systems and are sometimes used as stand-alone systems. Very often, however, a computer is part of a larger system. In this case, the computer itself is only a small subsystem. It may provide an input to a larger system. It may be a part of the larger system's process, feedback, or comparison mechanisms. It may even provide the command input to a larger system. Because a computer can be made in many different sizes and can be programmed to perform a very wide variety of functions, it has become one of the most useful and widespread tools of the technological age.

Computers can be used as systems, or can be small parts of larger systems.

SUMMARY

The use of electronics has revolutionized all aspects of technology during this century. Electricity has been made to represent, store, change, and communicate information.

All materials are made out of atoms. Atoms contain smaller particles called protons, neutrons, and electrons. Protons have a positive electric charge, while electrons have a negative electric charge. Particles with similar charges repel each other. Particles with opposite charges attract each other.

Materials whose atoms give up electrons easily are called conductors. Materials whose atoms hold tightly to their electrons are called insulators. Electric current flows when electrons move through a material. The force that makes electric current flow is called electromotive force, and it is measured in volts. The measure of how strongly a conductor opposes current flow is called resistance.

Simple electronic components include resistors, diodes, transistors, and capacitors. Electronic components connected together in various combinations to perform specific functions are called circuits. Because of printed circuit and integrated circuit technologies, more and more circuits have been fit into smaller spaces. Integrated circuits implement entire circuit functions in areas less than 1/4 inch by 1/4 inch. Very dense integrated circuits have over one hundred thousand transistors on them.

Analog circuits deal with information in a continuously varying manner. Digital circuits deal with information that is converted into binary codes (1s and 0s). Information stored, processed, or communicated in digital form is less sensitive to noise than if it were in analog form.

Computers are general-purpose tools of the technological age. They can be programmed or instructed to do a specific task. Computers have processors, main memory, secondary storage, and input/output components. Computing power has been increasing exponentially for forty years, while the cost of many computer components has been falling exponentially.

Computers can be roughly categorized as microcomputers, minicomputers, mainframe computers, or supercomputers. The definitions of these four categories are constantly changing as computers become more capable. What was considered a supercomputer only twenty years ago is now available as a desk-top computer.

Computer software may be divided into operating systems and applications programs. Applications programs may be purchased "off the shelf" or may be custom-written in a programming language.

Like other technological systems, the computer may be described by the general system model. Computer terminology is very similar to system terminology, making it easy to model a computer as a system.

Computers may be configured in a variety of sizes. They can be programmed to perform specialized tasks. Because of this and because their cost has been going down at an exponential rate for decades, they have become very widely used. Today, they are used in appliances, automobiles, and entertainment equipment, as well as in the more traditional areas of word processing, engineering, and business.

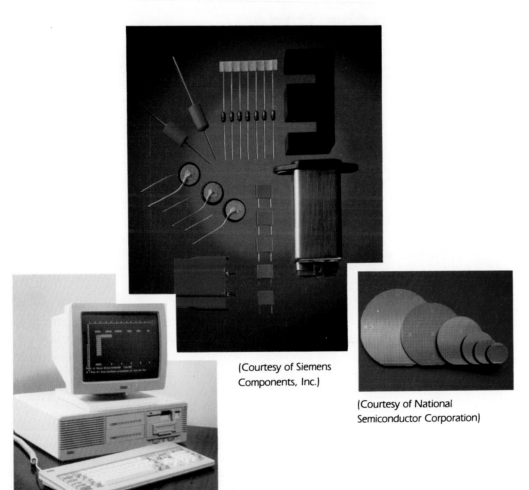

(Courtesy of Siemens Components, Inc.)

(Courtesy of National Semiconductor Corporation)

(Courtesy of NCR Corporation)

COMPUTERS IN INDUSTRY

Setting the Stage

Computers are an important tool for industry. They allow us to simulate real-life situations and to test ideas. They are also frequently used for computer-aided drafting. As a member of a design team, for example, you might be involved with computer designing of a new car. How can a computer help you develop and test your design before it is modeled?

Your Challenge

Operate a computer, disk drive, printer, and joystick or mouse. Use programs that require you to input information to make a more informed decision concerning technology projects. Examples include "The Factory," "Car Builder," and "Blazing Paddles."

Procedure

1. Insert "The Factory" disk into the disk drive and load the program. Use all three levels or files. The experience you gain from learning how to operate the machines, design a product, and understand how a product was built at the factory will help you in other technology activities.

   File 1: Test a Machine
   a. A punching, rotating, or striping machine is available for testing.
   b. Use the arrow keys to choose the process and input your choices with the RETURN or ENTER key.
   c. Once you have learned how to use each machine you will then be able to build a factory using the second file.

   File 2: Build a Factory
   a. Using the arrow keys, put up to eight machines together in an assembly line to produce a finished product.
   b. Once your factory is assembled, erase the assembly line, leaving the finished product. Then, you or a partner should attempt to reestablish the assembly line that produces the product.

Suggested Resources

Computer
Disk drive
Printer
Joystick or mouse
"The Factory"
"Car Builder"
"Blazing Paddles"

File 3: Make a Product

You will be shown a product that has been made by several machines. Reconstruct the sequence of machines and processes used in the construction of the project.

2. You have just experienced how a computer can help people in choosing processing techniques and sequence of operations. Now try to use a program for computer-aided drafting (CAD) and for testing a product. Load "Car Builder" into the computer.

File 1: By following the design sequence, design and construct a car. Chassis length, engine placement and size, type of steering, body shape, and types of tires are a sampling of decisions you will have to make.

File 2: The next file on "Car Builder" will test your car in a wind tunnel. The drag form factor will also be displayed for the car you designed.

File 3: Using the final file, road test the vehicle you designed. By following the directions on the disk, save and/or print your CAD vehicle.

3. Now try a program that will allow the printer to become a graphics plotter and instant drafting machine. Load "Blazing Paddles" into the computer.

By using a joy stick or mouse and following the directions, you will be able to produce quality drawings in a short time. There are hundreds of programs on the market, each with their own system of commands and operations to be performed. Many of them are complex, so stick with the ones that meet your requirements. "Blazing Paddles" will allow you to draw any plans you might need for the activities in later chapters.

Technology Connections

1. After using "Car Builder," model the body you designed. Test it in the lab wind tunnel.
2. Name and properly hook up all the computer accessories that you used for the above programs.
3. List some uses for the computer in our technological world.
4. What other communication systems are used in combination with the computer?

Science and Math Concepts

▶ One disk can hold more than one million magnetic charges or bits of information. There are eight bits in a byte of information. The middle section and two other small areas of the disk are exposed so the computer can read and write to the disk.

119

LOW-POWER RADIO TRANSMITTER

Suggested Resources

Safety glasses and lab apron

1 copper clad circuit board—2" × 3"

Direct etching dry transfers (Radio Shack #276-1577)

Etchant

Shallow pan for etching

Plastic funnel

Carbon paper

Steel wool (000)

Wire cutters

Soldering pencil and solder

Drill and drill bits

Aluminum or plastic case

The following electronic components:

T1—2N3906 PNP transistor

R1— 130 K ohm resistor

C3 & C4—.022 $\mu f d$ capacitor

C2—.0047 $\mu f d$ capacitor

C1—100 pfd ($\mu \mu f$d) capacitor

L1—Adjustable tapped loopstick antenna coil for broadcast band

L2—10 or 15 turns of #30 enameled wire around L1

B1—9 volt alkaline transistor battery with 9 volt battery snaps

M1—1000 ohm magnetic-type earphone or microphone with ⅛" phone plug

S1—Miniature spst toggle switch

ANT—30" telescoping antenna

⅛" miniature phone jack— open circuit

Setting the Stage

Electronics and radio communications were developed by people like Edison, Morse, DeForest, and Marconi. The devices used to transmit and receive data have since been much improved. Supersensitive receivers with huge antennas can now pick up weak radio signals from spacecraft millions of miles away.

Your Challenge

Construct a small radio transmitter that can send a message through an AM radio.

Procedure

1. Be sure to wear safety glasses and a lab coat.
2. Cut a 2" × 3" piece of single-sided copper clad circuit board with a squaring shear or fine-toothed saw. Clean with 000 steel wool.
3. Duplicate the printed circuit layout on the circuit board with direct etching dry transfers.
4. Etch the circuit. This takes about 20 minutes.
5. Remove the resist material and wash the circuit board in water. Again clean with 000 steel wool.
6. Drill ¹⁄₃₂" (or #52) holes in the board on the donut dots so that pigtail leads and components can be mounted. Also, drill ³⁄₁₆" diameter mounting holes in each corner of the board.
7. Wrap 10 or 15 turns of #30 enameled wire around L1.
8. Solder all components with a small soldering pencil. CAUTION: Excess heat will cause the copper to lift from the board and can destroy electronic components. Use a small alligator clip as a heatsink whenever possible.
9. When wiring the battery snaps in place, watch the polarity (+ or −). The transmitter will not work if the battery is wired incorrectly. DOUBLE CHECK your wiring.
10. Connect the antenna, microphone, and battery.
11. Tune an AM radio to a spot where there is no station.
12. Turn on your transmitter (S1) and adjust the tuning slug of L1 (the adjustable tapped loopstick antenna coil) until you hear a whistle in the AM radio.
13. Now speak directly into the microphone. You should hear your own voice.
14. If there is a problem: Recheck all wiring. Look for short circuits (a piece of steel wool across two points), open circuits (a small

crack in the copper foil), and poor solder connections. Also, make sure your battery isn't dead or its polarity reversed. Try a different microphone if possible.

15. If everything works fine, you may wish to construct a small aluminum box for the transmitter.

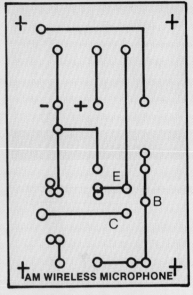

VIEW FROM THE FOIL SIDE
OF THE CIRCUIT BOARD

BLACK IS COPPER

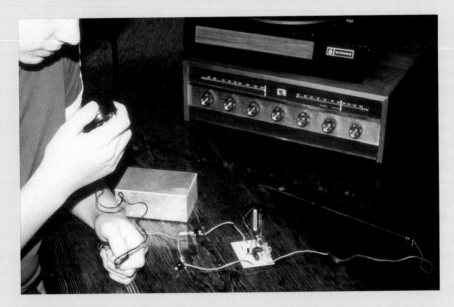

Technology Connections

1. Electronic circuits are made up of *components*. Each component has a specific function in the circuit. What is the function of the microphone in the transmitter? The battery?
2. The use of electronics has revolutionized all aspects of technology. What would your life be like without electricity? What things would you miss the most?
3. What is it called when many components of a circuit are miniaturized and produced on one piece of semiconductor material?

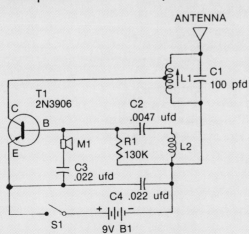

Science and Math Concepts

▶ A *capacitor* is made from a combination of conducting plates separated by an insulator. It can store an electric charge.

▶ Semiconductors are made from materials such as germanium or silicon. Their electrical resistance is somewhere between conductors and insulators.

121

REVIEW QUESTIONS

1. Name three particles found inside of atoms.
2. In an electronic circuit powered by a 9-volt battery, one ampere of current flows. If the 9-volt battery is replaced by a 20-volt battery, does more or less current flow? Why?
3. Name five electronic components that can be used to build circuits.
4. Why was the invention of the integrated circuit important in the history of technology?
5. Describe how the use of electronics has radically changed manufacturing, transportation, information, and health care technologies.
6. Should a telephone be an analog or a digital instrument? Why?
7. Name four parts of a computer.
8. Describe how a computer might be used in a production system, a transportation system, an information system, and a health care system.
9. Describe the difference between operating system software and applications software.
10. A computer can be used for mailing letters to thousands of people. List the major components that you would expect to find in such a computer system. Draw the system using a general system model.
11. Do you think it's a good idea to have a totally automated system, with no involvement by people? Why or why not? Give an example to support your answer.

KEY WORDS

Analog	Current	Operating system	Supercomputer
Bit	Digital	Processor	Transistor
Byte	Electron	Program	Voltage
Circuit	Insulator	RAM	
Component	Integrated circuit	Resistance	
Conductor	Keyboard	Semiconductor	

SEE YOUR TEACHER FOR THE CROSSTECH PUZZLE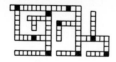

CHAPTER 5

COMMUNICATION SYSTEMS

MAJOR CONCEPTS

After reading this chapter, you will know that:

- Communication includes having a message sent, received, and understood.
- Humans, animals, and machines can all communicate.
- All communication systems have inputs, processes, and outputs.
- The communication process consists of a transmitter, a channel over which the message travels, and a receiver.
- Communication systems are used to inform, persuade, educate, and entertain.
- Communication systems require the use of the seven technological resources.
- Two categories of communication systems are graphic communication and electronic communication.

WHAT IS COMMUNICATION?

As long as humans have lived on the earth, they have needed to communicate. To get assistance in acquiring the things necessary for their comfort and well-being, people have had to make their needs and wants known to each other.

Every society has its own methods of communication. Most societies, whether advanced or primitive, use some sort of spoken language to communicate thoughts and ideas to others. Hundreds of different languages are spoken throughout the world, but each contains words that express the needs and desires of the people.

In the United States, Canada, the British Isles, and South Africa, many people speak English. Certain English words, however, may not mean the same thing in one country as they do in another country. For example, in American English, the word for a large vehicle with wheels that can transport a lot of material is "truck." In the English spoken in Great Britain, that word is "lorry." Likewise, what Americans call an "apartment," English-speaking people in South Africa call a "flat."

Sometimes, it is difficult for people to communicate with each other even when they are speaking the same language. When people from different countries need to communicate with each other, it is even more difficult, unless one person knows the other's language.

To have good communication with someone else, we must be sure that what we are saying is clearly understood. We must know that our words have the same meaning to the other person as they do to us. Just speaking to someone else doesn't mean that we have communicated with that person.

Communication includes having a message sent, received, and understood.

TYPES OF COMMUNICATION

Person-to-Person Communication

People can communicate with each other in many different ways. You can probably list many ways you communicate with your friends. We do most of our communication through speech. Sometimes, however, we communicate

124

People come from many different backgrounds, but they all need to communicate. (Courtesy of Southern California Edison)

without speaking. For example, when we are happy, we smile. When we are sad, we cry. If we are sharing a secret, we might wink. We express our feelings using gestures or facial expressions.

We also use body language to communicate our feelings. When we feel defensive, we might cross our arms in front of our bodies. When we are embarrassed, we might squirm. Sometimes, we don't even know that we are letting others know our feelings by our body language.

People communicate by using all their senses. Love is communicated through touch. Perfume or cologne communicates something about its wearer through the sense of smell. Our vision and hearing tell us a great deal about what is happening around us. They are very important communication "receiving devices."

Some people claim that they can transmit thoughts to another person or receive thoughts from another person by just thinking really hard about something. These people call themselves **psychics.** They claim to be able to predict the future or know things about the past that most people don't have knowledge of. It is not known whether these people do in fact have special powers. They may have just been making some lucky guesses.

Animal Communication

People can communicate with other people. People can also communicate with animals. If you have ever had a pet, or know a friend who did, you know that some kind of communication took place between the person and the animal. You might tell a dog to sit and expect the dog to understand. Certainly, when a dog is hungry or wants a treat, it lets you know in a hurry!

Animals communicate with each other in many ways. This bee is showing the direction and distance of a food source.

Animals can also communicate with other animals. When bees find a source of food, they come back to the hive. They fly in a pattern that looks like a figure eight. The direction they fly when making the line in the middle of the eight is the direction of the food source. The bees even communicate the distance the food source is from the hive by the length of time they take to fly that line.

One of the most interesting examples of animal communication is the story of the chimpanzee who was taught to use sign language. The chimpanzee was able to ask for things using sign language. Even more astoundingly, when the chimpanzee had babies of her own, she taught them to use sign language to communicate with her.

Machine Communication

In addition to being able to communicate with animals, people are also able to communicate with machines. One example of this is the use of a joystick on a video arcade game. When the joystick or paddle is moved by the player, the motion is converted into electrical signals in the game. An example of machine-to-human communication is the output of a computer. Computer output can be printed paper, a display on a video terminal, or sounds from a synthesizer.

Machines can also communicate with each other. In manufacturing plants, computers are connected to other machines to control the machines' operation. When automobiles are painted, the spraying is done by a robotic device. The length of time that the robot sprays paint on a particular part of the car is controlled by a computer. The computer and the robotic spray painter communicate with each other.

Another instance of machine-to-machine communication is a home heating system. A furnace heats the room to a temperature preset by a thermostatic control. When the

As automation becomes more widespread, machine-to-machine and machine-to-human communications become more important. (Courtesy of Allen-Bradley, a Rockwell International Company)

room reaches the proper temperature, the thermostat sends a message (an electrical signal) to the furnace. The signal turns the furnace off.

We have seen examples of how people communicate with other people, how people communicate with animals, and how people communicate with machines. We have also seen how animals communicate with other animals, and how machines communicate with other machines. Can you think of some examples of how an animal might communicate with a machine or how a machine might communicate with an animal?

Humans, animals, and machines can all communicate.

COMMUNICATION SYSTEMS

When people speak to each other, they form a **communication system.** In Chapter 3, a **system** was defined as a method of achieving the results that we desire. Every system has an **input,** a **process,** and an **output.**

A communication system could involve two people speaking to each other. In such a case, the **input** is the desired result (what it is you want to communicate). The **process** is how you will communicate the input. The **output** is the message that is actually received by the other person. **Feedback** tells you whether the other person has understood your message.

All communication systems have inputs, processes, and outputs.

When you are speaking to someone, how do you get feedback? How do you know that the message you wanted to communicate was actually received and understood? You might ask, "Did you understand what I said?" You might look at the person's facial expression to see if he or she looks confused.

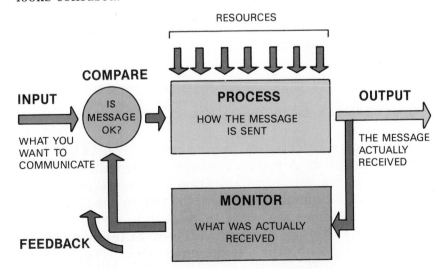

The general system diagram for a communication system

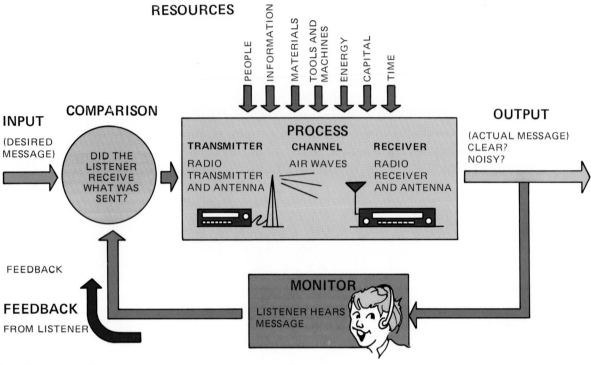

A radio communication system

THE PROCESS OF COMMUNICATION

All communication systems are alike in important ways. Since all are systems, all have inputs, processes, and outputs. To have effective communication, feedback is needed.

The communication **process** includes three parts. There is a means of **transmitting** (sending) the message. There is also a **channel,** or route the message takes. Finally, there is a **receiver,** which accepts the message. Transmitters, channels, and receivers are present in all communication processes.

The communication process consists of a transmitter, a channel over which the message travels, and a receiver.

DESIGNING THE MESSAGE

Before we communicate, we need to think about what we want to accomplish. A message must be designed for a specific purpose. We use communication systems to reach a variety of goals. Sometimes we need to **inform** people about an event. Sometimes we want to **persuade** people to do something. We also use communication to **entertain.** And we communicate with others to **educate** them.

The communication process consists of a transmitter, a channel, and a receiver.

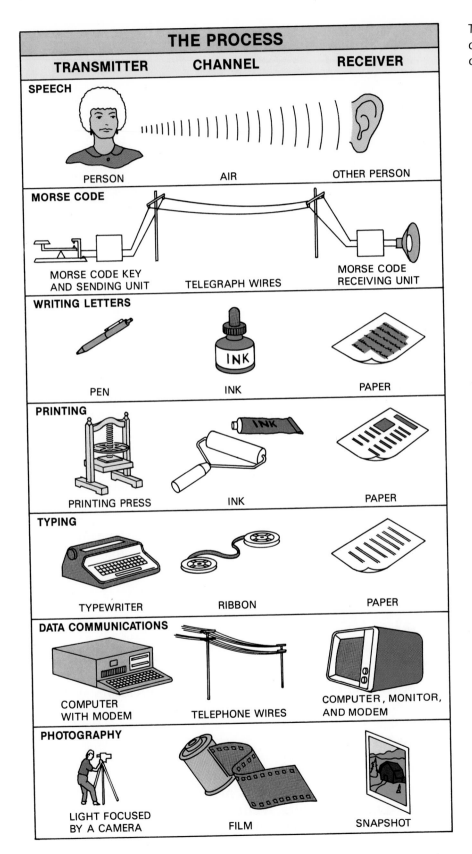

THE PROCESS

TRANSMITTER	CHANNEL	RECEIVER
SPEECH		
PERSON	AIR	OTHER PERSON
MORSE CODE		
MORSE CODE KEY AND SENDING UNIT	TELEGRAPH WIRES	MORSE CODE RECEIVING UNIT
WRITING LETTERS		
PEN	INK	PAPER
PRINTING		
PRINTING PRESS	INK	PAPER
TYPING		
TYPEWRITER	RIBBON	PAPER
DATA COMMUNICATIONS		
COMPUTER WITH MODEM	TELEPHONE WIRES	COMPUTER, MONITOR, AND MODEM
PHOTOGRAPHY		
LIGHT FOCUSED BY A CAMERA	FILM	SNAPSHOT

Television communication is often used to entertain. (Courtesy of National Broadcasting Company)

A communication system is chosen because it best satisfies a need. To inform people, you might choose a printed brochure. To persuade someone to do something, you might use a telephone and speak to the person. You might choose cartoon pictures on television to entertain an audience. To educate a group of people, you might use charts and drawings.

Often, there are several good ways to communicate a message. The choice of means will be based on the type and size of the audience you need to reach. If you are selling a new kind of toothpaste, you probably will want to reach the general public. You most likely will use one of the **mass media**, like radio, television, or newspapers. Mass media are those that reach large numbers of people (the masses). On the other hand, if you need to persuade a small group of business leaders, like the heads of particular companies, you might choose to produce a newsletter or a flyer to carry your message.

Communication systems are used to inform, persuade, educate, and entertain.

RESOURCES FOR COMMUNICATION SYSTEMS

Like other technological systems, communication systems make use of the seven types of resources.

People

People design the systems, conceive the message to be sent, receive the message, and help to deliver the message. Actors, camera operators, technicians, and set designers are needed for video productions. Writers develop radio and

A favorite form of entertainment provides communication careers for many people. (Courtesy of General Electric)

television scripts, books, and other printed materials. Artists illustrate brochures and flyers. Photographers take and print the photographs that often accompany the written text. Printers operate machines to produce multiple copies of books, newspapers, and magazines. People also maintain and repair communication systems.

Information

Information is needed about the audience to which the message is sent. The likes and dislikes of the audience will influence the way advertisers create the message. For example, teenagers and senior citizens like and respond to different kinds of music. Publishers also need information about the audience. What different groups need and want to know will influence what books are published.

All communicators need to know how to produce high-quality communications at the lowest cost. That involves knowing how to select everything from capable writers to good quality paper. Engineers and technicians need mathematical, scientific, and practical knowledge in order to design, operate, and maintain communication systems.

This LaserCard™ is a new credit card that can store the amount of information contained in two 400-page books. (Courtesy of Drexler Technical Corporation)

Materials

Materials such as paper, film, and tape are often used to transport the message. New kinds of materials are being developed constantly to improve communication. Photographic film is now available that can take pictures in very low light. Inks and paper surfaces have been developed that look like metals. Optical disks are now available that can store large amounts of information in a small space.

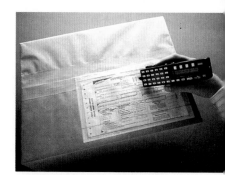

Federal Express keeps track of its packages with a Super Tracker. (Courtesy of Federal Express Corporation)

Tools like this 28-ounce portable terminal can be used to communicate with a central computer by radio. (Courtesy of Motorola Inc.)

Tools and Machines

Tools and machines used in communication systems are radio transmitters, printing presses, cameras, tape recorders, computers, and printers. These devices are used to design, transmit, receive, and store messages. Craftspeople like printers and graphic artists also use hand tools like airbrushes and special rulers, pens, and pencils.

Energy

Energy moves the message from one place to another. Electrical energy powers radio and television transmitters and receivers. Electricity also runs computer systems and word processors. Motors change electrical energy into mechanical energy. Motors run printing presses and electric typewriters. In computer printers, electrical energy is changed into magnetic energy by **electromagnets.** The electromagnets operate the print-head. Light energy is used to expose photographic film.

High capital expenses are required to finance a newspaper plant. (Courtesy of Gannett Co. Inc.)

Capital

Capital is needed to set up and operate the communication system. Equipment must be bought, and people must be hired. Facilities like television studios and print shops must either be built or rented. Light and heat must be provided.

Time

Moving the message from the transmitter to the receiver takes time. The time it takes depends on the length of the message and the rate at which the message is sent. Electrical signals travel at the speed of light (186,000 miles per second). Sound energy travels at a speed of 1,096 feet per second. Satellites are now used to relay data from one location to another. The satellites are in orbit about 22,500 miles above the earth. It takes about ¼ second for a message to be relayed from one point on the earth to another point on the earth via satellite. It takes another ¼ second for a response to be transmitted back in the other direction. Therefore, a person using a satellite to send a message can expect at least a half-second delay between the time the message is sent and the time the response is received.

Communication systems make use of the seven technological resources.

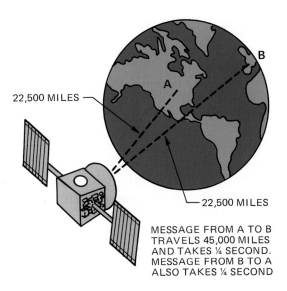

22,500 MILES

22,500 MILES

MESSAGE FROM A TO B
TRAVELS 45,000 MILES
AND TAKES ¼ SECOND.
MESSAGE FROM B TO A
ALSO TAKES ¼ SECOND

The person at *A* says "How are you?" It takes ¼ second for this message to reach the person at *B*. The person at *B* says "Fine, thanks." It takes another ¼ second for that response to reach the person at *A*. The total delay for the person at *A* is ½ second.

Interference—Noise in the System

In perfect systems, the channel does not affect the information sent by the transmitter. In real systems, however, the communication channel is not perfect. Often, when we try to communicate a message, some imperfection in the channel affects our ability to clearly understand the message.

One common kind of imperfection is **noise.** Noise is something that interferes with the communication process. It does not have to be audible. Some kinds of noise are very familiar, like the static you hear on the radio during a thunderstorm. Another common type of noise is caused by airplanes flying overhead. Planes interfere with a television picture, making it flutter. Dust on a record can cause noise. A smudge on a drawing is noise because it obscures the picture. When engineers and designers build communication systems, they try to reduce the system noise as much as possible.

Television "ghost" images are a form of noise.

CATEGORIES OF COMMUNICATION SYSTEMS

Many methods have been used by people for hundreds of years to communicate thoughts and ideas. These methods include speaking, writing, signaling, and drawing. Modern communication methods, however, can be divided into two categories: **graphic communication** and **electronic communication.** Graphic communication systems are those where the channel carries images or printed words. Electronic communication systems are those where the channel carries electrical signals. In the next chapter, you will learn about various types of graphic communication systems. Chapter 7 will discuss electronic communication systems.

Two categories of communication systems are graphic communication and electronic communication.

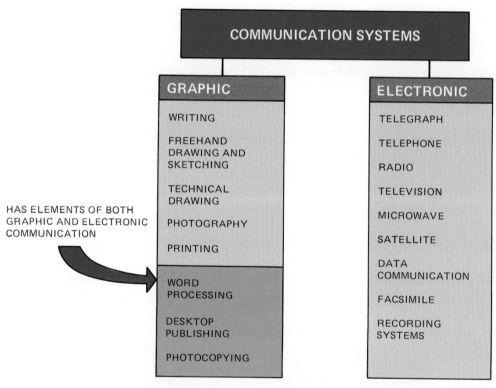

Categories of communication systems

SUMMARY

Humans, animals, and machines can communicate among themselves and with each other. Communication includes both sending a message and having it received and understood. In a communication system, the indication that the message was received and understood is the feedback.

All communication systems have inputs, processes, and outputs. The communication process includes a transmitter, a channel over which the message travels, and a receiver.

We use communication systems to inform, persuade, entertain, and educate. The choice of a communication system is based on the type and size of the audience we need to reach. Like other technological systems, communication systems make use of the seven types of resources.

Modern communication systems can be divided into two categories: graphic communication and electronic communication. In graphic communication systems, the channel carries images or printed words. In electronic communication systems, the channel carries electrical signals.

(Courtesy of International
Business Machines Corp.)

(Courtesy of Hewlett-Packard)

INTERNATIONAL LANGUAGE

Setting the Stage

Chin Lee is the business manager of a large firm that manufactures watches that are sold all over the world. A set of instructions printed in various languages is included with each watch. He could save the company a lot of money and possibly earn himself a pay raise if he could devise a small pamphlet containing pictured instructions to be understood by everyone. This pamphlet could be included with every watch, saving the company so many different printings.

Your Challenge

Using the international language of pictures, assemble a booklet of directions for an activity of your choosing.

Procedure

1. Study the example of picture directions provided by your teacher. Translate the directions by writing them down.
2. Look for directions that come with purchases and bring in examples of picture directions. For example, these might include how to install a VCR. Share these picture directions and build a file for later use in the classroom.
3. Assemble a list of examples of international symbols used to provide information. These will help you when you begin your task.
4. Choose a procedure that you are familiar with. Outline the procedure in writing. Use short sentences. Your procedure should reflect safety in each step.
5. Your outline might be long. Now you must begin to see how many ideas can be presented in one drawing. "One picture is worth 1,000 words." Group the ideas you would like presented in each drawing.
6. Produce your first picture using drawings, photography, computers, or clip art and ask fellow students to interpret it. Use feedback they give you to modify and change areas that are not clear. Color can be added to emphasize important points.
7. Repeat step #6 until you have finished describing how to do the activity.
8. Check your booklet by having another student write a set of directions while he or she looks at your picture directions.

Suggested Resources

Colored pencils
Markers
Stick figures
Stencils
Drawing equipment
Overhead projector
Opaque projector
Photographs
Computers
Clip art

Technology Connections

1. Communication includes having a message sent, received, and understood.
2. The communication process consists of a transmitter, a channel over which the message travels, and a receiver. In this activity, what is the transmitter? The channel? The receiver?
3. What purpose (inform, persuade, educate, or entertain) does your communication system accomplish?
4. How did you use each of the seven technological resources to create your communication system?

REVIEW QUESTIONS

1. Define communication in your own words. Then use your definition to tell which of the following are not communication.
   a. a baby crying for its mother
   b. a sound made by breaking glass
   c. a car horn
   d. a railroad car screeching on the tracks
   e. a bird chirping
   f. static on the radio
2. Give an example of animal-to-human communication.
3. Give an example of machine-to-human communication and an example of human-to-machine communication.
4. Pick one communication system and explain how it uses the seven technological resources.
5. What are four ways communication systems are used?
6. What three elements make up the process part of a communication system?
7. Give two examples of noise that is not audible.
8. Draw a labelled system diagram of a person-to-person communication system.
9. What kind of communication system would you use to do the following:
   a. sell a used stereo
   b. collect money for cancer research
   c. let people know you are looking for a part-time job
   d. entertain a large group in the school auditorium
10. Describe the difference between graphic and electronic communication.

KEY WORDS

Channel	**Educate**	**Noise**	**Receiver**
Communication system	**Electronic communication**	**Machine communication**	**Transmitter**
Graphic communication	**Entertain**	**Mass media**	
	Inform	**Persuade**	

SEE YOUR TEACHER FOR
THE CROSSTECH PUZZLE

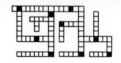

CHAPTER 6

GRAPHIC COMMUNICATION

MAJOR CONCEPTS

After reading this chapter, you will know that:

- In graphic communication systems, the channel carries images or printed words.
- Pictorial drawings show an object in three dimensions. Orthographic drawings generally show top, front, and side views of an object.
- The five elements needed for photography are light, film, a camera, chemicals, and a dark area.
- Four types of printing are relief, gravure, screen, and offset.
- Word processing has made office workers more productive.
- Desktop publishing systems combine words and pictures.

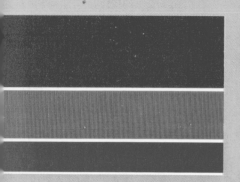

INTRODUCTION

The word **graph** means "to draw" or "to write." Graphic communication systems use images or printed words to convey a message.

To be successful, a graphic message must appeal to its intended audience. Once the audience is clearly identified, a communication process is chosen that will do the best job for the least cost. In other words, the most cost-effective process is chosen.

Let's assume that you want to advertise your new lawn-mowing business. Your audience consists of people in your neighborhood. Although you could advertise on television, handing out a printed flyer would be more cost-effective.

In graphic communication systems, the channel carries images or printed words.

Advertisements are designed to appeal to specific audiences.

PLANNING AND DESIGNING THE MESSAGE

Effective graphic communication requires planning and design. Designing a graphic message involves picking the right words and pictures and placing them creatively on the page. Designing the message also involves choosing ink colors and paper. Sizes and styles of lettering (type fonts) must also be chosen, along with **design elements** (like lines or bars across the page).

Graphic communication systems include writing, freehand drawing and sketching, technical drawing (drawing using special instruments), photography, printing, and photocopying.

Modern computerized word processors are electronic tools, but they are also considered graphic communication systems. In word processing systems, the message consists of graphic images and printed words on paper.

WRITING

The earliest writing known to us in **cuneiform writing.** Cuneiform writing was done by people who lived in the Middle East about 6,000 years ago (around 4000 B.C.). This kind of writing was done by gouging or chiseling wedge-shaped symbols into cave walls and trees. Later, the Egyptians developed their own method of writing called **hieroglyphics.** Hieroglyphics were made by pressing a tool into a soft clay tablet, leaving a series of impressions. Each impression had a different meaning.

The first alphabet was also developed by people living in the Middle East. The Hebrew alphabet started out with the letters "aleph" and "bet." Later, the first two letters of the Greek alphabet were "alpha" and "beta." These two letters together give us the word **alphabet.**

Hieroglyphics from the pyramid at Nur-Sudan.
(Courtesy of Egyptian Tourist Authority)

THE INVENTION OF PAPER

The earliest form of paper was made from the papyrus plant by the Egyptians around 2500 B.C. The fibers of the plant were soaked in water. They were then mashed together and matted to form thin sheets. Paper similar to that we know today was developed by the Chinese about 2,000 years ago. It was not until about A.D. 1400 that paper of really good quality was made. The invention of lightweight, low-cost paper made it much easier for people to record their ideas and share them with others. The sharing of ideas was largely responsible for technological development. Once people were able to learn what other people had done, they could build upon the experiences of others.

FREEHAND DRAWING AND SKETCHING

Prehistoric people recorded their experiences by drawing pictures on cave walls. Often, these cave drawings showed hunting scenes. They included people, animals, and tools

An early cave drawing

like spears and arrows. Primitive artists may have thought drawings had magical powers to make things happen. They may have drawn these hunting scenes to ensure a successful hunt.

Because people found drawings pleasing to look at, they began to use drawings for decoration. Artistic design was used in metalworking, pottery, and sculpture as an important means of communicating pleasant feelings about objects.

A drawing can be simple or complex. Generally, drawings begin as **sketches.** A sketch is a simplified view of an object or scene. It gives just the basic outline and a few details.

A simple sketch

Making realistic drawings became more and more important as people needed to communicate actual shapes and sizes of objects. Craftspeople, particularly, had to work from clear sketches and accurate drawings in order to reproduce objects exactly. During the period called the Renaissance (about A.D. 1350–1500), there were great artistic advances.

TECHNICAL DRAWING

Technical drawing was developed to communicate size and shape information accurately. Precise instruments helped architects, craftspeople, and mechanics draw products according to given specifications. Technical drawing is today a very important communications method. It is used by artists, engineers, architects, designers, and drafters.

Technical drawing normally involves the use of a **drawing board** and **instruments** like a T-square, a plastic triangle, paper, and a pencil.

The T-square is held tightly against the edge of the drawing board and is moved up or down. In this way, horizontal lines can be drawn that are exactly parallel to each other. The plastic triangle is placed against the long edge of the T-square and used as a guide to draw vertical lines or lines at an angle. By using these instruments, lines can be drawn in many directions, and objects can be drawn very realistically.

In technical drawing, a relatively small number of tools is used to produce complex drawings. (Courtesy of Keuffel & Esser Company, Parsippany, NJ 07054)

By moving the T-square and triangle on the drawing board, horizontal, vertical, and angled lines can be drawn.

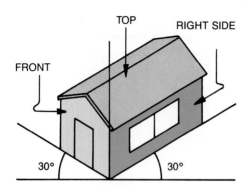

A three-dimensional view (an isometric drawing) of a house.

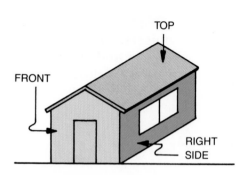

In an oblique drawing, the object is drawn with a straight-on view of one surface.

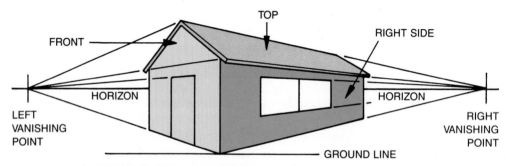

A perspective drawing makes an object look natural. If the horizontal lines on the drawing are extended, they will meet at two points called **vanishing points.** This is an example of two-point perspective.

Pictorial drawings show an object in three dimensions. Orthographic drawings generally show top, front, and side views of an object separately.

Technical drawings can show objects in two ways. One way is to draw a three-dimensional view of the object. This kind of a drawing is called a **pictorial** drawing. Three common types of pictorial drawings are isometric, oblique, and perspective drawings. An **isometric** drawing is drawn within a framework of three lines, called an isometric axis. These three lines represent the edges of a cube. The two base lines of the axis are drawn at an angle of 30 degrees to the horizontal. An **oblique** drawing also presents a three-dimensional picture. Oblique drawings show one surface as a straight-on view of the object. The other two surfaces are shown at an angle. A **perspective drawing** is the most realistic. In a perspective drawing, parts of the object that are further away appear smaller.

A second way is to draw several straight-on views of the object. Three views (top, front, and side) are usually enough to completely describe an object. This kind of a drawing is called an **orthographic** drawing.

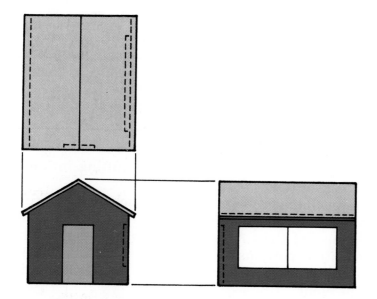

An orthographic drawing of a house. Three views are enough to completely describe its shape and size.

Computer-Aided Design

The computer has brought about many improvements in the fields of drawing and design. Computer-aided drafting (CAD) has become one of the most important modeling tools used by engineers, drafters, and designers. CAD stands for computer-aided design or computer-aided drafting. CADD stands for computer-aided design and drafting.

A CAD system consists of people, software, and hardware. The people who operate the CAD system may be drafters, CAD technicians, or engineers, depending on the tasks involved.

The software consists of computer programs. These programs cause the hardware to perform the tasks the operator desires. CAD software can be used to insert symbols, rotate drawings, or zoom in for a closer look at one portion of a drawing.

The hardware includes a keyboard, display screen, drawing tablet, and plotter. The operator can enter lines and symbols by keying in commands on the keyboard or drawing directly on the drawing tablet. A pen plotter is used to automatically draw the final drawing. The pen plotter is used instead of a display screen when hard copy (actual drawings) are desired.

Using a CAD workstation, a designer can create very precise electronic drawings. The drawings can then be stored in the memory of the CAD system. CAD drawings can represent mechanical parts, electronic circuits, architectural designs, or many other things.

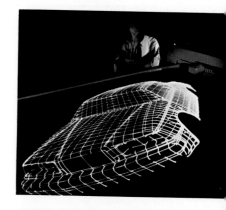

A CAD operator using a CAD system to model an automobile. (Courtesy of Ford Motor Company)

A CAD system can also calculate the strength of a mechanical part or the operation of an electronic circuit and point out potential flaws to the designer. Actual drawings may be obtained from the CAD system by means of a plotter. In addition, many CAD systems will generate parts lists and price information directly from the design.

The following are some of the advantages a CAD system has over hand-drawn designs.

1. The combination of design and drafting into one function saves time.
2. The combination of other tasks (such as generating parts lists) with the CAD function reduces the chance of error.
3. The use of the CAD system cuts out some of the most repetitive portions of the hand drawing. Time-saving estimates range from a 2:1 to an 8:1 saving in drawing time alone.
4. New designs and changes to existing designs are much faster using CAD systems than using hand drawings.
5. Accuracy and consistency are improved from one drawing to the next by the use of a CAD system.

CAD has made drafters, designers, and engineers much more productive. With CAD they can spend more time doing mental design work and less time doing actual drawings.

Computers are changing technical drawing. With computer-aided drafting (CAD) systems, drafters can create, change, and document more complex drawings than ever before. (Courtesy of Autodesk)

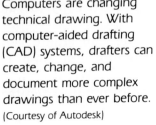

A plotter uses pens with different color inks to make drawings. (Courtesy of Hewlett-Packard)

CAD systems come in all sizes. This is a small system being used in a school. (Courtesy of Texas Instruments, Inc.)

PHOTOGRAPHY

Artists use their eyes to interpret what exists and place this interpretation on paper with paint. Nature herself duplicates her beauty onto photographic film, using light and silver rather than the eye and paint. The word **photography** means "to write with light."

All cameras have at least three common elements: a dark chamber, a lens with a mechanism for admitting light for a controlled amount of time, and film. The very earliest camera-like device was developed in Italy in the 1500s. It was called the **camera obscura.** In Italian, this meant "dark room." Its name came from the fact that it was a dark room with no windows, except for a small lens put in a wall facing the street. An inverted image of the view of the street was projected onto the opposite wall on a bright day. An artist, working with paint on a canvas, could paint the street scene by tracing over the image.

In 1839, a Frenchman named Louis Daguerre developed a method for making permanent photographs. He used a light-sensitive film made of silver-coated metal with an iodine solution on it. Since the film was not very sensitive, someone who wanted a picture taken had to sit still in front of the camera for about half an hour. The pictures taken were called **daguerreotypes.**

In the late 1800s George Eastman introduced the Kodak camera. Until then, most photographers were professionals who did their work in studios. The Kodak camera simplified picture taking and made photography available to just about anyone. The first Kodak camera included a roll of film that could take 100 pictures. It was so simple to use that the Kodak company advertised, "You press the button, we do the rest." Before long, picture taking became a fad. Almost every family owned a camera and a collection of snapshots.

A nineteenth-century daguerreotype of feminist Lucy Stone
(Courtesy of National Portrait Gallery, Smithsonian Institution, Washington, D.C.)

The Five Elements of Photography

There are five elements necessary for photography. They are

1. **light** (the sun, a light bulb, or a flash);
2. **film** (color or black and white);
3. a **camera** (large or small, with a lens);
4. **chemicals** (for developing film and printing pictures); and
5. a **dark area** to process the film in.

All photography requires a source of **light**. In photography, light is reflected by an object and captured as a permanent record on light-sensitive film. When photographers first began to experiment, photographs were taken in indoor studios. A bright source of artificial light was needed. Early photographers burned magnesium powder, which gave off a very intense light. The powder, however, was smoky and dangerous.

Today, **film** is made out of acetate (a plastic). Many types of films are available for special purposes. Besides color and black-and-white film, there are films that are sensitive to all colors except red. Such films can be developed in a darkroom that has red safelights. Some films are sensitive to ultraviolet light. Some are sensitive to heat. Some can be used with very low light levels.

Photographic film is coated with tiny grains of silver. These particles are so small that they are hardly visible. When the lens focuses an image upon the film and the camera's shutter is opened, the silver grains are exposed to light. When the exposed grains of silver are developed in chemicals, they turn black. The unexposed grains of silver are those that do not receive light. They are washed away during the developing process and leave clear areas on the film. This process forms a **negative.**

There are four major types of **cameras.** The **view camera** is the simplest. It has a lens on one end and a place for film and focusing on the other. There is a shutter in the front behind the lens and a dark chamber (called a bellows) made of flexible material. The camera is usually very large. It must be supported by a tripod.

In a view camera you look through the back of the camera to focus. In a **viewfinder camera,** you look through a separate viewfinder to compose the picture.

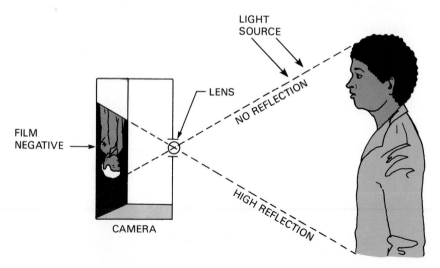

Formation of a negative
image on film

A single lens reflex (SLR) camera (Courtesy of Minolta Corp.)

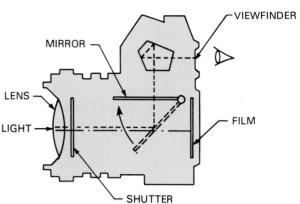

Diagram of a single lens reflex camera
(From Dennis, Applied Photography, © 1985 by
Delmar Publishers Inc. Used with permission)

Instamatic® cameras and disk cameras are viewfinder cameras.

The **twin lens reflex camera** uses two lenses. One is for viewing, and one is for focusing light on the film. This kind of camera also has a mirror that reflects the light from the viewing lens up to the top, for easy focusing. Hence the name "reflex camera."

A **single lens reflex camera** is very similar to the twin lens reflex camera. Its single lens, however, is used for both viewing and focusing. The mirror, which reflects the light, is movable. In its normal position, the mirror reflects the light from the lens to the viewing eyepiece. The photographer can thus see what he or she is taking a picture of. As the picture is taken, the mirror swings up out of the way. This allows the light that enters the lens to reach the film.

Chemicals develop the film and print the photographs. The three kinds of chemicals used are called **developer, stop bath,** and **fixer.** Developer turns the exposed silver particles black. Stop bath stops the developing action. Fixer washes away all the unexposed silver particles and clears those portions of the film that received no light.

Photographic paper and film are very sensitive to light. The chemical processing is usually done in a dark area. Many amateur photographers set up a **darkroom** in their homes, using very modest equipment. They are able to develop film and print their own pictures at low cost. A dark room can be very elaborate, with sophisticated equipment. A dark area can also be as simple as a small tank or a closet in one's home.

The five elements needed for photography are light, film, a camera, chemicals, and a dark area.

A darkroom is used to develop photographs. (Photo by Michael Hacker)

THE DEVELOPMENT OF PRINTING

Four types of printing are relief, gravure, screen, and offset.

One of the most significant developments in graphic communication technology took place in Germany about 1450 A.D. Johannes Gutenberg perfected a way to cast individual letters (movable type) by pouring hot metal into molds. Before Gutenberg's invention, type was made of wood. Each letter was carved individually, which took a great deal of time.

Since the days of Gutenberg, the printing industry has reached into every corner of the world. Millions of books and magazines are printed every year. Knowledge can be shared with people in even the most remote locations.

Newspapers were the first medium of communication for the masses. They became very popular toward the end of the nineteenth century. By that time, the Industrial Revolution had brought about the printing machinery necessary for large and rapid print runs. Newspapers were published in the American Colonies as early as 1704. The first newspaper in America was the *Boston News-Letter*. Most early newspapers were printed on hand-operated printing presses. About 150 years passed before most companies had adopted machine-driven presses. In 1837, a New York book publisher, Harper and Brothers, still used a press powered by a mule.

Newspapers brought people closer together because of improved communication. News from one place was known in other places as soon as the newspapers were delivered. Publishers hoped that the newspaper would improve understanding among people all over the world.

In 1840, the *New York Sun* had a circulation of 40,000 newspapers per day. This circulation was the highest in the country. Today, the *New York Times* prints 1.5 million Sunday papers. The news stories are composed on computers instead of typewriters. The pages are optically scanned by lasers and sent from one plant to another by microwave or satellite transmission. Huge newspaper presses continuously feed large spools of paper through their rollers.

Some newspapers, like the *Wall Street Journal* and *USA Today*, print their papers in several different locations. This cuts transportation costs. The information is transmitted to satellites, which then rebroadcast the information to receivers in different cities. The papers are then printed and distributed locally. The expense is far less than would be the case if the papers were transported by truck or airplane to the various locations.

With the help of satellite communications, newspapers like *USA Today* can be printed and distributed at several locations. (Courtesy of Scott Malay/Gannett Co. Inc.)

Johannes Gutenberg and the Printing of Books

Johannes Gutenberg, inventor of the mechanical printing process
(Courtesy of Inter Nationes)

In Mainz, Germany, about 50 years before Columbus discovered America, Johannes Gutenberg invented a process that would revolutionize society. That invention was the mechanical printing of books.

When Gutenberg was a young man, he studied from books that had been hand-written by scribes. He watched the monks as they labored for years to make copies of the Bible.

Gutenberg wanted books to be made available to everyone. He felt there must be a way to make the work of the scribes easier. He began to think about mechanical printing. He knew that wooden blocks were used to print playing cards. He also knew of metal stamps used to imprint coins.

He realized that printing books mechanically would require separating the letters and making them movable. He knew that he must not use soft wood but hard metal. For each letter, Gutenberg made a steel stamp,

which he then stamped into a block of softer copper. Into this mold he could pour molten metal. The letters produced by this casting process were made of an alloy of lead, tin, antimony, and a small amount of bismuth.

But movable type was not enough. Gutenberg also designed a method to hold individual letters in place. He used a screw press (like the kind used to squeeze grapes for wine making) to do the printing. In addition, good quality ink that adhered to the metal type had to be produced.

By 1455, after three years of hard work, Gutenberg had printed about

Gutenberg's printing press
(Courtesy of Inter Nationes)

One of the original Gutenberg Bibles
(Courtesy of Inter Nationes)

200 copies of the Bible. He printed thirty of these on vellum, a paper made from animal skin. For these, the hides of almost 10,000 calves were used.

Gutenberg's real contribution was that he created a printing system. The system was composed of a press, good paper, the proper ink, and movable metal type. After Gutenberg, people were able to reproduce books and manuscripts in large quantities. The average person could finally afford to buy books. Word of Gutenberg's printing system spread quickly. The ability to print books mechanically promoted the spread of knowledge to all corners of the earth.

Relief Printing

Gutenberg's printing method made use of raised surfaces. This method is called **relief printing.** In relief printing, only the raised surfaces of the letters receive ink. The lower surfaces do not. When the letter is pressed against a piece of paper, only the inked raised surface prints.

Relief printing is also known as **letterpress printing.** The method is still used today, with modern refinements, to print a wide variety of items, such as newspapers and greeting cards.

Most typewriters make use of the relief printing process. The letters on the typing elements are raised surfaces. When the raised portion of the letter strikes the ribbon, it transfers ink to the paper. Typewriting can be thought of as an intermediate step between writing by hand and printing.

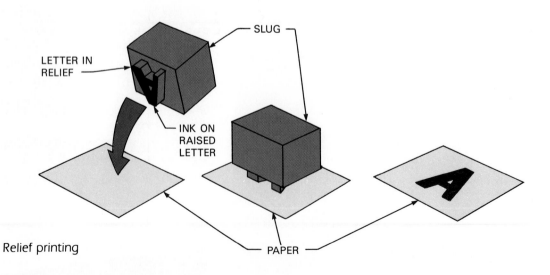

Relief printing

Gravure Printing

Besides printing from a raised surface, we can also print from a recessed (lowered) surface. First, we scratch a line into the surface of a piece of metal. We then apply ink to the entire surface and wipe it all off. The ink remains only in the scratch. When a piece of paper is pressed hard against the metal surface, the paper is forced into the scratch and pulls the ink out. This type of printing is called **gravure,** or **intaglio printing.** Gravure is used to print some magazines, such as the magazine section included with many Sunday papers.

Screen Printing

To print posters, decals, and designs on T-shirts, we use a stencil, or mask. The stencil permits ink to pass through only certain areas. It prevents ink from passing through other areas. Stencils, often made from plastic film, are attached to a fine mesh silk screen. The silk screen is stretched tightly across a wooden frame and placed in contact with the T-shirt. This method is known as screen printing.

Screen process printing

Offset Lithography

Today, most commercial printing is done by **offset lithography.** **Lithography** means writing (**graphy**) on stone (**litho**). It is based on the principle that oil and water do not mix. The process was first developed by a German artist, Alois Senefelder. Senefelder drew a line with a waxy crayon on a smooth, flat piece of limestone. He then wet the entire surface. He found that an oil-base ink applied to the limestone adhered only to the waxy crayon lines, and not to the wet part of the limestone. When the surface of the limestone was pressed up against a piece of paper, only the inked part (the crayon lines) printed.

The modern offset printing process uses flat metal sheets rather than pieces of limestone. The principle, however, is exactly the same. Through a photographic process, a greasy image is placed upon a sheet of aluminum. The aluminum, which has been chemically treated so that it is sensitive to light, is called an **offset plate.** The plate is wrapped around a metal cylinder. The cylinder is moistened by a water solution.

Ink is applied to the entire plate. It adheres only to the greasy image, and not to the moistened areas. As the plate revolves, it presses up against another cylinder, which is covered with a thin rubber blanket. The image is transferred (or offset) to the rubber blanket, but the image is in reverse. The rubber blanket rotates, and paper is fed through the press. The image is again offset from the rubber blanket to the paper. This time the image reads correctly.

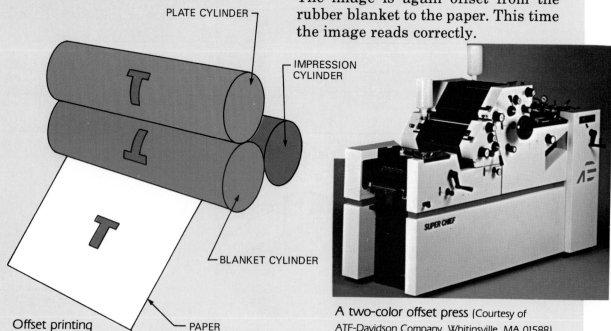

PLATE CYLINDER

IMPRESSION CYLINDER

BLANKET CYLINDER

Offset printing

PAPER

A two-color offset press (Courtesy of ATF-Davidson Company, Whitinsville, MA 01588)

Word processing is rapidly replacing typing in most offices. (Courtesy of Michael Hacker)

WORD PROCESSING

Word processors combine typewriter and computer technologies. When you type on the keyboard, your text is stored in the computer's memory. The text is displayed on a video display terminal instead of on a piece of paper. If you make a mistake, you can go over your text and correct it easily. Only after you have written everything exactly as you want it do you print the text on paper. All corrections are made electronically, before the text is printed.

Word processors offer many advantages to typists. If you have misspelled a word, you can use the **search and replace** command. Then each time the word appears in the text, the word processor will automatically replace it with the correct spelling. If you wish to rearrange paragraphs in a letter, you can do so with a **move** command. If you want to copy a chart from one page to another, you can do that with a **copy** command. Most offices are replacing standard typewriters with word processors. They save typists a lot of time.

Word processing has made office workers more productive.

DESKTOP PUBLISHING

In addition to straight text, computer systems are now able to produce entire pages, including headlines and pictures. **Desktop publishing** describes the linkup of a personal computer, special software, a mouse (a device used to point to things on the screen), and a laser printer. The result of this combination is a page of text and graphics of very high quality.

Desktop publishing lets you arrange the words and the pictures on the computer screen exactly the way you want them to appear on paper. The image on the screen is called a **WYSIWYG image.** In computer jargon, WYSIWYG is pronounced whizzy-wig and stands for "what you see is what you get!"

Many companies are now using desktop publishing programs to prepare their own newsletters and advertising brochures. The software is easy to use. Generally, the operator selects the column width and the type size and style (font). Graphics are then placed on the page. Even photographs can be included. With a device called a **scanner,** photographs will appear right on the screen. Finally, the text is inserted from a word processor and made to flow around the pictures. The entire page can be re-arranged into a different format at the touch of a key.

Most of the time, many copies of printed materials are needed. However, only one piece of copy is produced as output from a desktop publishing system. That original must then be reproduced. This is usually done by transferring the image photographically to an offset plate. Then an offset printing press is used to produce large numbers of copies. For this reason, desktop publishing is commonly referred to as a pre-press operation. If only a few copies are needed, a copying machine can be used.

Desktop publishing systems combine words and pictures.

Desktop publishing permits you to compose a page layout using a computer. The finished layout is then output by a printer.
(Courtesy of Aldus Corporation)

Samples of materials generated by the desktop publishing process.
(Courtesy of Aldus Corporation)

A computerized phototypesetting machine (Courtesy of Compugraphic Association)

Today, some newspapers use a publishing process called **pagination** to make up their pages. The entire page can be viewed, or the operator can zoom in on a particular column or phrase. Once the page is composed, it is sent electronically to a **phototypesetter.** The phototypesetter is like a computerized printer. It produces a high-quality printout of the newspaper page. An offset plate is then made from the printout, and copies are run.

COMPUTER PRINTERS

Mechanical typewriters gave way to electric machines in the 1940s. The IBM Selectric typewriters needed about 1,500 mechanical adjustments, as many as some automobiles.

The newest typewriters are **electronic.** Many of the mechanical functions have been replaced by integrated circuits. (For a discussion of integrated circuits, see pages 97-99.) Therefore, electronic machines are easier to adjust and maintain. The capabilities of the machines have been improved as well. For example, many electronic typewriters have a memory and can store pages of information. Electronic typewriters can also do things like center text and underline words automatically.

Much of the new typewriter technology has been incorporated in the printers used by computers. Three types of printers are **daisy wheel, dot matrix** and **laser** printers. The daisy wheel printer uses a daisy-shaped print-head made of plastic. It contains an entire set of characters (letters, numbers, punctuation and symbols). When a typewriter key is depressed, the daisy wheel rapidly spins into position. The correct letter then strikes against the ribbon to print the character. The quality of this type of printing is excellent. It is called "letter quality," because it is good enough to use on business letters.

A daisy wheel (Courtesy of Michael Hacker)

Dot matrix printers are typically less expensive than daisy wheel printers and use an entirely different technology. The print-head on a dot matrix printer consists of a group of pins arranged in a rectangular format. The rectangular format is called a matrix. When a particular letter is struck on the keyboard, a combination of pins sticks out to form that letter. These press into the ribbon, printing the letter onto the paper. (See Chapter 7.) An advantage of dot matrix printers is that the pins can form all kinds of shapes, not just letters. Therefore, these printers are used for computer graphics.

Laser printers combine the good quality of daisy wheel printers and the flexibility of dot matrix printers. They can print letter quality text and graphics very rapidly.

Laser Printers

Laser printers are often used by businesses that have desktop publishing systems. High quality originals which include text and graphics can be rapidly produced.

At the heart of a laser printer is a **photosensitive drum.** The drum is made from aluminum. It looks like a beverage can. On the surface of the drum is a layer of material that is sensitive to light.

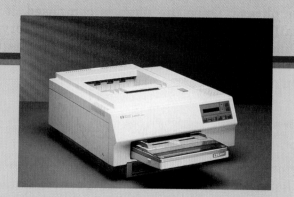

Laserjet II printer (Courtesy of Hewlett-Packard)

The drum rotates. As it does, it passes under a wire that is as long as the drum and fixed in a position above it. The wire carries high voltage electricity. It is called a **corona wire.** (A corona is the bright light you see when lightning flashes. The air around the lightning ionizes which means it becomes electrically charged and can conduct electricity.) The air space between the corona wire and the drum also ionizes. The airspace then conducts electricity. Negative charges from the corona wire flow to the surface of the drum.

After rotating past the corona wire, the drum has a voltage of −600 volts of static electricity on its surface. This is a high negative charge.

A laser beam then focuses laser light on portions of the drum. The drum becomes positively charged where the laser light strikes it. The drum now has negative charges where there was no light and positive charges where the laser exposed it.

The laser beam first shines onto a mirror with six sides. The mirror is shaped like a hexagon. As the mirror rotates, the beam is caused to move in an arc. The beam is focused by a lens, bounces off another mirror, and

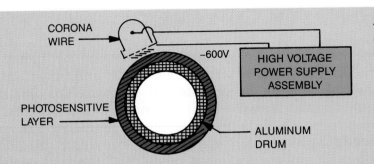

CORONA WIRE

−600V

HIGH VOLTAGE POWER SUPPLY ASSEMBLY

PHOTOSENSITIVE LAYER

ALUMINUM DRUM

A negative static charge is applied to the surface of the drum.

sweeps across the drum in a horizontal line. The laser **pulses** (is turned on or off) in response to data (letters and graphics) that come from the computer.

Since the beam is sweeping horizontally while the drum is rotating, the entire drum can be covered by the laser beam. In this way, an electrostatic image is placed on the drum. This image is made up of positive and negative charges and is not yet visible.

Toner is then applied to the drum. Toner is a powder made of iron particles and plastic resin. The toner receives a negative charge. Opposite charges attract each other. The negatively charged particles of toner are attracted to the positive areas of the drum that have been exposed to laser light. The electrostatic image is thus turned into a visible image.

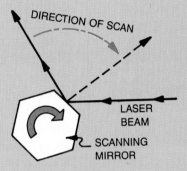

DIRECTION OF SCAN

LASER BEAM

SCANNING MIRROR

The mirror rotates and causes the laser beam to scan in an arc.

The toner image on the drum is then transferred to the paper. The paper is charged positively by another corona wire. The positive charges on the paper pull the negatively charged toner off the drum. Finally, the toner is melted and forced into the paper by heat and pressure. This fuses the image on to the paper.

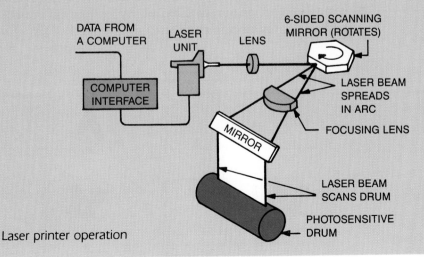

DATA FROM A COMPUTER

LASER UNIT

LENS

6-SIDED SCANNING MIRROR (ROTATES)

COMPUTER INTERFACE

LASER BEAM SPREADS IN ARC

FOCUSING LENS

MIRROR

LASER BEAM SCANS DRUM

PHOTOSENSITIVE DRUM

Laser printer operation

PHOTOCOPYING

An early form of copying

In the Middle Ages, copying was done by hand. Since monks knew how to write, the difficult and time-consuming task of copying documents fell to them. In 1937, a law student from New York City named Chester Carlson developed a method to make copies. He called his process **xerography,** which means "dry copying." In 1959, the Xerox Company developed model 914, the first copier that could be easily used in an office. It could make copies without using messy inks or fluids.

Copying machines make use of photography and static electricity. Static electricity is what is created when you walk across a rug on a cold, dry day and then touch a doorknob. You build up an electrical charge on your body. The charge is discharged when you touch an object with a different charge. Like charges (two positives or two negatives) repel each other, while unlike charges (a positive and a negative) attract each other.

Copying machines use a metal plate coated with a light-sensitive material (see part 1 of the diagram below). The metal plate is given a positive charge as it passes under a wire in the copying machine (2). The paper to be copied is exposed to a very bright light (3). The light is reflected by the white areas and not reflected by the areas dark with printing or writing. The light that is reflected by the white areas destroys the positive charge on the metal plate. The rest of the positive charge (where the black image is) remains. A black powder called **toner,** which has a negative charge, is dusted over the plate. Toner is attracted to the positively charged image area (4). Another sheet of paper (5) is charged positively. It attracts the negative toner to it (6). Finally, the paper is heated by a fusing roller. The heat permanently bonds the toner and makes a finished copy (7).

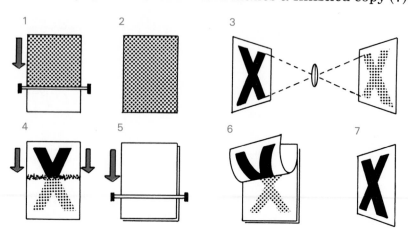

Diagram of the xerographic process (Courtesy of Xerox Corp.)

SUMMARY

In graphic communication systems, the channel carries images or printed words. Examples of graphic communication systems are writing, freehand drawing and sketching, technical drawing, photography, printing, word processing, and photocopying.

The design of a graphic message involves picking the right words, arranging them creatively with pictures, selecting design elements, choosing appealing colors, and specifying sizes and styles of lettering.

Cuneiform writing dates back to around 4000 B.C. Later, the Egyptians developed hieroglyphics. They also made paper from the papyrus plant.

Generally, drawings begin as sketches. A sketch is a simplified view of an object or scene. Perspective drawing is a way of making things look realistic by making the distances in the drawing look correct to the viewer.

Technical drawing is an important method of recording and passing along ideas and information. While many technical drawings are made by hand with instruments, computer-aided drafting (CAD) has become an important tool for engineers, drafters, and designers.

Pictorial drawings show an object in three dimensions. Orthographic drawings generally show top, front, and side views of an object separately.

The five elements required for photography are light, film, a camera, chemicals, and a dark area. Four types of cameras are the view camera, the viewfinder camera, the twin lens reflex camera, and the single lens reflex camera.

Since the days of Gutenberg, the printing industry has grown tremendously. Relief printing is printing from a raised surface. Gravure printing is printing from a recessed surface. Screen printing uses a stencil attached to a fine mesh silk screen. Offset printing involves printing from a flat sheet of photosensitive material.

Word processing combines typewriter and computer technologies. Electronic publishing systems can combine words and pictures. Desktop publishing allows you to see WYSIWYG images. You can compose entire pages of text and graphics on the screen, then make high-quality copies.

Three kinds of computer printers are daisy wheel, dot matrix, and laser printers. Laser printers print high-quality graphics and text at high speed.

Photocopiers make use of photography and static electricity. Negatively charged toner is attracted to a positively charged piece of paper and fused to the paper by heat.

TEAM PICTURE

Setting the Stage

It's only the first day of track practice and already the team is looking depressed. Every other team got brand new uniforms. Because of a budgeting mistake, the track team will have to wait until next year.

"Isn't there something we can do to get uniforms by ourselves?" asked Marty. "How about making our own!" said Amy.

Your Challenge

Design and make a stencil or iron-on transfer using a computer and graphics software. Use it to personalize a T-shirt or sweatshirt.

Procedure

1. Be sure to wear safety glasses and a lab coat.
2. Look through books, magazines, etc. for design ideas.
3. Use a computer to draw the image you want printed on the T-shirt.
4. If you are using a *graphics tablet,* it may be possible to digitize your design by 'tracing' it into the computer memory.
5. If you are using a *video camera* as the input device, it may be possible to digitize your own portrait! This requires special equipment and software such as Computereyes® or MacVision®.
6. If you are using software such as Print Shop®, many graphics are already included in a Graphics Library.
7. Once you are satisfied with your graphic, print it out on a dot matrix printer.
8. If you are using the Underware® direct transfer iron-on printer ribbon, and text is included in your design, FLIP YOUR DESIGN HORIZONTALLY to produce a mirror image *before* printing it out.
9. Iron the image directly onto a T-shirt. The transfer is safe and washable.
10. If you are using thermal screens to produce a stencil with the thermofax machine, do *not* flip your design into a mirror image. Instead, print it out using a dark printer ribbon.
11. To make a stencil using the thermal screens, simply pass the computer printout, together with a piece of thermal screen, through the thermofax machine.

Suggested Resources

Safety glasses and lab aprons

Computer with software for graphics and printing

Graphics tablet, light pen, mouse, or other input device

MacVision® digitizer (requires a video camera)

Computereyes®/2 (requires a video camera)

Dot matrix printer

Thermal screens—9" × 11⅜"

Plastic frames—7¾" × 10½"

Squeegees—7⅜" wide

Screen mounting tape, double-faced, ¼" × 60 yds.

Hinged frame

Versatex textile paints— assorted colors

Thermofax machine

Underware® ribbons—direct iron-on transfer ribbons (available for most printers)

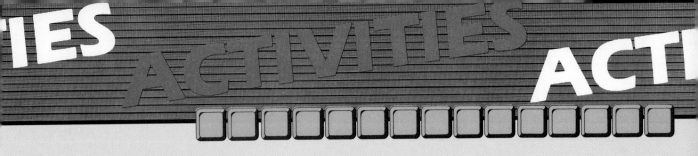

12. Now mount the thermal screen stencil on a hinged frame using double-sided tape.
13. Using a squeegee, press textile paint through the thermal screen stencil directly onto a T-shirt.

Technology Connections

1. Computer-generated graphics, including animation, are important in today's communications. How is this technology used by the entertainment industry? What about the advertising industry?
2. Scientists and engineers use computer-generated graphics and computer aided drawing (CAD) to help plan and test new automobile and aircraft designs. How does this help to keep design costs down?
3. Computer graphics and text are often combined in *desktop publishing*. What is desktop publishing? Do you think it will play an important part in the communications field?

Science and Math Concepts

▶ Graphic communication has been around since prehistoric times. Prehistoric people recorded their experiences on cave walls using natural pigments for color.

▶ Computer animation is done by drawing and erasing the object over and over. Each time the object is redrawn it is moved to a slightly different location.

163

THE NEW COMPUTER

Setting the Stage

The entire class is buzzing with excitement. The long-awaited computer for the class has arrived. Not only does the class have a new computer, but an entire CAD system as well. The class now has a computer, a monitor, a drawing tablet, a pen plotter, and numerous pieces of software.

With everything set up on the computer table, there is a problem. Where will all of the manuals, guides, and instruction books be kept? These items cannot be piled on the table. They might fall off or be ruined. Besides, the class could never find anything or keep track of the materials if they were just in a pile.

Your Challenge

Design a set of holders for the manuals, guides, and books so that they stand neatly on edge. The holders should "fit" the design of the computer and other hardware.

Procedure

1. Look through magazines, catalogs, etc. for design ideas.
2. Sketch at least six designs.
3. Combine good ideas or select the design you like best and make a final sketch of your design.
4. Carefully measure the size requirements of your design and add them to your sketch. This is called *dimensioning*.
5. Using the technical drawing reference books, make an orthographic sketch of your design and add dimensions.
6. Using the manuals for your CAD software and your hardware, make an orthographic drawing of your design. Dimension the drawing. This is called a *working drawing* and can be used to make the object(s) you have just designed.
7. Print out your drawing with the pen plotter to make a hard copy.

Suggested Resources

Computer with CAD software
Graphics tablet or mouse
Plotter (dot matrix printer will work)
Sketch pad and soft lead pencils or felt tip pens
Ruler
Three reference books on sketching and technical drawing
Software and hardware manuals

164

Technology Connections

1. Computer-aided design has become one of the most important drafting tools for scientists and engineers. How has CAD changed the way people work?
2. Aircraft and space vehicles are designed and tested using CAD and computer-generated graphics. How has this made them safer and more efficient?
3. Sometimes, computers using similar CAD programs communicate with each other by telephone. They send drawings hundreds or even thousands of miles. How are they able to do this?

Science and Math Concepts

▶ Accurate measurements must be made if a design is to work, whether it is drawn by hand, with drafting equipment, or with CAD.

PHOTOGRAMS

Setting the Stage

A photogram is a silhouette picture made on a piece of photographic paper. Objects are placed on the paper and exposed to white light. When the picture is developed, we see that the place where the objects were laid is white. The other areas are black. This is because light cannot penetrate these objects, thus producing a white image. The areas exposed to light turn black.

Your Challenge

Create a unique photogram. Use your photogram to convey a message or describe an emotion.

Procedure

1. Locate all materials in your darkroom.
2. If not already mixed, mix the developer, stop bath, and fixer according to the instructions on the package. Wear rubber gloves, eyeglasses, and apron. Place each chemical in a separate tray. Be sure to remember what is in each tray! Your instructor can help you out in mixing the chemicals if you are having problems.
3. Place water in the fourth tray.
4. Turn the safelight on, and turn off all white lights in the room. It will take a little while for your eyes to get used to the dark.
5. Take one piece of photographic paper out of its protective cover. Remember that this paper is sensitive to all light except your safelight. *Do not* expose the paper to the white room lights at this time!
6. Place the photographic paper with the shiny side up on a flat surface under your light source. This could be an enlarger or some other white light source.
7. Place your items for the photogram on top of the photographic paper. Arrange them in a unique way. Try to describe an emotion (love, hate, anger, etc.) or convey a message.
8. Turn on the enlarger or other light source and make the exposure. You may have to experiment with the amount of time the light is on, depending on the intensity of the light hitting the photographic paper.
9. Take the exposed piece of photographic paper and place it in the paper developer tray exposed side up. Rock the tray gently. You should see an image appear on the paper. Depending on the type of developer used, the time will vary. Typical time would be

Suggested Resources

Darkroom or equivalent with running water

Photographic paper—black and white 8.5" × 11" (resin coated)

Enlarger (or some other light source)

Photo chemicals
—paper developer
—stop bath
—paper fixer

Four chemical trays (to fit the paper size)

Safelight (for black and white photo paper)

Squeegee

Timer (wall clock or watch)

Objects to create photogram, such as rings, string, nails, etc.

Rubber gloves, safety glasses, and apron

around 1 minute and 20 seconds in the tray. *Note:* If the print turned out too dark, you need less exposure time. If the print turned out very light, you need more exposure time.

10. Next place the print in the stop bath. This is normally for around 10–15 seconds. This stops the development.

11. Place the print in the fixer. Fixing time is typically 5 minutes. The fixer preserves the image by hardening it on the paper. If not fixed properly, the print will turn brown in a period of time.

12. Next, place the print in the water tray. It is good if the water is running slowly in the tray. This removes any excess chemicals from your print. Normal wash time for resin coated paper is 4 minutes.

13. Place your photogram on a flat surface and squeegee off any excess water. Be careful not to scratch the surface of the print.

14. Place your newly created photogram in a print dryer, or just let it air dry.

Technology Connections

1. What are some commercial and industrial uses of photography?
2. How has current technology altered photography? Hint: What about camera types, styles, and uses? Instant photography?
3. How do you think large photographic processing companies process your film and prints?
4. Why is photography a unique communication process as compared to other methods of communication?
5. Why did some objects cause the paper, when processed, to appear gray, not really black or white?

Science and Math Concepts

▶ The photographic process requires certain chemicals to process film and photographic paper.

▶ Silver particles on the surface of photographic paper or film are sensitive to light. The chemical developer turns exposed silver particles black. The stop bath stops developing action. The fixer washes away unexposed silver particles.

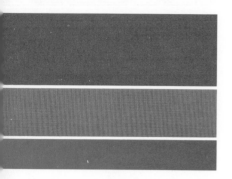

REVIEW QUESTIONS

1. What is meant by graphic communication?
2. Draw an isometric and an orthographic sketch of a 2″ × 2″ × 1″ block.
3. How has CAD affected the field of technical drawing?
4. Using the library as a resource, name five careers that require skill in technical drawing.
5. What five elements are needed for photography?
6. What is the main difference between a single lens reflex camera and a viewfinder camera?
7. What contribution did Johannes Gutenberg make to communication technology?
8. Give an example of something printed by each of the following processes:
   a. relief printing
   b. gravure printing
   c. offset printing
   d. screen printing
9. How do desktop publishing systems differ from word processors?
10. How is static electricity used in laser printers and copying machines?

KEY WORDS

CAD	Gravure printing	Orthographic	Screen printing
Daisy wheel	Isometric	Perspective	Technical
Desktop	Laser printer	Pictorial	drawing
publishing	Negative	Phototypesetter	
Dot matrix	Offset printing	Relief printing	

SEE YOUR TEACHER FOR
THE CROSSTECH PUZZLE

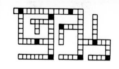

CHAPTER 7

ELECTRONIC COMMUNICATION

MAJOR CONCEPTS

After reading this chapter, you will know that:

- In an electronic communication system, the channel carries an electronic or electromagnetic signal.
- Electricity and electronics have revolutionized communication technology.
- The telephone is the most commonly used instrument for electronic communication.
- In radio communication, information is sent through the air from a transmitting antenna to a receiving antenna located some distance away.
- Computing and communication technologies are merging to produce powerful tools for information processing.
- A number of computers or computer devices joined together is called a network.

WHAT IS ELECTRONIC COMMUNICATION?

In Chapter 5, you learned that all communication systems have a transmitter, a receiver, and a channel that carries the message from the transmitter to the receiver. In some systems studied in Chapter 6, the channel was ink or photographic film. If the channel uses electrical energy to carry information, then the system is said to be an **electronic communication system**.

Many communication systems and devices that are not electronic communication systems use electronics to help make them work better. For example, a camera may use a microcomputer to help it to focus properly or to select the correct shutter speed. However, that does not make the camera an electronic communication system. Similarly, a copying machine is not an electronic communication system, even though it also uses a microprocessor and other electronic circuits to make it work correctly. Cameras and copying machines are not electronic communication systems because their channels do not carry electronic signals. In a camera, the channel is film. Copying machines use toner as the channel.

The use of electricity and electronics in communication systems has created the information age, just as steam engines fueled the industrial age. Today, we can watch live TV programs coming from other continents. We often speak on the phone with friends and relatives thousands of miles away. We use automated bank teller machines to bank after closing hours. All of these activities have been made possible by the application of electricity and electronics to communication systems.

Electronic communications systems help organizations handle large amounts of information quickly.
(Photo by Jay Fries with permission, Tandem Computers Inc.)

In electronic communication systems, the message can be sent through cables (as in a telephone system) or through the air (as in a radio system). A message can also be stored in some medium (as on recording tapes or compact disks) and physically moved. Systems that send messages immediately are called **transmitting and receiving systems**. Systems that store messages electronically are called **recording systems**.

TRANSMITTING AND RECEIVING SYSTEMS

The first electronic communication system to come into widespread use was the **telegraph**. During the first half of the 1800s, small telegraph systems were set up in Europe and the United States. In 1843, Samuel F.B. Morse obtained $30,000 to construct a telegraph line from Washington, D.C. to Baltimore. Morse's first system used a pen that made marks on a piece of paper. This soon gave way to the telegraph sounder. Operators could easily learn to interpret the clacking sounds produced by this machine. The Morse code, although changed somewhat over time, is still used today.

Soon, telegraph wires sprang up all over Europe and the United States. In 1858, the first transatlantic telegraph cable was laid. Instant communication was achieved between England and the United States for the first time.

The Telephone

With nearly 150 million telephones in the United States today, the telephone may well be the most commonly recognizable instrument for electronic communication. Telephones have evolved from the crank-box phones of the late 1800s to sleek, TouchTone® phones with many convenient features.

Alexander Graham Bell, the inventor of the telephone, was a teacher of the deaf. He knew a great deal about the way the human ear works. He applied his knowledge of speech and hearing to the invention of the telephone. Another inventor, Elisha Gray, patented a device very similar to Bell's telephone only a few hours later than Bell did. Bell and Gray were involved in a legal dispute for

Using a Telegraph System

A simple telegraph system contains a telegraph key, a power source, a buzzer or sounder, and wires to connect the sending station to the receiving station. When the sending operator presses the key down, the contacts conduct electricity, sending power to the buzzer. The sending operator holds the key down for short and long periods of time, forming a code in which different groups of shorts and longs (dots and dashes) stand for each letter, number, and punctuation mark. The Morse code is the most commonly used code of this type. It can also be used for signaling with lights.

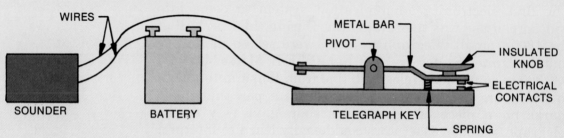

Diagram of a telegraph system

	The Morse Code				
A	• —	K	— • —	U	• • —
B	— • • •	L	• — • •	V	• • • —
C	— • — •	M	— —	W	• — —
D	— • •	N	— •	X	— • • —
E	•	O	— — —	Y	— • — —
F	• • — •	P	• — — •	Z	— — • •
G	— — •	Q	— — • —	.	• — • — • —
H	• • • •	R	• — •	/	— • • — •
I	• •	S	• • •	?	• • — — • •
J	• — — —	T	—	—	— • • • • —
1	• — — — —	4	• • • • —	8	— — — • •
2	• • — — —	5	• • • • •	9	— — — — •
3	• • • — —	6	— • • • •	0	— — — — —
		7	— — • • •		

several years. Finally, Bell was awarded the rights to the patent.

When you speak into a telephone mouthpiece, the sound of your voice creates pressure on tiny carbon particles in the telephone microphone. As the pressure on the carbon granules increases (when you speak loudly), they become more tightly packed. This permits more electricity to flow through the telephone circuit. When you speak softly, the carbon granules are loosely packed. Less electricity can flow. The electric current that flows through the telephone line changes in response to your voice.

As electricity flows through the telephone earpiece on the receiving end, the current makes a thin piece of metal vibrate. These vibrations are directly related to the voice that is causing the current to flow. Therefore, the original speech is exactly reproduced.

The telephone is the most commonly used instrument for electronic communication.

Sometimes very long distances separate the two connected parties. In such cases, amplifiers or repeaters boost the electric current. Many modern telephone circuits convert the varying electric currents to digital pulses. This makes it easier to switch and relay the signals over long distances.

The connection of your telephone to another telephone is called **telephone switching.** Telephone switching is now done by special computers. Such computers are built to perform only that function. You can understand the importance of telephone switching if you consider that any one of the 150 million telephones in the United States today must be able to call any other telephone anywhere in the world.

In the modern telephone network, a smoothly varying electric current is produced by the mouthpiece microphone. This current, an analog signal, is converted into digital pulses similar to computer data. The pulses are then sent to the central office telephone switch. The modern telephone switch is actually a computer designed especially for

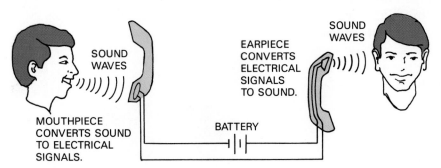

A telephone converts sound to electrical signals and electrical signals back to sound.

telephone switching. At the receiving end, the digital pulses are converted back into analog (smoothly varying) electric current. They are then sent to the earpiece of the receiving telephone.

Since many telephone conversations take place at the same time, switches need to be connected with a large number of wires. In the modern telephone network, microwave and fiber optic cables connect a switch in one town to a switch in another town. Microwave radios can carry more than one thousand simultaneous conversations on one channel, while a single fiber optic cable can carry almost ten thousand conversations at the same time.

The digital switch makes special telephone services like call forwarding and call waiting affordable. Another ad-

The Automatic Telephone Switch: An Invention Born of Necessity

Inventions are often made by people who require their use. In the late 1800s, Almon Strowger was a funeral home director in a small town in the Midwest. The town's telephone operator was the wife of the owner of a competing funeral parlor. When people called the operator and asked to be connected to a funeral parlor, she connected them with her husband's funeral parlor. Mr. Strowger's business declined.

In order to save his business, Strowger designed a device that made telephone connections automatically, bypassing the operator. The switch he invented was the first automatic system to be used in a public exchange. Strowger switches are still in use today in some telephone offices.

(Courtesy of Science Museum Library, London)

Almon Strowger and his automatic telephone switch

(Courtesy of AT&T Bell Laboratories)

vance due to digital technology is the use of voice synthesizers to provide directory information, time of day, and other routine telephone inquiries.

Improvements in telephone technology are occurring at a rapid pace. One new device is the cordless telephone. The cordless telephone includes a small radio transmitter in the part that you carry around with you. This small transmitter sends a signal to the base of the telephone, which is connected by wire to the telephone lines.

Automobile telephones are becoming increasingly popular. A relatively new technology called **cellular radio** is responsible for the growth of automobile telephone systems. Like cordless phones, automobile telephones transmit radio signals. Cordless phone signals, however, can be transmitted only over a short distance. Cellular radio provides reliable communication over a wide area, such as an entire city.

In the future, the telephone will be an even more useful device than it is today. The average home telephone will be connected to a computer. You will therefore be able to find people's telephone numbers without looking them up in a telephone book. The telephone will also call for help automatically when we need medical or security aid.

A cordless telephone
(Courtesy of Tandy/Radio Shack)

Radio

During the mid-1800s, James Maxwell, a Scottish physicist, mathematically predicted that it was possible to send signals through the air to remote locations. Not until the turn of the century, however, were practical devices built that actually transmitted and received radio waves.

On December 12, 1901, in a deserted old hospital in Newfoundland, Canada, the three dots of the letter "S" were heard in Morse code by a young man named Guglielmo Marconi. The Morse signals had crossed the Atlantic by wireless. Marconi had achieved radio communication across the Atlantic. Can you imagine the excitement when people realized that they could communicate over great distances without wires?

The great advantages of radio communication were first recognized by sailors. They finally had a method of rapid communication while at sea, both for routine communications and for emergency distress calls. New inventions, including the vacuum tubes of Fleming and deForest, allowed voice and music to be carried directly on radio waves.

Wavelength and Frequency

All electromagnetic waves travel through free space at the same speed. You may already know what this speed is, because light is a form of electromagnetic wave. Radio waves and light waves travel at about 186,000 miles per second, which is about 300 million meters per second. Thus, a wave sent by a radio transmitter (or a light source) is 186,000 miles (or about 300 million meters) away one second later. This is true no matter what the frequency of the wave is.

During that same second, the radio transmitter generates the number of wave cycles defined by its frequency. Thus, a 10 MHz (megahertz) transmitter will generate 10 million cycles in one second. These 10 million cycles cover 300 million meters in one second. Therefore, the **wavelength,** or distance covered by one cycle, is:

$$\text{Wavelength} = \frac{300 \text{ million meters per second}}{10 \text{ million cycles per second}}$$

$$\text{Wavelength} = 30 \text{ meters}$$

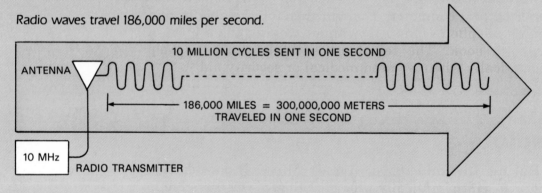

Radio waves travel 186,000 miles per second.

10 MILLION CYCLES SENT IN ONE SECOND

ANTENNA

186,000 MILES = 300,000,000 METERS
TRAVELED IN ONE SECOND

10 MHz

RADIO TRANSMITTER

Regular radio broadcasts began in the 1920s at KDKA in Pittsburgh. Today, you can listen to any one of dozens of radio stations in almost any area of the country.

Modern radios allow us to listen to music, news, sporting events, and other programming. Radios allow us to have two-way communications between our homes and cars, boats, or airplanes. We can even communicate with people who are on the space shuttle or exploring the moon. Most radios use the same basic principles.

Radio transmitters send messages into the air at a specific frequency in the spectrum. All radio frequency electromagnetic signals change polarity (go from positive to negative) on a regular basis. This polarity change defines their **frequency.** Frequency is measured in **cycles per second,** or changes per second. One cycle per second is called one hertz. Radio signals with low frequencies have

long wavelengths. Radio signals with high frequencies have **short wavelengths.**

Different wavelength signals travel differently through the atmosphere. The correct wavelength needs to be chosen for each type of radio communication. For example, high frequency (HF) radio waves bounce off the atmosphere's upper layer, the **ionosphere.** This permits long-distance communication between two points on the earth. Super high frequency (SHF) radio waves are used for satellite communication. They travel straight through the ionosphere.

In radio communication, information is sent through the air from a transmitting antenna to a receiving antenna located some distance away. The channel is the portion of the electromagnetic spectrum used in the air between the transmitter and the receiver. Because each transmitter uses a different frequency, many transmitters can send messages at the same time. This is why you can select one of a number of TV or radio stations.

In radio communication, information is sent through the air from a transmitting antenna to a receiving antenna located some distance away.

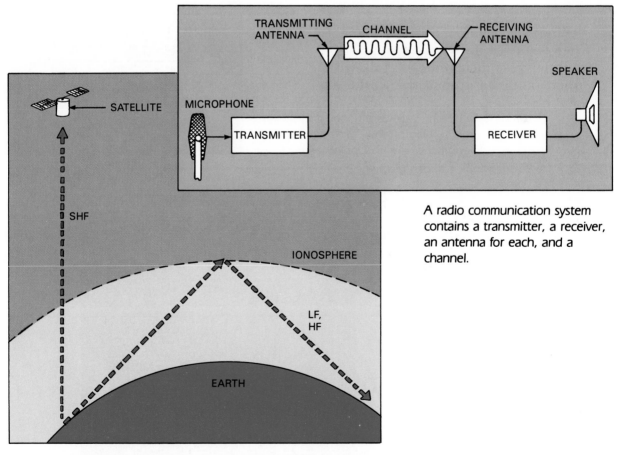

A radio communication system contains a transmitter, a receiver, an antenna for each, and a channel.

Radio signals at low frequencies (LF) and high frequencies (HF) are reflected at the ionosphere. Signals at super high frequencies (SHF) pass directly through.

Amateur Radio

Early in the development of radio technology, experimenters called **radio amateurs**, or "ham" radio operators, made many of the discoveries and advances that led to the use of short wave radio and to improvements in equipment. Today, amateur radio operators often still build their own equipment. They can also purchase a wide variety of commercially-made equipment. In order to use this equipment, they must first pass a written test and be issued a license by the Federal Communications Commission.

Today's amateurs range in age from teenagers to grandparents. They routinely use their radios to talk with other amateurs around the world, as well as to those across town. They perform many public service functions, and often provide the only means of communicating with people in an area hit by a natural disaster, such as an earthquake or a tornado.

Amateur radio ("ham" radio) is a hobby that allows private operators to talk to other hams around the world. This is another example of point-to-point communication. (Photo A courtesy of American Radio Relay League, Photo B courtesy of QST October 1986)

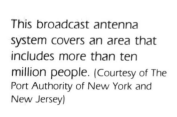

This broadcast antenna system covers an area that includes more than ten million people. (Courtesy of The Port Authority of New York and New Jersey)

Radio communication systems can be used in different ways. For example, a single transmitter can send out a signal intended for many listeners. This kind of operation is called **broadcasting.** You are familiar with AM and FM radio broadcasting. If a transmission is intended for only one receiver, the service is called **point-to-point.** Point-to-point transmissions can provide communications for remote locations (such as fire-ranger towers) or for mobile users (such as a traveling appliance repair truck).

Television

In television, visual information is converted into electrical signals. The signals are transported across a distance. Finally, the signals are changed back to visual images on a screen. Television is used in broadcast communication systems. It is much like radio communication except the kind of information being transported is different.

In a television system, the camera is part of the transmitter. The camera converts the visual scene to electrical signals (**video**). A mechanism then either stores or transports the video signals. This is the channel. Lastly, a monitor converts the video into visual images on a screen. This is the receiver, which may be watched by the viewer. Video signals may be stored in video recorders (for example, a video cassette recorder or VCR). Video may also be transported directly by cable (as in closed circuit TV), sent by cable TV, or sent over the air by using TV transmitters.

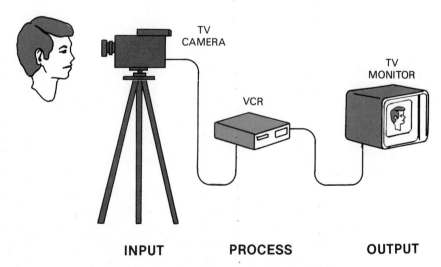

A TV system converts a visual image into an electrical video signal. The signal can be stored in a video recorder or broadcast live.

Using a closed circuit TV system, security guards can watch many areas at once.
(Courtesy of ADT, Inc.)

Cameras change the visual scene into video signals. The signals can be either black and white or color. Color signals break down light into red, blue, and green components for transmission and reassembly at the monitor.

Closed circuit TV requires simple cable connections from camera to monitor. Closed circuit TV is often used to allow overflow audiences at lectures or concerts to watch the event at a nearby location. With closed circuit TV, security guards can watch several areas at the same time.

Broadcast TV is the most common method of transporting video signals. In broadcast TV, transmitters send the video signals on carrier frequencies. The carrier frequencies are determined by the TV channel assigned. TV receivers (TV sets) are actually monitors with tuners. The tuners select the channel. TV transmitters send sound at the same time as the picture.

Television transmissions have a limited useful range from the transmitter to the receiver. Many people live beyond that useful range. To provide good quality reception in outlying areas, cable TV systems use expensive, sophisticated antennas and/or relay links to obtain good TV reception. The system then delivers all of the channels received to subscriber homes over a single cable.

Programming other than broadcast TV can be added to the cable. This programming includes movies, sporting events, concerts, and community news. Because of this extra programming, cable TV systems have been built in most major metropolitan areas, even where regular television reception is good.

Microwave Communication

FM radio, AM radio, TV, microwave radio, and fiber optic communications all have something in common. They all use **electromagnetic waves** to send a message from the transmitter to the receiver. Radio waves, microwaves, and light waves are all electromagnetic waves that occur at different frequencies. As the frequency increases, the wavelength of the signal decreases.

Microwave radios are radios that operate at higher frequencies than FM radios or TV broadcasts. Because their frequencies are higher, their wavelengths are lower. They are therefore called microwave or "little wave." The small wavelength allows the use of small antennas to focus the radio waves into a narrow beam from one microwave station to another. Microwave radios are used to send many telephone conversations from one telephone central office to another and to send TV signals to remote locations.

Each of these antennas is used in a point-to-point microwave radio link.
(Courtesy of Andrew Corporation)

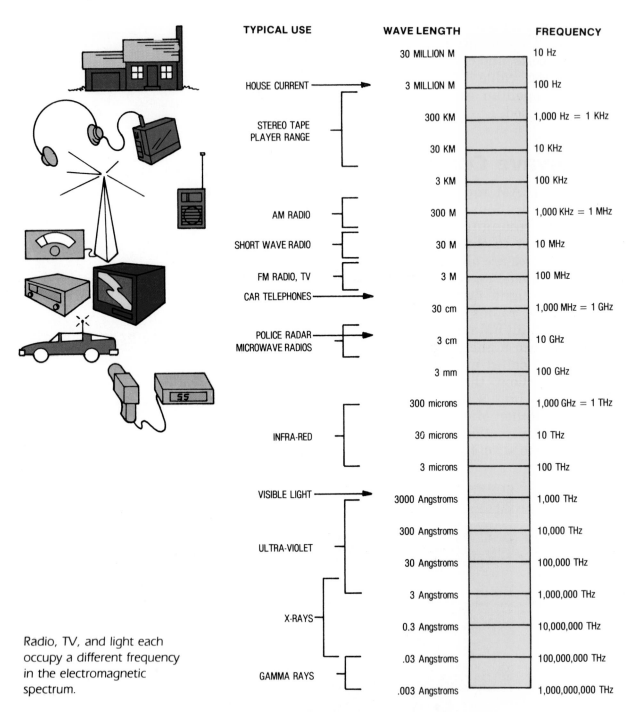

TYPICAL USE	WAVE LENGTH	FREQUENCY
	30 MILLION M	10 Hz
HOUSE CURRENT	3 MILLION M	100 Hz
STEREO TAPE PLAYER RANGE	300 KM	1,000 Hz = 1 KHz
	30 KM	10 KHz
	3 KM	100 KHz
AM RADIO	300 M	1,000 KHz = 1 MHz
SHORT WAVE RADIO	30 M	10 MHz
FM RADIO, TV	3 M	100 MHz
CAR TELEPHONES	30 cm	1,000 MHz = 1 GHz
POLICE RADAR MICROWAVE RADIOS	3 cm	10 GHz
	3 mm	100 GHz
	300 microns	1,000 GHz = 1 THz
INFRA-RED	30 microns	10 THz
	3 microns	100 THz
VISIBLE LIGHT	3000 Angstroms	1,000 THz
ULTRA-VIOLET	300 Angstroms	10,000 THz
	30 Angstroms	100,000 THz
	3 Angstroms	1,000,000 THz
X-RAYS	0.3 Angstroms	10,000,000 THz
GAMMA RAYS	.03 Angstroms	100,000,000 THz
	.003 Angstroms	1,000,000,000 THz

Radio, TV, and light each occupy a different frequency in the electromagnetic spectrum.

Satellite Communication

Some writers have said that satellite communication, more than any other single electronic development, has contributed to the "shrinking" of the globe. Thanks to satellites, we now have instant communication with the most remote parts of the world.

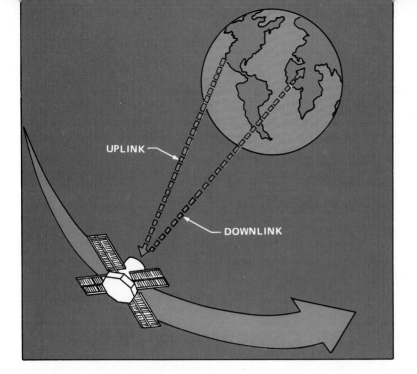

Both the satellite and the earth make a complete revolution in the same time (24 hours). Viewed from the earth, the satellite appears not to move.

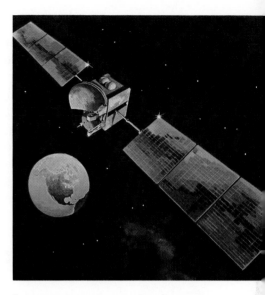

Spacenet I (Copyright 1984 GTE Spacenet Corporation. Spacenet is a registered service mark of GTE Spacenet Corporation)

Communication satellites are really radio relay stations, called **repeaters.** They are located high above the earth and give coverage to a very large area. Several technologies combine to provide satellite communication. Communication satellites are very sophisticated electronic systems. They are housed in specialized space vehicles that must first be manufactured. Then they are transported into space. They must then remain in orbit around the earth.

Perhaps you have tried tying a rope to a bucket of water and swinging the bucket around your head. If you swing the bucket too slowly, the water will fall out as the bucket is traveling over your head. If you swing the bucket faster, the water stays, seemingly stuck to the bottom of the bucket.

A satellite can be thought of as swinging around the earth. Gravity is the "rope" holding the satellite to the earth. At low altitudes, the satellite must go around the earth very quickly to stay up. At an altitude of about 22,500 miles, the satellite needs to move once around the earth every 24 hours to stay up.

Suppose a satellite above the equator circles the earth once in 24 hours. To an observer on the earth, it appears to stay in the same place in the sky because the earth is also turning once in 24 hours. When a satellite appears stationary, it is said to be in **geosynchronous orbit** (**geo** means earth and **synchronous** means locked together).

Since the satellite doesn't appear to move, an antenna pointed at it doesn't have to move. Because the satellite is very high, its signals can be picked up over a very large area of the earth.

A transmitting station on the earth will send a transmission to the satellite on one frequency (the **uplink**). In the satellite, a receiver accepts the signal and shifts it to a second frequency. The satellite transmitter then retransmits the signal at the second frequency (the **downlink**) toward the earth. Anyone with a satellite receiver in the coverage area can receive the transmission. Some systems now allow inexpensive direct home reception of satellite broadcasts.

Fiber Optic Communication

Telephone switches are commonly connected by wires that run for miles on telephone poles or underground. Because of the many conversations that may be going on at the same time, a large number of wires have to be installed on these long routes. The wires are made out of copper, which is heavy, relatively expensive, and sometimes large in diameter. A new technology, using light for communications, is reducing the size, weight, and cost of these long distance cables.

Communication using light as a carrier has come into widespread use in telephone and data networks. A flexible cable, made out of a very thin strand of glass fibers with protective coatings, can guide light around corners. This **fiber optic cable** can carry light for long distances (up to 60 miles) without needing to amplify the light. Usually, many fibers are contained in one bundle.

A controllable, intense light source, such as a **laser** or a **light emitting diode (LED),** is the transmitter in this communication system. Information is put onto the light wave by varying the voltage driving the laser or LED. The fiber optic cable is the channel. The receiver is a light-sensitive semiconductor device called a **photodiode.** The photodiode converts light energy into electricity.

Light is actually very high frequency electromagnetic radiation. It has a frequency far above radio frequencies in the spectrum. (This means that its wavelength is very much shorter.) It has an extremely large bandwidth. It can therefore carry very large volumes of information. A single fiber in a multiple-fiber cable often carries 1,344 telephone circuits or more. Cables are often built with 144 or more such fibers in a single cable. Most major cities now have fiber

A cross section of a fiber-optic cable (Courtesy of AT&T Bell Laboratories)

optic cable carrying a large amount of telephone traffic. Fiber optic cables between cities are becoming more and more common.

DATA COMMUNICATION SYSTEMS

Recent advances in communications have had a lasting effect on how computers are built and used. Communication between computers or from a computer to a terminal, printer, or other "peripheral" (outside) device is called **data communication**. During the early development of the computer, data communication was relatively rare and was often confined to the computer room. All of the peripheral equipment that supported the computer was located in the same room as the computer. People who wanted to use the computer went to the computer room. This kind of arrangement is called **centralized computing**.

Today, new communication facilities are available and smaller and less expensive computers are able to perform sophisticated tasks. It is now common to have small computers located where people live and work, connected to larger computers at other locations. Some computing can be done on the small computers, while the small computers send large, difficult tasks to the large computer to perform. This arrangement is called **distributed computing**.

Computing and communication technologies are merging to produce powerful tools for information processing.

Computer Codes

Computers and other digital equipment send messages back and forth using codes. In digital codes, a group of bits (called a character) stands for a letter, number, or punctuation mark. Characters are usually five to eight bits long.

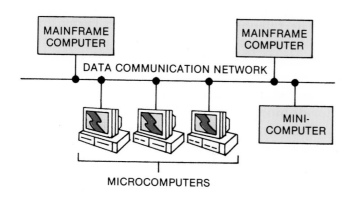

Advances in communications and low-cost computers have made distributed computing possible.

Character	ASCII Code
<	0 111 100
=	0 111 101
>	0 111 110
?	0 111 111
@	1 000 000
A	1 000 001
B	1 000 010
C	1 000 011
D	1 000 100
E	1 000 101
F	1 000 110
G	1 000 111
H	1 001 000

Part of the ASCII code

ASCII (pronounced *as-key*) is a common code that computers and computer devices use to communicate with each other. Another code commonly used in computer communication is **EBCDIC** (pronounced *ebb-sa-dick*), which stands for **E**xtended **B**inary **C**oded **D**ecimal **I**nterchange **C**ode. Both ASCII and EBCDIC are used to represent letters, numbers, and punctuation, but the two codes are different. When a computer device that uses ASCII is required to communicate with a computer device that uses EBCDIC, a code conversion has to be made.

Modems

Telephone lines are often used to carry data communication. Digital equipment cannot use telephone lines directly, however, because data signals are very different from telephone voice signals. To use telephone lines, a device called a **modem** is placed between the digital equipment and the telephone line. A modem turns data signals into audio tones that the telephone line can carry. This process is called **modulation**. At the other end, a similar modem turns the audio tones back into data signals. This process is called **demodulation**. The word "modem" means **mo**dulator-**dem**odulator.

Computers can be connected to each other using modems and telephone lines. In this way, we can communicate data from home to home, home to office, and office to office. Many people are using data communications to obtain

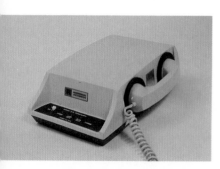

Acoustic couplers are modems that connect computers to the telephone system through a telephone handset. (Courtesy of Anderson-Jacobson Inc.)

Direct-connect modems link computers via the telephone system. (Courtesy of Hayes Microcomputer)

A traveling executive receives messages from his office electronic mail system by means of a hand-held computer and modem at a public telephone.
(Courtesy of RCA)

information from **electronic bulletin board** services. An electronic bulletin board is really a computer that you can call, using your telephone, computer, and modem. You can store messages on the bulletin board for others to read and read messages that they have stored there.

A more personalized form of data communication is **electronic mail**. In an electronic mail system, you are given an "electronic mailbox" that others can put messages in just for you. These messages can be made so that certain other people get copies, just like paper memos in an office. Workers can sort through the mail in their electronic mailboxes very quickly. They can throw away (erase) those messages they don't need, save (store electronically) those they want to keep, and reply (by another piece of electronic mail) to those that need an answer. While no paper ever needs to be used, a paper copy of a message can be made using a printer. Electronic mail systems are used in large companies (private mail systems). Using modems and small computers, individuals at home and at small companies can use public electronic mail systems to communicate with each other.

Ticket agents can make reservations on an airline's computerized reservation system from anywhere in the world through data networks that serve thousands of agents at a time. (Courtesy of United Airlines)

Data Networks

Some computers have many remote terminals or smaller computers connected to them using telephone lines. A number of computers or computer devices joined together is called a **network**. Computer networks exist that have thousands of terminals attached to many large computers. Computer networks allow us to make airline reservations instantly from anywhere in the world and to use automatic banking machines late at night after the banks have closed.

In many offices today, small data networks, called **Local Area Networks**, allow small computers on each desk to share data. Many small computers can also share a single printer, a large amount of memory, or a modem. In addition,

A number of computers or computer devices joined together is called a network.

Many office functions are now automated through a combination of computer and communication technologies. (Courtesy of NCR Corp.)

modern typewriters, copiers, and other office tools contain microprocessors and data communications connections. By connecting these machines together and to the small computers in the office, all of these machines can work together and expand their capabilities and effectiveness. For example, a typewriter connected to an office computer can act either as a typewriter or as a computer printer. The term **office automation** refers to the effective use of computers and communications in an office setting.

Facsimile

Facsimile (FAX) is a method of sending pictures electronically. Newspapers use FAX to send news photos from one location to another. Weather maps have been sent this way since the 1920s. Recently, facsimile transmission has become a very important communication method. Corporations like Federal Express® are using FAX (they call it ZAP Mail) to send pictures and words electronically over long distances. Such companies are now in direct competition with the U.S. Postal Service.

Facsimile transmission involves breaking up an image into patterns of white and black. First, an optical scanning device moves across the page. The white and black areas are converted to electrical impulses. The impulses are transmitted over wires to another location. A facsimile machine at the second location then reverses the process. The electrical impulses are turned back into a black-and-white picture.

Courier services can pick up a document and deliver a copy anywhere in the country within two hours.
(Photo by Dana Duke, courtesy of Federal Express Corporation. All rights reserved.)

RECORDING SYSTEMS

Communication systems that store the message on a magnetic, optical, or other medium are called **recording systems**. The medium, which can be a magnetic recording tape, magnetic disk, phonograph record, optical disk, or some other physical device, can be taken from one place to another. The message can then be played back, or reproduced, at the second location. The message can thus be played many times, even though it was recorded ("sent") only once. Recording systems can store sound (audio), pictures (video), or digital information (data). Data storage has already been described in Chapter 4. We will focus here on sound and video recording.

Stereo

Most record players today are **stereophonic**, which means that they have two sound channels. Two different microphones in two places are used to pick up the original sound when the recording is first made. Each sound channel is amplified separately and imbedded in the record grooves. The stereo phonograph that plays the record also amplifies each channel separately. Each of the two speakers reproduces what the two microphones originally "heard." Because people have two ears, stereo sounds much more natural and complete.

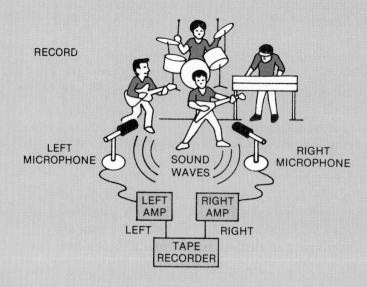

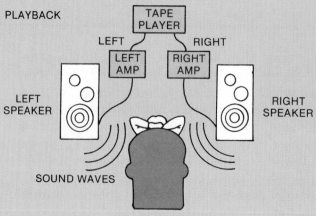

How stereo sound is recorded and played back

Phonograph Records

Sound recording has come a long way since Thomas Edison invented the phonograph in 1877. His device could both record and play back. To record, a metal needle scratched a groove into a rotating cylinder coated with a thin sheet of tin. The depth of the groove depended upon the loudness of the voice or music being recorded. The metal needle could also play back the recording.

Edison's phonograph was able to record and play back sound. (Courtesy of RCA)

Today, phonograph records are made out of durable vinyl plastic. They provide high quality sound when played on modern record players. Records are made by pressing grooves into a flat round plastic disk. When the recording was made, a microphone turned sound waves into electrical signals. The electrical signals were made larger in a device called an **amplifier**. The amplifier then drove a recording stylus (needle) that made grooves (cuts) of varying sizes that corresponded to the sound waves.

When the record is played on your home phonograph, the small pointed needle that rides in the grooves is part of a cartridge that turns the variations in the grooves into small electrical signals. The signals are then made larger in the amplifier, and turned back into sound waves by the speakers.

Compact Disks and Video Disks

A record with grooves in it can pick up dust and dirt. It therefore might be noisy when it is played. A newer kind of record, called the **laser disk**, has no grooves at all. Laser disks are sometimes called **compact disks** because they are smaller than the 12-inch LP phonograph disks.

Before sound is recorded on a laser disk, it is converted into digital bits (see Chapter 4). Laser disks are made of spiral tracks that have been recorded by a laser beam. The

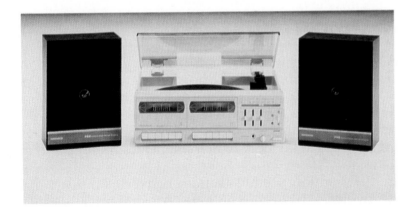

Modern phonographs produce high quality sound reproduction. (Courtesy of N.A.P. Consumer Electronics Corp.)

Compact disks combine optical and digital technologies to produce extremely high quality audio and video recordings. (Courtesy of Sony Corporation of America)

laser beam burns millions of tiny pits into the surface of the disk. Deep pits are made to represent "0s," while shallow pits represent "1s." Even the deep pits are so shallow that you can't tell by eye that they are below the surface. A protective coating is then placed over the entire surface.

When a compact disk is played, a laser beam shines light on the disk. A light-sensing device "reads" the reflection from the pits. A deep pit reflects only a little bit of light. One that is shallow reflects more light. These reflections are converted into digital pulses of electricity, which are then turned into sound.

While compact disks are widely used for stereo sound systems, larger disks using optical storage are also used for video systems. These disks, called **video disks**, also use lasers to read the information stored on the disk. To date, video disks have not been as popular as video cassette recorders. This is because current video disks are used for playback only; that is, home equipment cannot record on video disks. If home equipment becomes available to record on video disks at a relatively low price, they will probably become more popular.

Tape Recordings

A very popular way of recording sound, video, and data messages is on **magnetic tape**. Magnetic tape is a long, thin piece of mylar plastic that has been coated with a metal oxide that can be magnetized. The tape is pulled past an electromagnet called a **tape head**. The head creates a changing magnetic field when the voltage supplied to it changes.

In the recording studio, sound is converted into a varying electric voltage by a microphone. An amplifier boosts the

Modern tape players provide high quality speech and music reproduction.
(Courtesy of N.A.P. Consumer Electronics Corp.)

small electric voltage from the microphone and sends the varying electrical signal to the recording head. As the tape moves past the varying magnetic field in the tape head, the metal oxide coating on the tape is magnetized to a varying degree.

On playback, the process is reversed. The tape is moved past a playback head. The playback head senses the changing magnetic field on the tape and converts it to a voltage. An amplifier boosts the voltage so that it is large enough to drive a speaker. The speaker converts the voltage back to sound waves.

Several different sounds can be recorded on one tape at the same time, if each one is recorded on a different part of the tape. These are called **tracks**. Four track and eight track tapes are common.

In order to store enough tape to last for an hour or more, reels of some kind are used. Tapes wound on open reels, called **reel to reel** tapes, are no longer popular in home use, but are still used widely in studios. Two reels enclosed in a plastic container, called a **cassette**, are very popular for home use. Both audio cassettes and video cassettes (often called video tapes) are in widespread use.

Cartridges use one reel and a continuous loop of tape. In a cartridge, tape is drawn out of the center of the reel, past the heads, and wound back on the outside of the reel. A cartridge can play continuously, but cannot be rewound or backed up as reel to reel tapes and cassettes can.

Recording tape comes packaged in different ways. One familiar package is the cartridge.
(Courtesy of Ampex Corporation, Magnetic Tape Division)

MODERN TELECOMMUNICATION SERVICES

Teleconferencing

A **teleconference** is a conference held with the participants at a distance. With travel costs increasing, it has become desirable for some meetings to be held without the participants meeting face to face. A teleconference connects the participants by telecommunication lines. Several kinds of teleconferences are possible, depending upon the type of communication lines used.

In the simplest form of teleconference, only voice lines are used. The participants hear each other as if they were all sitting in a group. This is sometimes called a conference call. It can be set up by using normal telephone lines. Each participant may use his or her regular telephone. If a telephone equipped with a loudspeaker is used, a small group of people sitting in one room can take part in the teleconference together.

Television can make the conference even more effective. This kind of teleconferencing is often called **videoconferencing.** Videoconferencing is an expensive technology. It is most cost effective over long distances, where travel costs for a normal meeting would be high.

Telecommuting

Using inexpensive, very powerful personal computers, some workers are able to do their normal work at home. Programmers, word processor operators, and data-entry clerks, for example, can work at home. The completed work can be brought to the office on a floppy disk at the end of the day. The work can also be transmitted in digital form to the company's computer by telephone lines and modems.

Telecommuting is attractive to some employees because it allows them to keep flexible work hours. It also reduces the number of times they have to travel to the office. It is particularly attractive for those parents who must watch their children during the normal workday. Telecommuting is attractive to employers because they do not have to provide desks, workspace, and electricity.

As with any new system, there may be undesirable outputs. Employers are concerned that the quality of work will suffer because the benefits of normal daily socializing with fellow employees are gone. Telecommuting employees who work at home may also be distracted by telephone calls and visitors.

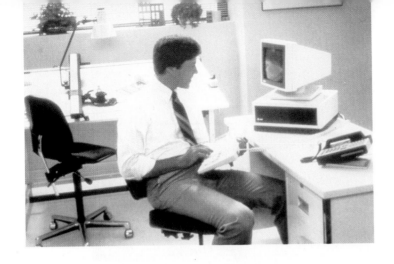

Effective data communication allows some information workers to work at home rather than in an office.
(Courtesy of AT&T Bell Laboratories)

Electronic Banking

The combination of computing and communications has changed the way banking is done. In most parts of the country, automated teller machines (ATMs) are springing up. They are even outnumbering bank branches in some areas. ATMs allow customers to deposit money, withdraw money, determine account balances, and pay bills at any time of the day or night.

ATMs are computer terminals connected by telephone lines to the bank's main computer. A customer may get his or her most current balances or make a transaction at any convenient time. In many areas, groups of banks use common ATMs. This gives customers easier access.

Automated teller machines allow customers to bank at any time of the day or night.
(Photo by Jay Fries with permission, Tandem Computers, Inc.)

Two-way Cable TV

The original cable TV systems could only send signals from the control center (the **head end**) to the subscribers. Many of the newer cable TV systems, however, can carry signals back from the subscribers to the head end. These systems are called two-way cable systems.

Two-way cable TV has made several new services possible. With **pay-per-view,** customers are charged only for the programs actually watched rather than paying a flat monthly fee. Sometimes, the billing is done automatically whenever the pay channel is selected. In other systems, the viewer must request the programming from the cable operator. This kind of system is now available at many hotels.

Shop-at-home allows customers to obtain information on many kinds of products. A keyboard mounted near the TV is used to request information. When a request is received, the products are demonstrated on a TV channel. The demonstrations may be on videotape or videodisk. The customer may then order the product.

Using **data base inquiry,** subscribers can browse through information contained in data bases. The data

Communications are changing the way we shop. (Courtesy of AT&T Bell Laboratories)

bases are maintained by major newspapers, reference organizations, and financial institutions. A viewer could use such a service to scan the latest issue of *The New York Times,* the *Washington Post,* or the *Wall Street Journal.*

SUMMARY

The use of electricity and electronics in communication systems has created the information age, just as steam engines fueled the industrial age. A communication system is called an electronic communication system if the channel uses electrical energy to carry information. In electronic communication systems, messages are sent immediately, or they are stored for transportation or later transmission.

One of the earliest and simplest electronic communication systems was the telegraph. In telegraph systems, codes of long and short sounds (dots and dashes) are used to represent letters, numbers, and punctuation. While the telegraph was a very fast means of communication compared with sending letters by stagecoach or boat, it is slow by electronic standards. Also, only people who knew the Morse code could use it directly.

Today, telephone is the most commonly used electronic communication system. Most modern telephone systems convert voice signals into digital data for switching and transmission. Telephone switches, which automatically connect parties, are special purpose computers. Conventional telephone technology has been combined with radio technology to produce new services such as cordless phones and mobile telephones.

People first used radios to communicate with ships at sea. Radios send information through the air from a transmitting antenna to a receiving antenna located some distance away. Radio, TV, microwave radio, and light are all forms of electromagnetic energy, but each occurs at a different frequency.

In television, visual information is converted into electrical signals that are sent through the air to a receiver that converts them back to a visual picture. Closed circuit TV systems transmit video over cables instead of through the air.

Microwave radios are used to transmit thousands of voice conversations from one telephone office to another. They are also used to relay TV signals from one location to another. Satellite communication uses microwave radios that communicate through a satellite. The satellite is a radio repeater that circles the earth at the same rate that the earth turns, so that it appears not to move when seen from the earth.

Fiber optic cable guides light from one end of the cable to the other. A light source at one end can transmit information that can be recovered at the other end of the cable. Fiber optic cables can carry the same amount of information as many larger, heavier, more expensive copper cables.

Data communication is the communication between computers or from a computer to a terminal, printer, or other peripheral device. Improvements in data communication and the reduction in price of computers have made distributed computing more common than centralized computing. Computer communication is done using standard data codes. Modems are used to convert computer digital signals to audio tones, and vice versa. Data communication over telephone lines usually requires a modem to turn the data signals into audio tones.

A number of computers or computer devices joined together is called a network. Networks can cover a very large area and use telephone lines to communicate, or they can cover only one office.

Sound is commonly recorded on magnetic tape, phonograph records, and laser disks (often called compact disks). Video is commonly recorded on magnetic tape and video disks.

Rapidly evolving communication technologies have brought us new services. We now have teleconferencing, electronic banking, and various services provided by two-way cable TV.

MORSE CODE COMMUNICATIONS

Setting the Stage

A *code* can be defined as a set of signals or symbols that have a specific meaning to both the sender and the receiver of the message. Satellites in earth orbit use coded radio signals to send data back to earth. Computers with modems, teletype machines, and many other devices use coded signals to transmit data.

Your Challenge

Design and build a telegraphy device. Using this device and the Morse code, send a message to another student.

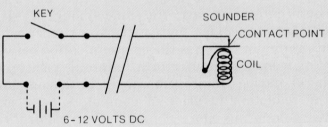

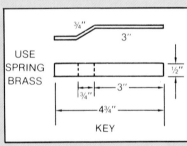

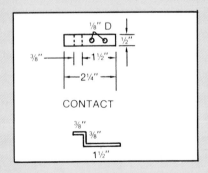

Suggested Resources

Safety glasses and lab apron

Copper magnet wire, enameled (single strand, 22 gauge, about 30' long)

Twin lead stranded wire

Electrical tape

Pine—¾"

Wire cutters

Hot glue

Spring brass

Drill, drill bits, countersink

6 flat head machine screws with hex nuts and wing nuts—1½" × ¼ #20

6 self-tapping screws—½" #6

25 gauge tin plate

35 mm film container cap

1 hex head, soft iron machine bolt with two nuts—3" long, ⅜ × 16 NC

Assorted woodworking and sheetmetal tools and equipment

6–12 volt DC power supply

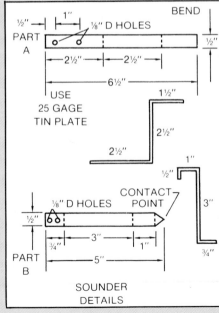

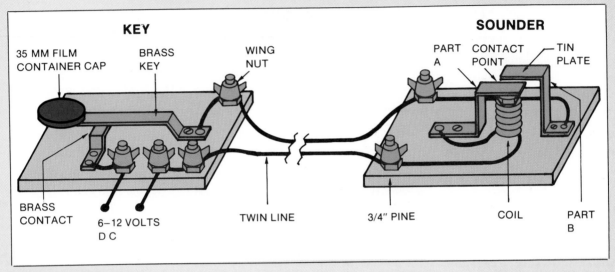

KEY

35 MM FILM CONTAINER CAP

BRASS KEY

WING NUT

BRASS CONTACT

6–12 VOLTS D C

TWIN LINE

SOUNDER

PART A

CONTACT POINT

TIN PLATE

3/4" PINE

COIL

PART B

Procedure

1. Be sure to wear safety glasses and a lab coat.
2. Cut and sand 4" × 6" pine bases.
3. Install the 3" machine bolt end nuts, and the six flat head machine screws and nuts.
4. Using the copper magnet wire, wrap a coil around the length of the 3" machine screw. Four to six layers of wire are okay. Start and finish the coil at the bottom of the bolt. Tape the bottom of the coil in place leaving 8–10" of extra wire at both ends.
5. Make Parts A and B of the sounder as shown and mount to base.
6. Make brass key and contact as shown. Mount to base. Hot glue film container cover to end of key.
7. Wire all parts as shown. The sounder pieces may have to be *slightly* bent for best results. You should hear a buzzing sound if everything is adjusted properly.
8. Send a message to another student.

Technology Connections

1. The communication process consists of a transmitter, a channel, and a receiver. In telegraphy, the Morse code key is the transmitter and the sounder is the receiver. What part of a telegraph circuit represents the channel, or route the message takes?
2. When telegraph lines are run long distances, relay stations are needed. Why?

Science and Math Concepts

▶ A wire carrying an electric current has a magnetic field around it. If the wire is turned into a coil, the magnetic field inside the coil becomes stronger than that formed by a straight wire.

▶ If a soft iron core is placed inside the coil, an *electromagnet* is formed.

▶ The strength of a magnetic field is measured in a unit called the *tesla*.

199

BEAM THAT SIGNAL

Setting the Stage

During the interview with your supervisor, you are informed that you must take a training session to learn about the manufacturing of semiconductors before you begin your public relations job. You are aware that a semiconductor laser is the smallest laser produced today. But how is it produced and why is it so useful in the field of communications?

Your Challenge

Simulate point-to-point transmission of radio communication systems with a helium-neon laser. The signal will be bounced off a satellite and received on the ground.

Procedure

1. Draw a horizontal line across the board 5" up from the bottom of the white board. This line will represent the earth.
2. In each corner of the board, project a line up to the top of the board near the center. This will form a triangle.
3. Measure each base angle and write the degrees near each angle. Use math to find the third angle. (The total number of degrees in a triangle equals 180°.) Check your work by measuring the third angle with a protractor.
4. Place a mirror at the top of the triangle.
5. Place the laser at the bottom of the board in one corner and the receiver in the other corner.
6. Obtain permission to turn on the laser. Adjust the mirror so the beam lines up on the sides of the triangle and hits the receiver.
7. Using the mirror for a guide, draw a line. Measure one angle between the mirror and the triangle. The other angle should be the same.
8. Play music from your tape recorder through the input jack of the laser. This will modulate the laser beam. Adjust the volume on the tape recorder for a clear sound.

Suggested Resources

4—4" × 4" mirrors mounted on ¾" thick plywood
Protractor
Ruler
Framing square
3' × 4' white board
Assorted markers
Audio-modulated helium-neon laser
Laser receiver and amplifier
Tape recorder

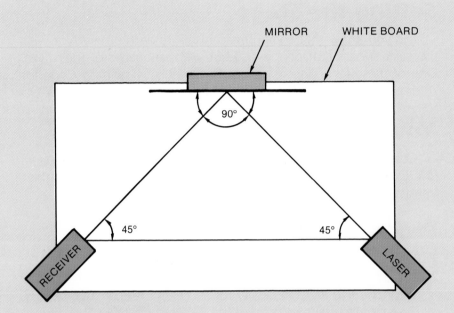

MIRROR   WHITE BOARD

90°

45°   45°

RECEIVER   LASER

Technology Connections

1. Hook a piece of fiber-optic cable between the laser and the receiver. Plug the tape recorder into the input jack of the laser. Again play the tape recorder. This will demonstrate fiber-optic communication.
2. Try the same experiment over a greater distance, such as 30 feet. Omit finding the angles. Keep all the equipment on the floor for safety. The laser beam must not hit your eyes.
3. How does a compact disc make use of a laser to record and play back information?
4. Fiber-optic cable is becoming increasingly common for what type of communication? How is the laser used in this type of communication?

Science and Math Concepts

▶ The angle of incidence equals the angle of reflection.
▶ Laser light is monochromatic (made up of only one color).
▶ The angle at which light spreads is called **divergence.** Laser light does decrease in its intensity as it travels further from the laser, but its divergence is much less than that of other light sources.

THE MISSING LINK

Suggested Resources

Computer with monitor
  and disk drive
Printer
Modem
Telephone line
Telecommunications
  software
Connecting cables

Setting the Stage

Did you know that computers can talk to each other? Not with the same words that people use but with languages of their own. One of these languages is called the *American Standard Code* for *Information Interchange,* or ASCII (pronounced as-key).

Your Challenge

Connect your computer to another computer using a *modem* (*MOD*ulator + *DEM*odulator = MODEM) and telephone lines. Then transfer some data back and forth between the computers.

Procedure

1. Read the instructions that came with the modem.
2. Proper modem connections between the computer and the telephone line are a necessity for successful telecommunications. The manufacturer's recommendations for connecting (interfacing) your hardware should be *carefully* followed.
3. Make sure the telecommunications *software* you are using is compatible with your computer and modem.
4. Get familiar with your software *before* you try to call another computer. Read the directions carefully! Watch for key words like: *baud rate* (transmission and reception speed of the modem), *dialing* a number (touch-tone or pulse dialing), *uploading* (sending a message), *downloading* (receiving a message), *full-duplex* (for connecting to a mainframe computer), *half-duplex* (for calling another microcomputer), disconnecting or hanging up, etc.
5. Use the computer, modem, and communications software to dial the phone number of another telecommunications system. If possible, call a student in another classroom or school.
6. Make sure both computers are set at the same *baud rate* (i.e., 300 or 1200) and in the *half-duplex* mode.
7. Once the connection has been established, take turns typing messages. When you are finished typing your message, send the word *GO.* This tells the other student to begin typing.
8. After your telecommunication is over, be sure to follow the *disconnect* procedure for your modem. *Don't just turn the computer off and walk away!*

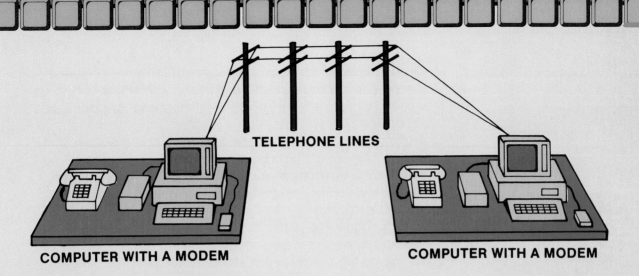

TELEPHONE LINES

COMPUTER WITH A MODEM   **COMPUTER WITH A MODEM**

Technology Connections

1. The communication process consists of a transmitter, a channel, and a receiver. In data communications, a computer with a modem is the sender, and another computer with a modem is the receiver. What part of the system represents the channel, or route the message takes?

2. A number of computers can be connected by telephone lines to form a *network.* What device is needed to connect a computer to a telephone line?

3. Why is it important for all computer manufacturers to use a standard code (such as ASCII) when designing new computers? What would happen if every computer used a different code?

Science and Math Concepts

▶ *Acoustics* is the science of sound.
▶ *Baud rate* refers to the speed at which a modem sends and receives data.

REVIEW QUESTIONS

1. Define electronic communication system. Give three examples.
2. A newspaper uses an electronic typesetting machine to help set up its print. Is the newspaper an electronic communication system? Why or why not?
3. Using the Morse code chart provided in the text, write the Morse code equivalent of "Technology Education."
4. Are telephone circuits analog or digital? Explain your answer.
5. What new products resulted from merging radio technology with telephone technology?
6. A radio signal has a frequency of 30 MHz (million cycles per second). What is its wavelength in meters?
7. Do microwave radios operate at a higher or lower frequency than FM radios?
8. Why is fiber optic cable better than copper wire for large numbers of telephone circuits?
9. How has communication technology changed the way computing is done?
10. What is a modem? How is it used?
11. Why does stereo sound more real and natural than single speaker sound?
12. Would you like to be a telecommuter? Why or why not?

KEY WORDS

Amplifier	Distributed	Local Area	Telecommute
ASCII	computing	Network	Teleconference
Broadcast	Downlink	Modem	Telegraph
Cassette	Electromagnetic	Network	Telephone
Compact disk	wave	Office automation	Tracks
Data	Fiber optic	Stereo	Uplink
communication	Frequency	Tape head	Wavelength

SEE YOUR TEACHER FOR THE CROSSTECH PUZZLE

CAREERS IN COMMUNICATION

The communication field includes a broad range of occupations having to do with research, writing, editing and production. They may be in the areas of education, journalism, publishing, television, business, advertising, public relations, photography, and speech.

OCCUPATION	FORMAL EDUCATION OR TRAINING	SKILLS NEEDED	EMPLOYMENT OPPORTUNITIES
ELECTRICAL ENGINEER AND TECHNICIAN— Electrical engineer designs, develops, tests, and supervises the manufacture of electrical and electronic equipment, such as radios, televisions, radar, industrial measuring and control devices, computers, and navigational equipment. Technicians build and service this equipment.	**Engineer:** Four-year college degree. **Technician:** Specialized training at technical institutes, junior and community colleges, and vocational and technical high schools.	**Engineer:** Excellent math skills, an analytical mind, and a capacity for detail. **Technician:** An aptitude for mathematics and science, and an enjoyment of technical work.	Much greater than average. There will be a strong demand for computers, robots, communications equipment, and electrical products for military, industrial and consumer use.
PHOTOGRAPHER— Takes pictures of a wide variety of subjects. Still photographers specialize in portrait, fashion, or advertising. Industrial photographers provide illustrations for scientific publications.	Entry-level jobs for photographers have no formal educational requirements. Special courses are available in universities, junior colleges, and high schools. Over 100 colleges offer a bachelor's degree.	Good eyesight and color vision, artistic ability and manual dexterity. Photographers should be patient, enjoy working with detail, and have some knowledge of chemistry, physics and mathematics.	Above average. The demand for photographers will be stimulated as business and industry place greater importance upon visual aids in meetings, stockholders reports, and public relations work.
PRINTING PRESS OPERATOR— Prepares and operates the printing presses in a pressroom or print shop.	Apprenticeship or training on the job. Courses in printing, communications technology, chemistry, electronics, and physics are helpful.	Mechanical aptitude is important in making press adjustments and repairs. An ability to visualize color is essential for work on color presses.	Average. Increased use of color printing will contribute to the growth of new jobs, computer-operated equipment will reduce labor requirements.

Data from *Occupational Outlook Handbook, 1986-87*, U.S. Department of Labor

COMMUNICATING A MESSAGE

Objectives

When you have finished this activity, you should be able to:
- Use a computer or word processor to input, store, and retrieve data.
- Assemble text and graphic images for reproduction using manual paste-up or desktop publishing procedures.
- Reproduce a sufficient number of copies using a photocopying process.
- Communicate factual information using an audio medium.
- Use an audio medium for persuasion (advertisement).
- Communicate factual information using a video medium.
- Use a video medium for persuasion (advertisement).
- Identify strengths and weaknesses in communication using print, audio, and video media.

Concepts and Information

Computers and electronics are affecting all areas of our lives. Nowhere has this been greater than in communication technology. Computers have impacted the graphic communication and printing industry as tools for data storage, retrieval, and transmission. Computers are used to create and arrange both text (words) and graphics (drawings and photographs).

Equipment and Supplies

Newsletter

Rough layout
Computer or word processor
Blank diskettes
Software
  word processing
  graphics
  desktop publishing
Modem (optional)
Mouse
Scanner/digitizer
Output device
Four paste-up boards
  11" × 14"
T-square and triangle
  (or equivalent)
X-Acto® knife
Pencil, non-photo blue
Pen, black ballpoint
Ruler
Wax or rubber cement or
  glue stick
Photocopier

Audio Recording

Audio recorder
Microphone
Audio tape

Video Recording

Video recorder
Video tape cassette
Script blanks

Major newspapers and magazines throughout the country use this technology every day. Writers input their stories into computers using keyboards, modems, and optical character readers. Stories are then modified by editors and other staff writers. Photographs are included to help communicate the story to the reader. Many photographs are electronically transmitted. The wording and graphics for advertisements also may be electronically input and manipulated. Entire pages are then electronically assembled by editors and staff using large monitors, and sent to the presses.

Much of our communication uses the audio medium. The records and radio broadcasts we listen to are examples of this. Radio provides us entertainment, news and information. It involves us by allowing our imaginations to create "pictures."

(Courtesy of WRGB-Newscenter 6)

Television, movies, and video have become major parts of our lives. The combining of audio and visual images can help us get a clear understanding of the ideas being communicated. Television and videos entertain us. Advertisements persuade us to buy products. News programs allow us to see what's happening all over the world.

Activity

This project will provide opportunities to use various technical communication systems. You will be able to use current technology in the production of a newsletter, an audio presentation (simulating radio), and a video presentation (simulating television). You will design and communicate a message to inform and to persuade.

Procedure

Newsletter

1. The newsletter will consist of four 8½" × 11" pages. A rough layout may be secured from your instructor. It suggests the content and space available on each page for text and graphics: page 1—News, page 2—Features, page 3—Editorial and Advertisement, page 4—Sports. You may wish to develop your own layout.
2. Write articles. Keep in mind the appropriateness of the subject matter and approximate length (based upon the rough layout).
3. Completed articles should be carefully checked for content, correct spelling and grammar.
4. Write headlines.
5. Drawings may be created on your computer or hand drawn.
6. Each photograph must be converted into a halftone (made up of a series of dots). If you have the facilities you may do this yourself using a photographic process. If you have access to a scanner, photographs may be entered into a computer and digitized to produce a halftone.
7. Select a product or concept to advertise. Write copy and make appropriate graphics.
8. Input and store articles on a disk. (Note: The line length will be determined by examination of the rough layout.)
9. Set headlines and copy for your ad. Keep in mind the line length.
10. Output articles, heads, and graphics in galley form (columns).
11. Proofread the text and heads, carefully checking content, spelling, and grammar. Add or delete copy to fit the article and head to space provided on the rough layout. Examine graphics for appearance and size.
12. Make appropriate corrections.
13. Using electronic publishing, for each page:
    a. Arrange text and graphics. (Examine the rough layout carefully to determine the position of text, graphics, and advertisements.)
    b. Output pages.
    c. Adhere any graphics not generated by a computer following manual paste-up procedures (**step d**).
    d. Add corner marks on each page (see **step f** in "manual paste-up procedures"). If there is not enough space on the ouput to add these corner marks, the output must be adhered to a paste-up board to allow more space.

(Courtesy of Aldus Corporation)

14. Using manual paste-up procedures, for each page:
    a. Output the corrected text, heads, and graphics.
    b. Using a T-square, align the paste-up board onto a drafting board (or light table). Use masking tape to secure each corner.
    c. Carefully measure and draw lines as indicated on the rough layout (paper lines, image lines, locations of articles, heads, and graphics) using a non-reproducing blue pencil, a T-square, and a triangle.
    d. Paste articles and heads in their proper positions using wax, rubber cement, or a glue stick. Use a T-square and triangle to carefully align elements according to rough layout. Make sure all lines are straight and be especially careful to smooth down elements without smearing and keep all work surfaces clean.
    e. Paste graphics (drawings and halftones) and advertisements as described in **step d**.
    f. Use a black ballpoint pen, a T-square, and a triangle to add corner marks (see diagram).
    g. Carefully double-check the alignment of all elements using a T-square and a triangle.
15. Photocopy each page and proofread it. This will be the last chance to correct errors in content, spelling, grammar, and position.
16. Tape a clean sheet of paper over the completed page for protection.
17. Photocopy as many copies of the completed page as desired.
18. Finish the newsletter by stapling the sheets together.

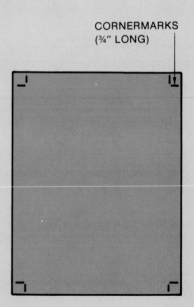

CORNERMARKS
(¾" LONG)

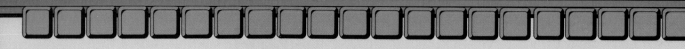

Audio Tape

The audio tape will simulate a radio broadcast. It will consist of two parts: (1) a factual or descriptive component and (2) an advertisement (for the same product as presented in the newsletter).

1. Identify the target population of the "broadcast" (age, gender, interests, etc.).
2. Decide the style to be used in the presentation (news, interview, dramatic, etc.).
3. Select the factual or descriptive content to be included. One of the stories from the newsletter could be used.
4. Examine the content of the advertisement used in the newsletter. How might it best be communicated in a radio broadcast?
5. Write a script. Keep in mind that an audio tape allows each listener to create "pictures" from the script.
    a. Outline the key topics.
    b. Write a specific narration (wording).
    c. Indicate the use of sound effects (music, special effects, etc.).
    d. Edit the script for appropriate content, grammar, and sequence.
6. Rehearse.
7. Produce a preliminary tape. Check your tape for background noise (unwanted sounds picked up by the microphone).
8. Review the preliminary tape. Revise the tape based on your review.
9. Record the "broadcast."
10. Review the final tape.

Video Tape

This medium will allow you to present a clear picture to the viewer, not just use words to help the "listener" create those images.

1. Identify the target population of the "broadcast" (age, gender, interests, etc.).
2. Determine the style to be used in the presentation (news, interview, dramatic, etc.). Think of television broadcasts you have seen to help get ideas.
3. Examine the content of the advertisement. How might it best be communicated using the video medium?
4. Outline the script.
5. Produce a preliminary storyboard. This will be an idea for each video image with accompanying script ideas.
6. Produce the script. Use script blanks (available from your teacher) to indicate each video image and accompanying audio.
    a. Each video image is represented by a rough sketch in the frame provided. This may also indicate camera angle and the depth of the picture.
    b. Exact wording should accompany each picture.
    c. Additional audio (music, sound effects, etc.) should be indicated.
    d. Review the script for appropriate content, grammar, and sequence.

7. Rehearse (check timing, script, and sequence).
8. Produce the preliminary video tape. Check for background "noise" (unwanted sounds or images).
9. Review the preliminary tape. Revise the tape on your review.
10. Record the "broadcast."
11. Review the final tape.

Review Questions

1. Discuss the roles of the seven technological resources in the completion of this activity.
2. How did the three parts of this activity attempt to inform, persuade, educate, and entertain?
3. What information is provided by a rough layout?
4. Discuss the advantages of desktop publishing and manual paste-up procedures.
5. In producing your newsletter, what methods were used to create graphics and to convert photographs to halftones?
6. List three methods that could have been used to input data for your newsletter.
7. What roles did electronic technology play in completion of the following activities: newsletter, audio tape, video tape?
8. What is "noise" in relation to audio and video? Why is it an important factor to consider?
9. What are the similarities and differences between a rough layout (newsletter) and a script (audio and video)?
10. What advantages and/or disadvantages did you discover in communicating using each of the following three media: (1) print, (2) audio, and (3) video?

SECTION

3

PRODUCTION

CHAPTER 8

PROCESSING MATERIALS

MAJOR CONCEPTS

After reading this chapter, you will know that:

- Processing materials involves changing their form to make them more useful to people.
- The changing, or conversion, of resources occurs within the technological process.
- Technological systems convert raw materials into basic industrial materials. Basic industrial materials are then converted into end products.
- There are several ways of processing materials. These processes are forming, separating, combining, and conditioning.
- Computers can control machines used to process materials.
- Materials are chosen on the basis of their mechanical, electrical, magnetic, thermal, or optical properties.

INTRODUCTION

To make material resources more useful, we process them. **Processing resources** means changing their form. For example, **raw materials** such as wood can be made into furniture. Cotton fiber can be processed into thread and then made into clothing by spinning and weaving. Paper can be made from wood chips. Animal hides can be made into shoes, handbags, and coats. Plants and animals can also be processed to provide food.

Processing materials involves changing their form to make them more useful to people.

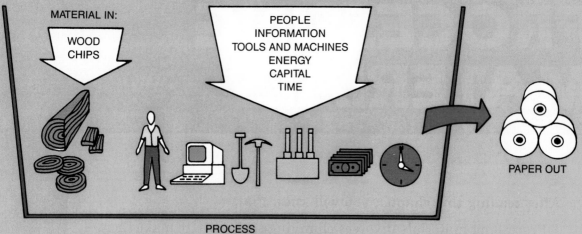

OTHER TECHNOLOGICAL RESOURCES IN:

MATERIAL IN:

WOOD CHIPS

PEOPLE
INFORMATION
TOOLS AND MACHINES
ENERGY
CAPITAL
TIME

PAPER OUT

PROCESS
(CRUSH WOOD CHIPS, MIX WITH WATER, BEAT, PRESS, AND DRY)

In a material conversion process, other technological resources are needed to convert materials from one form to another.

PROCESSING MATERIAL RESOURCES

The changing, or conversion, of resources occurs within the technological process.

Industries such as the timber and mining industries are called **primary industries** because they produce our **primary,** or most basic, **raw materials.** Primary raw materials are taken from the earth and then processed into **basic industrial materials.** Timber (from trees) is an example of a primary raw material. Timber is processed into standard-size wooden planks. The wooden planks are examples of industrial materials. They are used by industry to make end products like furniture and houses.

Another raw material, natural rubber (latex), is processed into a basic industrial material by heating it and adding sulfur dust. The process is called **vulcanization.** This process makes the rubber able to withstand extreme temperature changes. Likewise, iron ore, limestone, and coke are processed into standard-size steel slabs, sheets, rods, or beams. Crude oil is processed into basic chemicals, from which plastic is made.

Most of our products are made from these basic industrial materials. We need to use all the other technological resources whenever we convert material from one form to another. We generally add value to a material when we process it.

Technological systems convert raw materials into basic industrial materials. Basic industrial materials are then converted into end products.

The aluminum industry processes bauxite ore into many different products. (Courtesy of Aluminum Company of America)

215

TYPES OF INDUSTRIAL MATERIALS

Among the great variety of industrial materials are wood, metal, ceramics, and plastic. Industrial materials are chosen on the basis of how well they suit a particular use. For example, plastic does not conduct electricity. Therefore it can be used to cover (insulate) bare wires. Because certain woods are strong and beautiful, they are used for furniture.

Wood is a material often chosen by craftspeople and builders. (Courtesy of Boise Cascade Corp.)

Wood

Wood is classified as **hardwood, softwood,** and **manufactured board.** Hardwoods come from trees like maples, oaks, and poplars. These trees have broad leaves and are called **deciduous** trees. Softwoods come from trees with needle-like leaves, such as pine and fir. These are called **coniferous** trees. Most of the time, hardwoods are hard and softwoods are soft. However, strictly speaking, the terms "hardwood" and "softwood" refer only to the kinds of leaves trees have. Some woods, like balsa and basswood, are classified as hardwoods even though their woods are soft.

Manufactured board like plywood and particleboard is made from wood chips and sawdust. This is often stronger then the original wood and will not warp or twist.

Wood is a renewable resource. It is available, affordable, and can be quite strong. Most kinds of wood can be easily cut and shaped. Wood has a variety of beautiful colors and grain patterns. Therefore, it is a favorite material for craftspeople and builders.

Metal

Iron and the steel made from it are probably the most important metals used in modern industry. They are turned into end products like automobiles, skyscrapers, and machine tools. Metals made from more than 50 percent iron (chemical symbol Fe) are called **ferrous** metals. Steel is a ferrous metal. Metals without a large amount of iron are called **nonferrous** metals. Aluminum, copper, magnesium, nickel, tin, and zinc are nonferrous metals. **Alloys** are combinations of metals. Brass is an alloy of copper and zinc. Bronze is an alloy of copper and tin. Pewter is an alloy of tin, antimony, and copper. Steel alloys can be made with special properties. For example, **stainless steel** includes some chromium and resists rust.

Electromagnetic ingot casting is being used to produce these aluminum ingots. (Courtesy of Alcoa Aluminum Corp.)

How Steel is Made

Steel is the world's most useful metal. It is used to make everything from tiny paper clips to gigantic bridges. Steel costs little to produce. It is made from iron ore, which is a common material. Iron ore is found in the earth's crust mixed with rock and earth.

To make steel, iron ore (iron oxide) is melted with limestone and coke in a **blast furnace.** The intense heat burns the oxygen away from the iron oxide. The result is molten iron. Some of the molten iron (called **hot metal**) is cast into molds. It solidifies into **pig iron.** The impurities rise to the top of the molten iron and are removed as **slag.**

Most of the hot metal is refined into steel in **basic oxygen furnaces (BOFs).** Oxygen is blown into the top of the BOF at very high speed. There it combines with the carbon in the molten iron and converts the iron to steel. Steel is also made in **open hearth furnaces** and **electric furnaces.** An open hearth furnace is a shallow furnace heated by burning gases. An electric furnace produces heat using an electric arc. Some of the new, smaller steel mills, called **mini-mills,** use electric furnaces to produce specialty steels.

From the furnace, the molten steel is processed in two ways. First, it can be cast into huge blocks called **ingots.** The ingots are then formed into basic steel shapes. These are rolled into finished products, like bars, rods, structural shapes, and sheets.

The molten steel can also be converted directly into basic steel shapes by means of continuous casters. In this way, ingot production is bypassed.

Coke is made by burning coal. When the coal is burned, gases, oils, and tar are removed. The coke that remains serves as fuel for the blast furnace.

In the blast furnace, a blast of air burns the coke. The resulting heat and gases remove oxygen from the iron ore. Pure molten iron collects in the bottom and is drawn off every few hours.

Huge ladles are used to pour molten steel into an ingot mold.

Electric furnaces like this one used to produce only special steels like tool steel and stainless steel. They now produce high volumes of regular (carbon) steel as well.

In this continuous casting process, an even flow of steel falls vertically down through rollers. The path of the steel becomes horizontal at the base. The steel slab is then cut into pieces.

Rolling mills convert ingots into basic shapes called slabs, billets, and blooms.

Slabs are further processed into steel sheets.

Fiberglass is a ceramic material used in the manufacturing of cars.
(Courtesy Ferrari, S.P.A.)

Ceramics

Ceramic objects are made from clay or similar inorganic (nonliving) materials. In addition to clay, ceramics include plaster, limestone, cement, and glass. To produce clay products, the clay is mined and combined with other ingredients. It is then fired in an oven called a **kiln** to around 2,000° Fahrenheit. Firing makes the clay very hard. The clay can then be coated with a glass-like material called **glaze.** The glaze protects its surface and gives it color. Ceramic dishes are common because clay is an abundant material and therefore costs very little. When glazed, ceramic dishes can be very attractive and are easily cleaned.

Ceramics generally do not conduct electricity well. Therefore, they are used as insulators by the electrical industry. Wires carrying high-voltage electricity from power stations to your home use ceramic insulators. Ceramics are also used in electronic components like light bulb sockets, switches, and capacitors.

Ceramic materials can withstand high temperatures and still retain their strength. For this reason, special blends of ceramic and metal are used inside automobile and rocket engines.

Glass is a ceramic material. Most glass is made by melting sand, lime, and sodium oxide together at a temperature of around 2,500° Fahrenheit. This type of glass is called **soda-lime glass.** It is used to make containers, bottles, light bulbs, and window panes.

Fiber glass, used for insulation, is made by dropping molten glass onto a rapidly rotating steel dish. The dish has

Ceramics are common industrial materials. (Courtesy of Corning Glass Works)

hundreds of tiny holes in it. Tiny glass fibers are spun out of these holes by centrifugal force.

Fiber optics is a technology that uses thin glass threads to carry light, as wires carry electricity. Information can be transmitted in the form of light. Thin, lightweight glass fibers can carry as much information as thick, heavy copper cables.

Plastic

Plastic is different in structure from materials like wood, metals, and ceramics. Most materials are made of small groups of molecules connected together in patterns. Plastics consist of long chains of molecules. These chains are called **polymers.** In the Greek language, "poly" means many, and "mers" means parts. Most polymers are synthetic materials (made by humans rather than found in nature).

Thermoplastics are plastics that soften when heated. Like wax, they can be melted and formed into shapes. When they are allowed to cool, thermoplastics regain their original hardness. Acrylic fibers (like nylon and Orlon), polyethylene (used for plastic bags), polyvinylchloride or PVC (used for plumbing pipe and electrical insulation), and vinyl are all thermoplastics.

Thermoset plastics, like bakelite and Formica, do not soften when heated but char and burn. They are less common than thermoplastics. "Unbreakable" plastic cups and dishes are made from melamine, a thermoset plastic.

FORMING PROCESSES

Forming a material means changing its shape without cutting it. For example, metal can be bent into shape. It can also be formed by other methods.

There are several ways of processing materials. These processes are forming, separating, combining, and conditioning.

Casting

One way of forming material is called **casting.** Castings are made from molds. When you walk on the beach and press your feet into the moist sand, you make a mold of your foot. If you melted metal, poured it into the sand mold, and let it cool, you would have a casting of your foot. We use the casting process to make ice cubes. We pour water into an ice cube tray. When the water freezes, we have made solid ice cube castings.

Molds can be one-piece molds or can be made from several pieces of material. The footprint is a one-piece mold. Another example of a one-piece mold is a cake pan. We pour the cake mix into the pan. Because the cake takes the shape of the pan and hardens, we have performed a casting process.

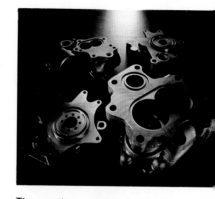

The casting process is used to make pump and transmission housings.
(Courtesy of Cross and Trecker Corp.)

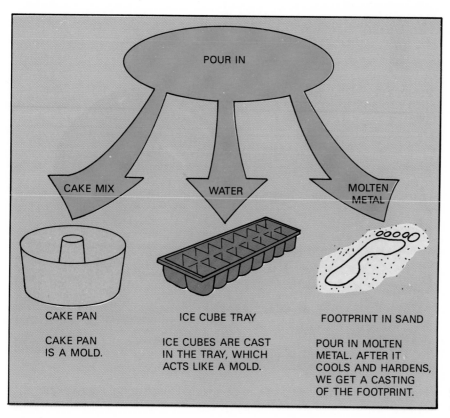

POUR IN

CAKE MIX   WATER   MOLTEN METAL

CAKE PAN   ICE CUBE TRAY   FOOTPRINT IN SAND

CAKE PAN IS A MOLD.

ICE CUBES ARE CAST IN THE TRAY, WHICH ACTS LIKE A MOLD.

POUR IN MOLTEN METAL. AFTER IT COOLS AND HARDENS, WE GET A CASTING OF THE FOOTPRINT.

Some common one-piece molds

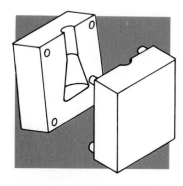

A two-piece mold for casting ceramic objects

We can use a two-piece mold to cast ceramic (clay) objects. First we make a two-piece mold from plaster. We then pour a liquid clay called **slip** into the mold. After allowing the clay to set for a few minutes, we pour the excess slip out. The plaster absorbs water from the slip and leaves a thin wall of clay inside the mold. When the clay dries, the mold is opened and the finished casting is removed.

Pressing

An important industrial forming process very similar to casting is **pressing.** In pressing, an exactly measured amount of material is poured into a mold. Then a plunger with its own shape is lowered down, forcing the material to spread out and fill the mold. This presses the material into the shape of the mold at the bottom and into the shape of

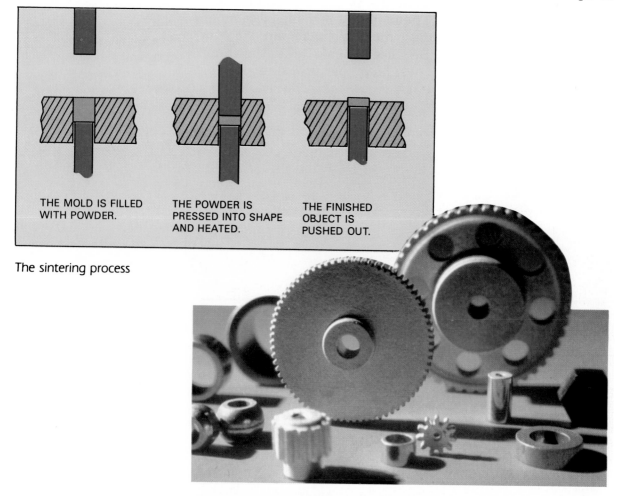

THE MOLD IS FILLED WITH POWDER.

THE POWDER IS PRESSED INTO SHAPE AND HEATED.

THE FINISHED OBJECT IS PUSHED OUT.

The sintering process

Finely ground metal powders can be sintered into strong, lightweight objects, like these gears and bearings. (Courtesy of Aluminum Company of America)

the plunger at the top. The plunger is then raised, and the object is removed. When we make hamburger patties, we press the meat into shape with our hands. Waffles are pressed by waffle irons.

Sometimes powdered metal is pressed to produce objects. The powder is placed into the bottom of a mold. The top of the mold is lowered, and great pressure forms the powder into a solid mass. The object is then heated to make the particles bond together. This pressing process is called **sintering.**

Forging

A **forging** is a metal part that has been heated (not melted) and hammered into shape. Many metal parts are made by the forging process. In earlier times, forging was done by hand. Blacksmiths used to heat up pieces of metal and hammer them into shape for horseshoes.

Today, metal forging is usually done by large machines. A piece of heated metal is placed in the lower half of a mold. A powerful downward ram presses the metal into shape with as much as 2,500 tons of force.

Extruding

Extruding is another method of forming. In this method, softened material is squeezed through an opening, like when we squeeze a tube of toothpaste. The material

These truck wheels are made from forged metal. (Courtesy of Alcoa Aluminum Corp.)

In days past, hand forging was a common method of forming tools and other useful items.
(Courtesy of The Citizens & Southern National Bank of South Carolina)

A large forge press (Courtesy of Cerro Metal Products)

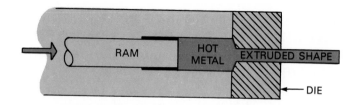

In extrusion, hot metal is forced through a hole called a die. The metal takes the shape of the die.

takes the shape of the opening. Suppose, for example, the hole at the end of the toothpaste tube is square. The ribbon of toothpaste would come out in a rectangular block. Extruding objects can save labor because the finished product does not require much additional shaping and machining.

Blow Molding and Vacuum Forming

Blow molding and **vacuum forming** are processes used to form plastic. In these processes, a thin plastic sheet is heated until it is softened. In the blow-molding process, air blows the softened plastic into a mold. Plastic bottles are made by blow molding. In the vacuum-forming process, a vacuum draws heated and softened plastic down. The plastic forms itself perfectly against whatever it is being drawn toward. Vacuum forming is used to package products that hang on cardboard cards in supermarkets and toy stores. A plastic sheet is drawn down against the object being packaged. This is known as blister packaging.

OTHER RESOURCES IN

MATERIAL RESOURCE IN

- CASTING
- PRESSING
- EXTRUDING
- FORGING
- BLOW MOLDING
- VACUUM FORMING

FORMED MATERIAL OUT

Some forming processes

SEPARATING PROCESSES

When we use knives, saws, or scissors we are **separating** one piece of material from another. We use a separating process every day when we cut our food with kitchen knives. Cutting is a separating process. There are many different kinds of tools and machines used for cutting. Some have been designed to cut specific materials. For example, there are special cutting tools for wood, metal, plastic, clay, leather, and paper. There are tools and machines that cut materials by **shearing, sawing, drilling, grinding, shaping,** and **turning.**

Shearing

Shearing is using a knife-like blade to separate material. We can either use one blade, like a knife, or two blades, like a pair of scissors. In fact, another name for scissors is shears.

When we cut material by shearing it, the knife-like edge actually compresses the material being cut. When the force on the material is high enough, the material breaks along the line of the cut. The sharper the edge on a knife, the greater the force on the actual area of the material being cut.

Sawing

Sawing involves separating material with a blade that has teeth. Each tooth actually chips away tiny bits of material as the saw cuts. Two kinds of sawing processes are used to cut wood. **Ripping** is a process used to cut wood in

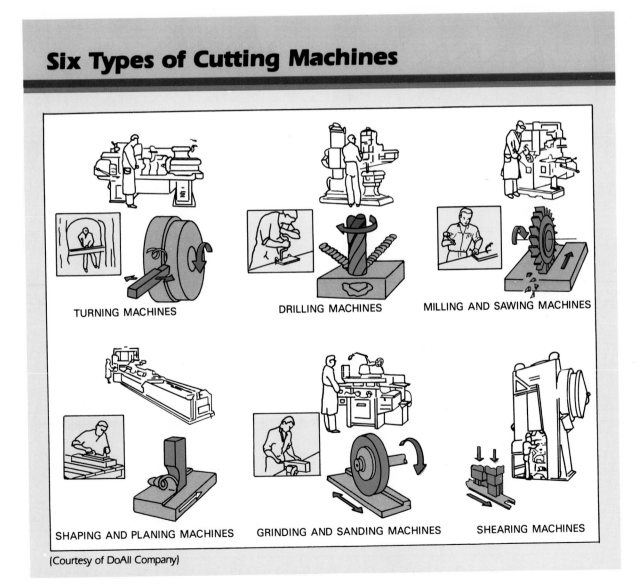

Six Types of Cutting Machines

TURNING MACHINES

DRILLING MACHINES

MILLING AND SAWING MACHINES

SHAPING AND PLANING MACHINES

GRINDING AND SANDING MACHINES

SHEARING MACHINES

(Courtesy of DoAll Company)

Saws and Sawing

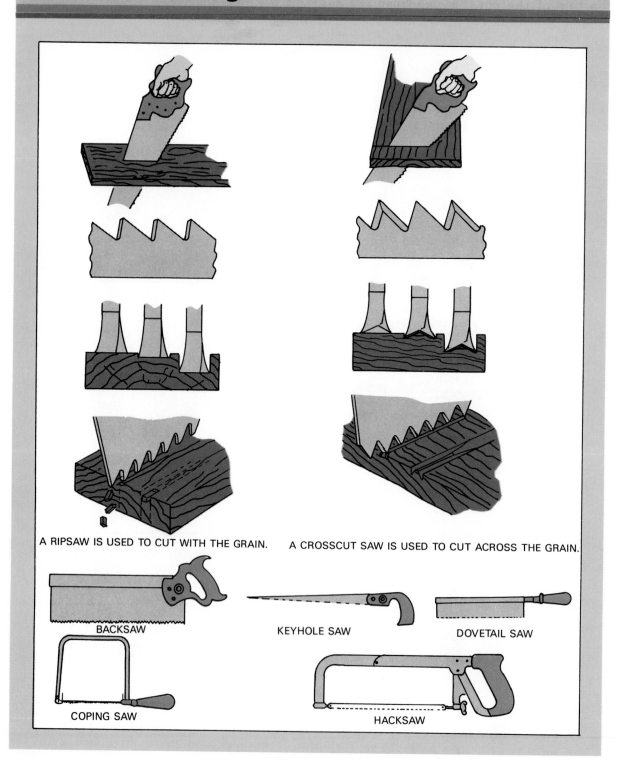

A RIPSAW IS USED TO CUT WITH THE GRAIN.

A CROSSCUT SAW IS USED TO CUT ACROSS THE GRAIN.

BACKSAW

KEYHOLE SAW

DOVETAIL SAW

COPING SAW

HACKSAW

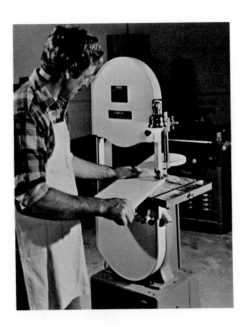

Operator using a band saw (Courtesy of Delta International Machine Corp.)

the direction of the grain. **Crosscutting** is a process used to cut across the grain. Hand rip and crosscut saws generally have from six to ten teeth per inch. Metal is cut by hand with a saw called a **hacksaw.** This type of saw has a very hard steel blade. The blade generally has about eighteen teeth per inch.

Saw blades can be made in circular shapes and in continuous bands. Some saws, such as radial arm saws and table saws, use circular saw blades. These blades are carefully made. Each tooth is very sharp. Since the blade spins rapidly, many cuts per minute are made on the material being sawed.

Drilling

Drilling is a separating process used to cut holes in materials. A pointed tool with a sharpened end (twist drill) is rotated very rapidly by a hand drill or an electric drill. Drilled holes can be made as small as 1/10,000 of an inch or as large as 3 1/2 inches.

Grinding

Grinding is done by tools like grinders and sanding machines. Grinding tools make use of pieces of very hard material called **abrasives.** Abrasives are crushed into very small particles. These small abrasive particles can then be

Drills and Drilling

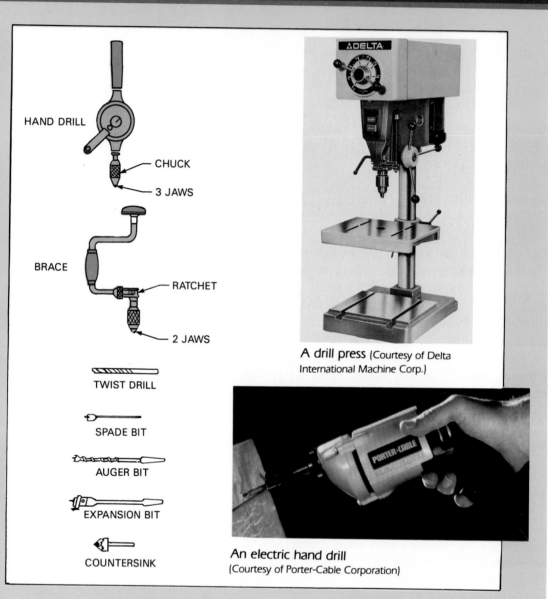

HAND DRILL

CHUCK

3 JAWS

BRACE

RATCHET

2 JAWS

TWIST DRILL

SPADE BIT

AUGER BIT

EXPANSION BIT

COUNTERSINK

A drill press (Courtesy of Delta International Machine Corp.)

An electric hand drill
(Courtesy of Porter-Cable Corporation)

glued onto a flat sheet to make sandpaper or emery cloth. The particles may also be formed into grinding wheels. As the tool operates, the abrasive particles rub up against and cut away tiny pieces of the material.

Often, we sharpen tools by grinding them. Grinding removes just a little bit of material at a time. With proper skill, we can control the cutting action very well.

Notice the safety shields over this grinding machine and the safety glasses worn by the operator. Sparks and tiny particles of material can fly into the eye unless proper safety precautions are taken. (Courtesy of Delta International Machine Corp.)

Polishing is really a form of grinding. Polishes usually contain some very fine abrasive powder. When you rub an object with a polish, you are actually removing microscopic pieces of the surface of the material. This makes the surface smooth, clean, and shiny. Toothpaste is really a very fine abrasive material. When we brush our teeth, we are using abrasive action to remove plaque.

Shaping

Shaping tools are used to change the shape, or contour, of a piece of material. Chisels and planes are examples of hand tools used for shaping. Shaping tools have cutters with chisel-like edges that chip away material. Jointers, planers, and shapers are examples of shaping machines.

Turning

Turning tools are also used to shape materials. A turning tool differs from other shaping tools in that the cutting tool doesn't move. Instead, the material to be cut (the workpiece)

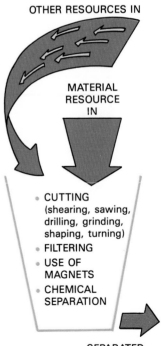

OTHER RESOURCES IN

MATERIAL RESOURCE IN

• CUTTING (shearing, sawing, drilling, grinding, shaping, turning)
• FILTERING
• USE OF MAGNETS
• CHEMICAL SEPARATION

SEPARATED MATERIAL OUT

Some separating processes

Wood lathes are used to make cylindrical objects, such as legs for tables and chairs. (Courtesy of Delta International Machine Corp.)

moves. A lathe spins the workpiece very rapidly. The cutting tool is held against the rotating material and cuts circular grooves. When the cutting tool is moved from left to right, it reduces the diameter of the workpiece. Very intricate and beautiful shapes can be made using lathes.

Other Separating Processes

Materials can be separated by means other than cutting. Materials may be separated **chemically,** like when water is separated into hydrogen and oxygen. We can obtain salt from salt water by letting the water evaporate. **Filtering** is a process that can separate liquids from solids in a mixture. For example, the vegetables in a can of vegetable soup can be separated from the liquid by pouring the contents through a filter or a strainer. **Magnets** can separate magnetic materials from nonmagnetic materials.

COMBINING PROCESSES

Sometimes materials must be combined with other materials. We might want to **fasten** materials to each other. We might want to **coat** materials with a protective finish. Or, we might want to develop a **composite** material that has improved properties.

Fastening Materials by Mechanical Means

By using nails, screws, and rivets, we can fasten materials together mechanically. Fastening devices are made specifically for certain materials. Nails are used to fasten two pieces of wood. We would not use nails to fasten two pieces of metal.

Years ago, nails were made individually, by hand. The ends of iron rods were heated red-hot. They were then hammered into a pointed shape and snapped off at the right length. Nails were very expensive because it took a long time to make them. Most houses were put together without nails. Today, automatic machines make nails by the thousands. Because the price of nails has dropped, it has become easier and cheaper for people to build homes.

For maximum holding power, nails should be driven into wood at right angles to the grain. When you are nailing two boards together, the nail should be long enough so that it goes two-thirds of the way through the second piece of wood. If the wood is very hard, the nail might split the wood. To avoid this, drill holes slightly smaller than the diameter of the nail in the wood. These holes, called **pilot holes,** make it easier for the nail to enter the wood.

There are five common types of nails: brads, finishing nails, casing nails, common nails, and box nails. Nails are purchased according to their size. The size of nails is measured in **pennyweight.** Pennyweight used to refer to the amount of money it cost to buy one hundred nails. An eight-penny nail (abbreviated 8d) used to cost eight cents per hundred. Now pennyweight refers only to nail length.

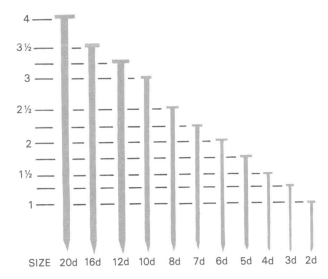

Relative sizes of nails

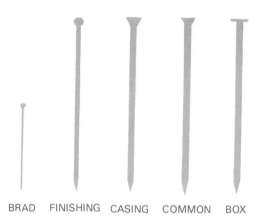

Types of nails

BRAD   FINISHING   CASING   COMMON   BOX

Screws can be used to draw one piece of material tightly against another. Screws provide more holding power than nails. Screws may also be removed easily. Different types of screws are available for use in metal and wood. The threads of wood screws start at the point and go about two-thirds of the way toward the head.

Screws used for fastening pieces of thin sheet metal are called **sheet metal screws.** These are threaded all the way to the head. Pilot holes are required for sheet metal screws because a sheet metal screw can't start itself in a piece of metal. The pilot hole should be drilled equal to the diameter of the body of the screw, without the threads. This is called the **root diameter** of the screw. Sheet metal screws make their own thread as they are threaded into a piece of sheet metal.

Machine screws and bolts do not have pointed ends. The thread is the same diameter from tip to head. Machine screws and bolts are held in place by a **nut** or a threaded hole. The nut has the same size thread as the machine screw or bolt. **Washers** are used between the nut and the material being fastened. Flat washers protect the material from being damaged by the nut. Lock washers keep the nut from loosening under vibration.

When you are driving screws, it is important to choose the correct size screwdriver. A screwdriver with too wide a blade will slip out of the screw slot and damage the material being worked on. A screwdriver with too narrow a blade could damage the screw slot and might not be powerful enough to drive the screw. **Never back up a screwdriver with your hand because the screwdriver may slip.**

Riveting is used frequently in the aircraft industry. Rivets hold pieces of sheet metal together. One end of a rivet is already formed. The other end is formed after it is placed through the two pieces of material to be fastened.

Types of Screws

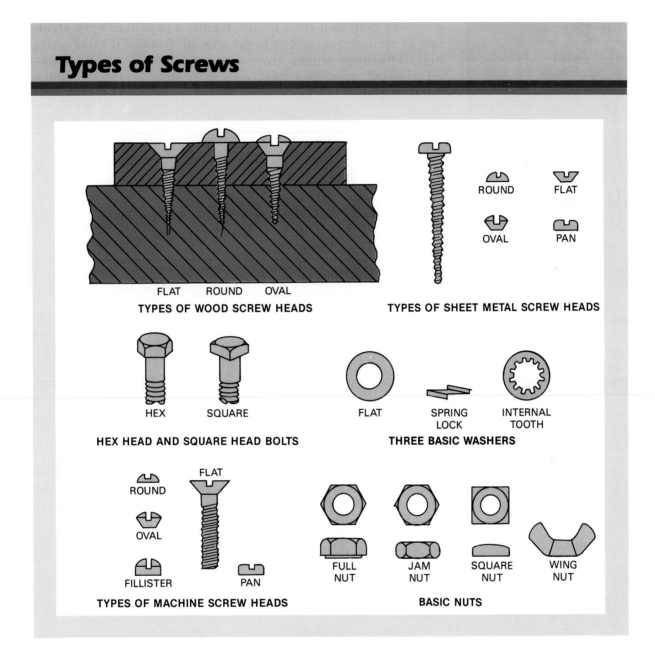

TYPES OF WOOD SCREW HEADS

ROUND   FLAT   OVAL   PAN

TYPES OF SHEET METAL SCREW HEADS

HEX   SQUARE

HEX HEAD AND SQUARE HEAD BOLTS

FLAT   SPRING LOCK   INTERNAL TOOTH

THREE BASIC WASHERS

ROUND   FLAT   OVAL   FILLISTER   PAN

TYPES OF MACHINE SCREW HEADS

FULL NUT   JAM NUT   SQUARE NUT   WING NUT

BASIC NUTS

Fastening Materials with Heat

Soldering is joining metals with heat and solder. Solder is an alloy made from lead and tin. It melts at about 450° Fahrenheit. Soldering irons or guns are used to melt the solder. Soldering is the most common method used to attach wires in electronic circuits.

Welding is another method of using heat to join metals. In welding, the metals to be welded are heated high enough to actually fuse together. Welding rod is used instead of solder to help join the metals. Welding produces very strong bonds and requires temperatures of 6,000–7,000° Fahrenheit. The heat comes from a welding torch that burns a mixture of gas and air (gas welding) or from welding machines that use high electrical currents (arc welding).

Gluing Materials

Gluing is another method of fastening materials. When we use glue, we make use of the chemical properties of materials. The glue creates chemical bonds between itself and the materials being glued. Special glues are made to glue wood, metal, ceramic, and plastic. For wood, white or yellow glues are recommended. For plastic, metal, and ceramic, use epoxy or SuperGlue®.

Today, there is a glue for fastening almost every type of material. For example, until recently copper pipe was used for most plumbing jobs. The pipe was fastened by soldering the pieces together. Today, plastic pipe called PVC (made from a chemical called polyvinyl chloride) is used. Special glues are used to join pieces of PVC pipe. The glues create a bond between the atoms of the two pieces of pipe.

A good glue joint is stronger than the material it is joining. Hot glues are applied with glue guns. These glues are very strong and set rapidly, allowing quick fastening. Such glues are now being used to hold airplane parts together.

Pennsylvania's Washington County courthouse has been restored to look as beautiful as it did when it was built at the turn of the century. The project required 450 gallons of paint, 15 gallons of stain, and 26 quarts of custom colorants. (Courtesy of PPG Industries)

Coating Materials

In order to beautify or protect the surface of a material, we can coat it with finishes. Paint, stain, and wax can be applied over the surface of a material. Ceramic dishes can be coated with glass-like glazes. The glazes make the dishes easy to clean. Metals can be coated with other metals by a process known as **electroplating.** Gold-plated jewelry and silver-plated tableware are made by this process. Aluminum is coated by a process called **anodizing.** Anodizing causes a thin oxide coating to appear on the surface of the aluminum. **Galvanizing** is a process that coats steel with zinc to keep the steel from rusting.

The Electroplating Process

Electroplating is an electrochemical process (a process that combines electricity and chemistry). The part to be plated is connected to the negative terminal of a battery. The pure metal, which does the plating, is connected to the positive terminal. The liquid plating solution for copper plating is a chemical called copper sulfate.

The copper sulfate solution breaks down into positive and negative charges (ions) when electric current flows through it. The copper ions are positive and are attracted to the negative terminal. They are replaced in the solution by ions from the pure copper. The electroplating process can be used to plate many different metals.

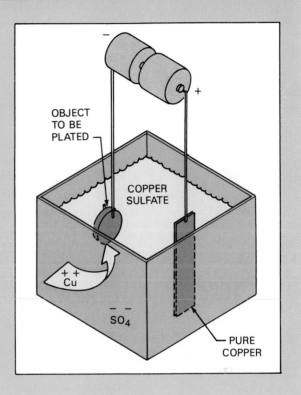

Making Composite Materials

Many years ago, the ancient Egyptians added straw to the clay they used for making bricks. The straw made the bricks stronger. Modern technology has since developed many new composite materials. A **composite material** is

The Pontiac Fiero, because of the latest fiberglass and coating technology, is lightweight and fuel efficient. (Courtesy of PPG Industries)

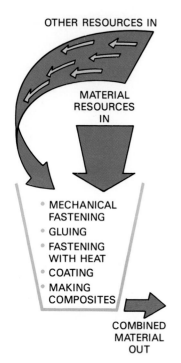

OTHER RESOURCES IN

MATERIAL
RESOURCES
IN

- MECHANICAL
  FASTENING
- GLUING
- FASTENING
  WITH HEAT
- COATING
- MAKING
  COMPOSITES

COMBINED
MATERIAL
OUT

Some combining processes

a material produced by combining several materials together. The new material has special properties. A good example of a composite material is ordinary plywood. Layers of wood panels are crisscrossed and glued together, and a much stronger wood is created.

Fiberglass is a composite material that can be made stronger than steel. Fiberglass, however, weighs much less than steel. Fiberglass is used to make the bodies of some automobiles.

CONDITIONING PROCESSES

Conditioning materials changes their internal properties. For example, we can **magnetize** a piece of steel. The magnetizing force rearranges the steel molecules so that they all line up in one direction.

Heat-treating can also cause changes within a material. When steel is heated to a cherry-red color and quickly cooled in water, the steel becomes harder. This process is called **hardening.** If we heat the steel again to an intermediate temperature and cool it quickly, the steel becomes very tough. We call this process **tempering.** If we heat steel to a cherry-red color and allow it to cool very slowly, the steel becomes softer. This process is called **annealing.**

When clay is fired in a kiln, it becomes much harder and stronger. When a piece of metal is hammered, it becomes harder because of **mechanical conditioning.** The metal's crystal structure changes and gets longer and thinner. When we mix plaster with water, heat is given off as a result of a chemical reaction. The plaster hardens because of this **chemical conditioning.** In each of these examples, the change takes place inside the material itself.

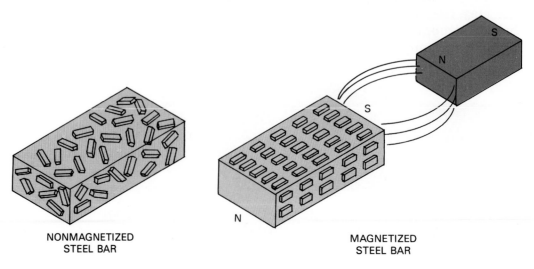

NONMAGNETIZED
STEEL BAR

MAGNETIZED
STEEL BAR

Other examples of conditioning include

- light exposing photographic film;
- chemicals developing photographic paper;
- baking a cake;
- cooking an egg;
- freezing water;
- boiling water;
- melting ice;
- immersing a metal in liquid nitrogen to make the metal a superconductor;
- using radiation on a tumor;
- making butter from cream;
- sending electricity through the filament wire of a light bulb; and
- making wine from grape juice.

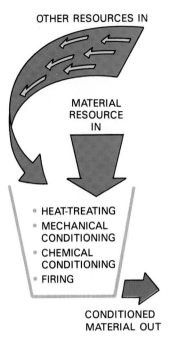

OTHER RESOURCES IN

MATERIAL RESOURCE IN

- HEAT-TREATING
- MECHANICAL CONDITIONING
- CHEMICAL CONDITIONING
- FIRING

CONDITIONED MATERIAL OUT

Some conditioning processes

USING COMPUTERS TO CONTROL PROCESSING OF MATERIALS

Computers can be used to control machines that process materials. For example, cooking is a conditioning process. When food is heated in a microwave oven, the oven temperature and cooking time are controlled by a small computer in the oven. In modern factories, machines that process industrial materials are also controlled by computers. Often, the computer is built right into the machine itself. The computer can be programmed to make a machine tool cut material along a specified path. For example, the machine might produce parts for a tractor. When the company decides to modify its tractor design, parts with a slightly different shape will be needed. The computer can then be reprogrammed to make the machine cut out the new parts. In the days before computer control, machine operators would spend considerable time setting up and readjusting machines by hand each time such changes were made.

Computer-numerical-controlled (CNC) machines and robots are examples of computer controlled machines used to process materials. Computers that direct the sequence of operations of a machine are often programmed through simplified computer languages. These languages use terms that are familiar to people who have worked with the machinery.

Computers can control machines used to process materials.

Computers can control machines used to process materials. In this two-machine/robot machining cell, computers control the machining of parts. (Courtesy of Cincinnati Milacron)

Robots are sometimes programmed by having a person "show" the robot the motions it is to make. With the robot in the "learn" mode of operation, the person uses a teach pendant to move the robot's arm through the motions needed in an operation. The robot records the motions in its memory. Then, when commanded, the robot will recreate the same motions again and again.

In this multiple exposure, a teach pendant is used to program a robot. (Courtesy of Microbot Inc.)

PROPERTIES OF MATERIALS

We choose our friends because they have characteristics that we like. Perhaps we like the way someone looks, or we enjoy a person's sense of humor. Materials are also chosen on the basis of their characteristics, or **properties.** Properties of materials include strength, hardness, appearance, ability to conduct electricity, and resistance to corrosion.

We use glass for windows because we can see through it. We use plastic for dishes because it is strong and can be cleaned easily. We make electrical wire out of copper because copper conducts electricity very well. We make phonograph needles out of small pieces of diamond because diamonds are very hard and will last for a long time. We make clothing out of nylon because it is lightweight and attractive.

Materials are chosen on the basis of their mechanical, electrical, magnetic, thermal, or optical properties.

Mechanical Properties

When force is applied to a material, the material will bend or break. Certain kinds of materials will bend more than others. A fishing rod made from fiberglass will bend a great deal without breaking; many kinds of wood will not. Materials that can be deformed without breaking are called **ductile materials.** When we make pots and pans, we start out with a flat sheet of metal. We then put pressure on part of the metal and form it into the shape we want. We may have to deform the metal a great deal. The metal must be able to stretch and get thinner without breaking. When we make wire, we start with a thick rod and pull it through a small hole to make it thinner. This process, called **drawing wire,** works because the metal rod is a ductile material. Any material that can be twisted, bent, or pressed into shape has high ductility.

The opposite of a ductile material is one that is brittle. A **brittle material** will not deform without breaking. Window glass is a good example of a brittle material.

There are two kinds of ductile materials: **elastic** and **plastic.** A material that can bend and then come back to its original shape and size is called an elastic material. It acts the way a piece of elastic does. Rubber bands, springs, and fishing rods are examples of materials with high elasticity. Elasticity refers to the stiffness of a material.

A property similar to elasticity is **plasticity.** Plastic materials can be bent and will stay bent. The material that we know as plastic got its name because of that characteristic. When we heat thermoplastic and bend it, it will stay bent after it cools. Other materials, however, are also considered to be plastic materials. Modeling clay is a plastic material. So are certain metals. They will stay deformed after the force is removed.

Some materials are stronger than others. Strength refers to a material's ability to retain its own shape while a force is being applied to it. Four kinds of force can be applied to materials. **Tension** is force that pulls on a piece of material. When we pull on a spring, it is under tension. **Compression** is the exact opposite of tension. It is a force that pushes or squeezes a material. Squeezing a sponge and walking on rubber-soled sneakers are examples of compression. **Torsion** occurs when a material is twisted. If we twist a piece of licorice candy, the material is in torsion. The twisting force itself is called **torque.** When we use a wrench to turn a bolt, we apply torque. A **shear** force acts on a material like the way a pair of shears (scissors) works. One part slides in one direction and the other part slides in an opposite direction.

Toughness is another of a material's mechanical properties. **Toughness** is the ability of a material to absorb energy without breaking. For example, leather animal hides are tough. Sometimes meat that we eat is tough. It takes a lot of chewing force to break down the fibers of the meat.

A very important mechanical property of materials is hardness. **Hardness** refers to a material's ability to resist being scratched or dented. Diamonds are the hardest materials. Certain metals are also very hard. Metal tools must be hardened in order to resist wear. Some synthetic materials, like tungsten carbide, are designed to be very hard. The teeth of some circular saw blades are made from tungsten carbide.

You Can Try This

To understand the difference between elasticity and plasticity, take a piece of steel rod, 1/4″ in diameter. Bend it slowly. If you don't exert too much force, the rod will spring back to its original position. This is an example of elasticity. Now bend the rod with more force. You will be able to bend it past its elastic limit. When you bend the rod past its elastic limit, it will stay bent. This is an example of plasticity.

Properties of Materials

MECHANICAL PROPERTIES

ELASTICITY
The rod returns to its original shape.

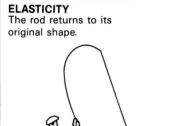

PLASTICITY
A coil of clay will stay bent.

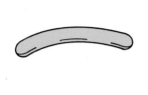

BRITTLENESS

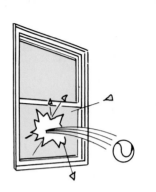

DUCTILITY

HARDNESS

TOUGHNESS
The frame of the car will withstand impact.

**STRENGTH
ABILITY TO WITHSTAND . . .**

Torsion

Compression

Tension

Shearing Action

OPTICAL PROPERTIES

OPTICAL CLARITY

REFLECTIVITY

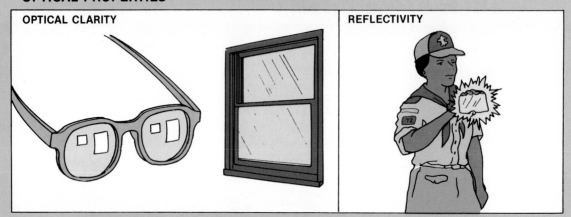

THERMAL PROPERTIES

CONDUCTION

INSULATION

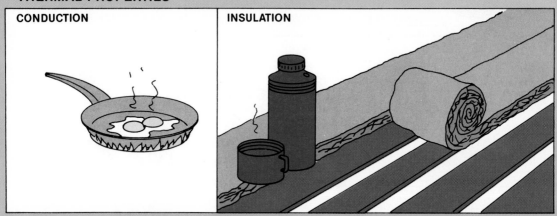

ELECTRICAL AND MAGNETIC PROPERTIES

CONDUCTION

MAGNETISM

INSULATION

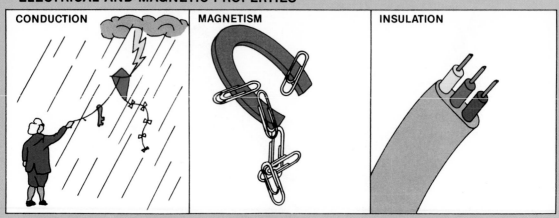

Electrical and Magnetic Properties

All materials resist the flow of electricity, but some materials, called **conductors,** offer very little electrical resistance. Most good conductors are metals. The very best conductor is silver. The next best is copper. Most wire is made from copper, however, because silver is so much more expensive. People make a trade-off (cost for performance) when they choose copper instead of silver.

Materials that do not conduct electricity well are called **insulators.** Sometimes we choose materials because they are good insulators. When we cover wire with plastic or rubber, we choose those materials for their insulating properties.

When electricity flows through wires, a magnetic field is produced. An electromagnet can be made by wrapping wire around a piece of iron. When an electric current (as from a battery) flows through the wire, the metal piece becomes a magnet. Materials that can be attracted to a magnet are **magnetic** materials. Iron, steel, nickel, and cobalt are examples of magnetic materials. Copper, wood, glass, and leather are examples of nonmagnetic materials.

Thermal Properties

In the Greek language, "therm" means heat. Words like thermostat and thermos bottle describe things that relate to heat. Thermal properties refer to a material's ability to conduct heat.

Metals are good conductors of heat. Copper and aluminum are two of the best. Other materials, such as rubber and fiberglass, do not conduct heat well. In building construction, we choose materials that do not conduct heat well. A layer of nonconducting material called **insulation** is placed inside the walls. The heat then stays inside the building.

Optical Properties

Optical properties refer to a material's ability to transmit or reflect light. Certain kinds of glass can transmit light very well. Window glass transmits light well enough to see through, but not well enough to be used for scientific devices like telescope lenses. Special optical glass is used for such

When metal was in short supply during World War II, people made bicycles out of wood. (Courtesy of Smithsonian Institution)

devices. Plastic is lightweight and will not shatter. It is therefore used for contact lenses and eyeglasses. Some pure types of glass can be made into glass fibers. Glass fibers can be used instead of wires for purposes of communication. Some metals reflect light very well and can be made into products like reflectors for headlights and flashlights.

Engineering the Properties of Materials

We use technology to create new materials with improved properties. Car bodies can be made from new kinds of plastics instead of metal. A new plastic called Arylon™ has recently been engineered. Arylon™ can be used for automobile parts like bumpers and body panels. It can also be used for circuit boards, sports equipment, and aircraft interiors.

Airplane parts can be made from special composites. Some aircraft are made from special purpose steels that are lightweight and resist the heat that builds up at very high speeds.

The Lockheed SR-71 spy plane is made out of titanium. (Courtesy of Lockheed-California Company)

DISPOSAL OF RESOURCES

We must also think about how we will dispose of resources when we are finished using them. Often, instead of repairing something, we throw it away. We live in a "throw-away" society. We use disposable paper cups, razors, and flashlights. Our soda comes in disposable metal cans or plastic bottles. Our food comes in disposable plastic or paper containers.

When we dispose of resources, we either store them in dumps, or burn them, or throw them into the oceans. But these resources can pollute the oceans, and burning the materials produces smoke, which pollutes the air. We must try to use resources that will decompose naturally (rot) and become part of the earth again. Those that will not decompose (like plastic or glass) should be **recycled** (reused).

SUMMARY

To make material resources more useful and more valuable, we process them. The other six technological resources are needed to process material resources.

Material Decomposition Time

Recycling is an important source of aluminum supply. (Courtesy of Aluminum Company of America)

MATERIAL	TIME TO DECOMPOSE
Orange peels	1 week to 6 months
Paper containers	2 weeks to 4 months
Paper containers with plastic coating	5 years
Plastic bags	10 to 20 years
Feathers	50 years
Plastic bottles	50 to 80 years
Aluminum cans with flip-top tabs	80 to 100 years
Plutonium	24,390 years (half-life)
Glass bottles	Indefinite

To dispose of radioactive waste, technologists cover it with lead and then bury it in salt formations far underground. (Courtesy of Rockwell International Corp.)

Waste disposal is a major technological problem. (Courtesy of Sperry Corp.)

Primary raw materials are processed into industrial materials. From industrial materials, end products are manufactured. Materials are processed by forming, separating, combining, and conditioning them.

Forming a material means changing its shape without cutting it. Forming processes include casting, pressing, forging, extruding, blow molding, and vacuum forming. Separating processes separate one piece of material from another.

Cutting is a separating process. We can cut materials by shearing, sawing, drilling, grinding, shaping, and turning. Combining materials means putting one material together with others. We can combine materials by fastening, coating, or making composite materials. Conditioning materials means changing their internal properties. Conditioning processes include magnetizing, heat-treating, mechanically working, and chemically processing a material. Computers can control machines used to process materials.

Materials are selected on the basis of their properties. Mechanical properties include: ductility (ability to bend without breaking); elasticity (stiffness); plasticity (ability to be bent and stay bent); strength (ability to withstand tension, compression, torsion, and shear); toughness (ability to absorb energy without breaking); and hardness (ability to resist scratching or denting). Other properties are: electrical (ability to conduct electricity); magnetic (ability to be attracted to a magnet); thermal (ability to conduct heat); and optical (ability to transmit or reflect light). We can use technology to create materials with new properties. How easy a resource is to dispose of or recycle also influences our selection.

A polishing and grinding line (Courtesy of Allegheny Ludlum Steel Corp.)

Laser measurement system (Courtesy of Hewlett Packard Company)

INVESTMENT CASTING

Setting the Stage

For the last few years, the senior class has been unhappy with the class rings they have purchased. The year and school logo have been fuzzy. You are on the committee to look into the matter. Thinking back to the metal-processing class you took last year, you remember how well a ring you made using the process of investment casting turned out. You offer to provide a die for repetitive castings of the senior class ring.

Your Challenge

Design and investment cast a ring.

PARTS OF A SPRUE SYSTEM

Suggested Resources

Assorted investment waxes
Sprue wax
Investment flasks
Alcohol lamp
Kerr investment
Metric scales and liquid
   measures
Enameling kiln
Centrifugal casting machine
Metal abrasives and
   polishing compounds
Sterling silver
Ring mandrel

Procedure

1. Determine the width of the ring. Draw two parallel lines 2 inches long separated by the desired width.
2. Select edge shapes and other repeating designs for your ring. Sketch the designs between the parallel lines. You may want to try several different designs.
3. Determine your ring size by using a ring sizer. Using a ring length chart, cut a piece of sheet wax the correct length and width.
4. Shape the wax ring with your fingers after warming the wax over an alcohol lamp.
5. Melt (weld) the ends of the wax ring together with a wax wire tool. Use a ¼" × 3" dowel with a 22-gauge music wire 3" long attached

248

to the end. Warm the wire over the alcohol lamp before using the tool.

6. Warm the wax ring and use the ring mandrel to round it.

7. Now use the various wax forming techniques that have been demonstrated by your teacher. Melt outs, adding more wax, shaping sheet wax, and using carving tools are a few of the processes you can use. Remember, investment casting reproduces exactly what you form.

8. Add sprues to the wax rings and weigh the rings in grams.

9. Find the amount of materials needed to investment cast your ring.

Volume of flask $= \pi\, r^2\, h$ (measure in cm)

Investment powder needed $=$ Volume of the flask (measure in cc) $\times$ 1.5

Water needed $=$ 18 cc per 50 grams of investment powder

Metal needed $=$ Specific gravity of metal $\times$ (ring + sprue wt.)

(The specific gravity of silver is 10.4.)

10. Invest the ring by pouring the slurry into the flask. Remove the base after the investment hardens. Label the top of your flask with your name. Store the flask in a jar with a damp paper towel. This will keep the flask moist until it is ready to be fired.

11. Bring the flask to the proper temperature in the kiln.

12. Cast the ring, using a centrifugal investment casting machine.

13. Use silicon carbide abrasive paper, files, tripoli, and white rouge to polish your ring.

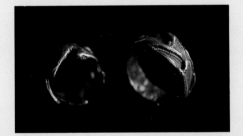

Technology Connections

1. Visit a local jewelry store. Compare the selling price of an industrially cast ring with the cost of materials for your ring. Compare the cost of your school class ring with the cost of your ring. Why is there such a difference?

2. Identify the forming, combining, and conditioning processes you used in making your ring.

Science and Math Concepts

▶ The specific gravity of a substance is the ratio of the weight of a given volume of the substance to the weight of the same volume of water at the same temperature.

CHECKERBOARD

Setting the Stage

You have received a new contract with a local restaurant. They want some unique and practical tables. The owner thinks more people would come to the restaurant if they could eat and also socialize. How might you build a tabletop that is easily cleaned, and also serves as a gameboard before or after food is served?

Your Challenge

Design and construct a checkerboard table top that will resist wear and be easily cleaned.

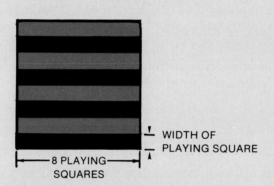

8 PLAYING SQUARES

WIDTH OF PLAYING SQUARE

|— LENGTH OF BOARD —|

CHECKERBOARD ARRANGEMENT

Procedure

1. Cut material of two contrasting colors into eight strips as shown.
2. Glue the strips together.
3. Cut across the strips to form eight pieces, with eight contrasting squares each. Then, flip every other strip to create contrasting squares in the other direction, forming a checkerboard.
4. Glue the strips back together. If you chose to use thin plastic or ceramic material, glue the squares to a solid backing.
5. To add a three-dimensional quality to the checkerboard, cast a ⅜" to ¾" clear layer of polyester resin on the checkerboard. (Steps 6–11 explain how to do this casting.)
6. Cut and assemble a frame for the completed checkerboard. The height of the frame should be the thickness of the checkerboard plus the desired thickness of the polyester resin layer.
7. Apply finish and polyester resin wax release to the frame and the bottom of the checkerboard. Polyester resin will spill onto the frame and bottom of the board during the casting procedure. The resin wax release will prevent the resin from sticking to the frame and bottom of the board.

Suggested Resources

Wood
Plastics
Ceramics
Adhesives
Clear polyester resin
Plate glass—¼" thick
Mylar or other resin wax release
Disposable mixing cups

8. Figure the volume of polyester resin needed to fill the board. Add three extra ounces. Follow the manufacturer's directions to mix the polyester resin.
9. Carefully pour the polyester resin into the frame. Fill the checkerboard to the top of the frame.
10. Pour extra polyester resin in the center of the checkerboard. Immediately place a piece of plate glass that has been waxed and polished with a release agent over the checkerboard and resin. This will force any surface air bubbles to the outside of the board and make the surface smooth. (*Note:* Some resin will run out from the sides of the plate glass, so be prepared.)
11. When the polyester resin has hardened, remove the plate glass with the help of your teacher. Clean up the resin on the outside edge of the checkerboard.

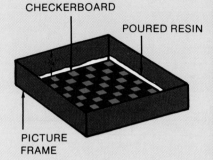

CHECKERBOARD

POURED RESIN

PICTURE FRAME

Technology Connections

1. Mylar will release from polyester resin. What other release agents might you use?
2. Identify all the tools and machines used in this lab.
3. Identify each material used as to whether it is renewable or nonrenewable.
4. List the procedures of material conversion that were used in this activity. These conversion processes can be listed under forming, separating, combining, and conditioning.

Science and Math Concepts

▶ Volume = length × width × height.
▶ Refraction of light is caused by a change in the speed of light as it leaves one medium and enters another. This is why the wood playing surface appears closer than it actually is when it is covered with the polyester resin.

251

REVIEW QUESTIONS

1. Why is it necessary to process materials?
2. Give an example of an edible material that is processed. Why is it necessary to process some foods?
3. What occurs within a technological process?
4. What is the role of computers in processing materials?
5. What are four ways materials may be processed?
6. What processes would you use to make hamburgers from raw meat?
7. How do sawing, drilling, and grinding differ?
8. How would you fasten a metal bracket to a wooden shelf?
9. Why would you want to weld two pieces of metal together rather than use screws or glue?
10. How does coating wood with wood finish differ from the way you would finish a metal surface?
11. What happens to the internal structure of a piece of steel when it is magnetized?
12. List five ways to condition materials.
13. Make a list of jobs that involve processing materials.
14. Explain why materials are chosen on the basis of their properties.
15. We want to manufacture a jigsaw puzzle suitable for four-year-old children. Our goal is to sell it for under two dollars. What material might we make it from?
16. You plan to purchase a screwdriver for electrical work. From what material should the handle be made? Why?
17. Why must people understand the uses and limitations of materials in order to choose them wisely?

KEY WORDS

Brittle	Elastic	Insulator	Shaping
Casting	Extruding	Plasticity	Shearing
Ceramics	Fastening	Polymers	Tension
Coating	Ferrous metals	Pressing	Thermal
Combining	Forging	Properties of	Thermoplastics
Composites	Forming	materials	Thermoset
Compression	Gluing	Raw materials	plastics
Conditioning	Grinding	Recycle	Torsion
Conductor	Heat-treating	Sawing	Toughness
Drilling	Industrial	Separating	Turning
Ductile	materials		

SEE YOUR TEACHER FOR
THE CROSSTECH PUZZLE

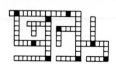

CHAPTER 9

████████████████████

MANUFACTURING

MAJOR CONCEPTS

After reading this chapter, you will know that:

- Production technologies satisfy many physical needs by means of manufacturing and construction systems.
- Manufacturing is producing goods in a workshop or factory. Construction is producing a structure on a site.
- Mass production and the factory system brought prices down. People were therefore able to improve their standard of living.
- Manufacturing systems make use of the seven types of technological resources.
- There are two subsystems within the manufacturing system: the material processing system and the business and management system.
- Automation has caused some jobs to be lost, but it has also created new jobs.
- Automation has greatly increased productivity in manufacturing.

PRODUCTION SYSTEMS

Look around you. We live in an environment that has been created by people. It is very different from the world we inherited from nature. Our homes, highways, and religious and recreational centers have been planned, designed, and constructed by humans employing technological means. People manufacture the furniture we use, the food we eat, and the vehicles that transport us. These items make our lives richer and more comfortable. It is hard to imagine what life would be like without the products of modern technology.

(Photo by Jonathan Plant)

(Courtesy of Maytag)

(Courtesy of Sony Corp. of America)

(Courtesy of Saab)

(Courtesy of Campbell's)

People use two kinds of production systems (manufacturing and construction) to satisfy many of their needs and wants.

MANUFACTURING SYSTEMS

People have used manufacturing technology for many thousands of years. During ancient times, products were produced by hand with primitive tools. People made use of natural materials to produce weapons, food, and clothing. Stone, bone, wood, and clay were the materials most often used by prehistoric people. Most tools were made by simply chopping or scraping a piece of available material into the desired shape and size.

Evidence shows that pottery was made from clay over 30,000 years ago. Firing clay was one of the earliest inventions. The earliest pieces of clay were probably fired by accident, when human beings made fires on ground rich in clay. People may have noticed that the clay became very hard after they cleared away the ash from the fire. They then began to put clay pieces into fires intentionally. This was one of the very first instances of manufacturing.

Natural metals, like copper, gold, and silver, were made into ornaments and tools. These tools were of better quality than those made from wood and stone. During the Bronze Age (about 3500 B.C.), people began casting metal.

Glass was perhaps the first synthetic (human-made) material. At one time, glass beads were considered to be very valuable. They were used as money in some places.

Manufacturing is producing goods in a workshop or factory. Construction is producing a structure on a site.

The Craft Approach

For centuries, objects were manufactured by individuals who produced items only for their own use and for the use of their families. After some time, it became clear that some people were better at making one thing than another. Soon, people began to specialize. Those who made shoes very well became shoemakers. They made shoes not only for their own needs, but for other people as well. In return, they received goods from other craftspeople. Manufacturing was done by such craftspeople as candlemakers, weavers, spinners, glassblowers, silversmiths (who made tableware), coopers (barrel makers), gunsmiths, musical instrument makers, and tailors.

These craftspeople generally made their products one at a time. Each item was made from start to finish by the master craftsperson. Most often, craftspeople did their manufacturing in their own homes or in small workshops. Craftspeople owned their own tools. Since tools were expensive, not everyone who wanted to become a craftsperson

Many musicians have their instruments custom-made.

(Photo by Michael Hacker)

could afford to do so. Often the master craftsperson hired young people who wanted to learn the trade. These young people were called apprentices. Instead of going to school to learn about the technology of the day, young people got their education on the job. The master craftsperson taught the apprentices how to perform all the processes necessary to create an object from beginning to end.

Our society still demands one-of-a-kind products, and we still have craft production today. Of course, we pay more for a product that is custom-made. Perhaps you have compared the price of clothing purchased at a department store with the price of clothing that is tailor-made for you. The custom-made clothing is much more expensive.

Products are custom-made for specialized markets. Handicapped people, for example, often have special needs. Products must be designed and produced specifically for the individual. Some people may want custom-made clothing, shoes, and jewelry. Craftspeople design and build cabinets, shelves, furniture, and many other items to suit the tastes of the individual purchaser.

Computers and Cowboy Boots

A modern-day industry that still employs craftspeople is the shoemaking industry. What is happening to the shoemaking craft in this modern technological age? It is an industry that is in the sunset of its life.

Because shoemaking requires skilled workers, labor costs are high. U.S. shoe manufacturers are finding it hard to compete with foreign competition. Today, most shoes that are purchased by people in the United States are made overseas. Between 1968 and 1986, the shoemaking industry in the United States lost about 100,000 jobs. Many companies have gone out of business.

Cowboy boots require fancy stitching. It would be very expensive for companies to hire U.S. workers to do this stitching by hand. The handwork is now done by workers in Spain and Portugal, who provide labor at a lower cost.

Some years ago, the Shoe Machinery Group of Emhart Corporation developed computer-controlled stitching machines. Computerized stitchers are bringing the manufacture of cowboy boots back to the United States.

Computer-controlled systems are being used for many other operations, including design. While the days of the cowboys and Indians may be over, it still looks good for cowboy boots in the United States.

This computerized stitcher is sewing a pattern on a cowboy boot. (Courtesy of Shoe Machinery Group—Emhart Corp.)

The Factory System

As the nation grew, railroads, canals, and highways opened up new markets. People demanded more and more goods. The masters hired more workers. Soon, instead of individual craftspeople producing goods, manufacturing was done by groups of people working for one boss in a factory.

During the Industrial Revolution in the late 1700s, there was an explosion of new inventions. Machines such as the steam engine, the cotton gin, and the sewing machine helped people accomplish tasks much more quickly than ever before. Many businesspeople used the new machinery to create improved methods of production. Individual craftspeople had made items from start to finish. Factory workers, on the other hand, were hired to do only one small part of the job. The factory workers were dependent upon each other to make one product. Also, the skills needed by factory workers to operate machines were different from those needed by craftspeople.

MASS PRODUCTION AND THE ASSEMBLY LINE

The factory system led to **mass production** of goods. Each worker was assigned only one job to do. The worker performed that job over and over. The worker became very skilled, but only at doing that one task. The production process was divided up into specialized tasks. Each worker did only his or her task on an **assembly line.** Through the use of mass-production methods and the assembly line, a much larger quantity of items could be produced in a given time period.

In 1798, Eli Whitney started one of the earliest mass-production assembly lines. Whitney signed a contract with the United States Army to make 10,000 rifles (an unheard-of number in those days) in two years. His method involved making a large number of each part at one time, with the parts being exactly alike (**standardized**). These parts could be kept in bins. Any one of the parts in a bin could be used in the assembly of a rifle. Since the parts were all alike, they were **interchangeable.** Before Whitney developed his method, rifles were made one at a time. The parts would fit into only the one rifle they were made for, and each part was a little different from the next.

Jigs and Fixtures

One of the ways manufacturers make standard-size parts is to use jigs and fixtures on the machinery used for production. A **jig** is used to hold and guide the item being processed. It also guides the tool that is doing the processing. A **fixture** is used to keep the item being processed in the proper position.

IMPACTS OF THE FACTORY SYSTEM

The factory system created a much larger supply of products for consumers. What used to be luxuries now were less expensive. More people could afford to buy the products of mass production. However, the factory system also created changes in society.

Many new industries were created as a result of mass production and the assembly line. The steel, automobile, appliance, and clothing industries provided millions of new

This student is using a jig. The jig holds a plastic strip in place so that the ends of the strip can be rounded evenly by a belt sander.
(Photo by Michael Hacker)

Mass Production and the Automobile Industry

One of the first moving assembly lines was at the Ford Motor Company in 1913. Auto parts were pushed from one worker to the next. This reduced production time by about one-half. By applying the same principle to the assembly of a total car, Henry Ford speeded up car production. A finished Model T Ford came off the assembly line every ten seconds.

The interchangeability of parts is one of the most important characteristics of a mass-production system. Modern automobiles, for example, are manufactured with interchangeable parts. Door latches for the Honda are alike. If one part breaks down, we can purchase another just like it from an automobile dealer.

A 1913 auto parts assembly line
(Courtesy of Ford Motor Company)

Workers making door latches for Honda automobiles
(Courtesy of Rockwell International)

jobs. Craftspeople and farmers became assembly-line workers in the nation's factories.

Factories created wealth by adding value to the resources that were processed into finished products. As a result, the standard of living improved for people who lived in industrialized nations.

Mass production and the factory system brought prices down. People were therefore able to improve their standard of living.

UNEXPECTED OUTPUTS OF THE FACTORY SYSTEM

Craft production and mass production were quite different. Craftspeople worked on an item, did some other chores around the house, and then went back to their craftwork. Factory workers, however, were expected to work regular shifts. Mass production caused time to become a much more important production factor. One manufacturing process now had to be followed immediately by another. The manufacturing operations were **synchronized.** If one process was delayed, the whole factory would slow down.

Factory owners often paid their workers by piecework. Workers were paid according to the number of items they completed in a day. It was not unusual for people to work twelve hours per day during the early part of the century. **Unions** were formed to protect the rights of workers. **Child labor laws** were passed to ensure that children were treated fairly.

During the craft era, families often worked together in the home workshop. There was always someone around to care for the children. When people began to work in factories, they had to leave the home. Society had to find a place for the children. Schools increased in number and importance as a result of the factory system.

Children were often used as factory workers in the late nineteenth and early twentieth centuries. They were often forced to work under harsh conditions and during evening hours.
(Courtesy of The Bettman Archive)

THE BUSINESS SIDE OF MANUFACTURING

Until the 1930s, most businesspeople thought mainly about technically improving production processes. They wanted their companies to be able to offer a standard product at a low price. A suggestion was once made to Henry Ford that he vary his product line by painting his Model Ts different colors. He responded, "Give it to them in any color, so long as it is black."

Competition began to occur among the many manufacturers. In the 1930s, General Motors started to produce automobile models that changed every year. Advertising, promoting, and selling the automobiles became as much a concern as actually producing them. Marketing and business management became important systems. These systems supported the technical side of production.

Differences Between Craft Manufacture and Mass Production

CRAFT MANUFACTURE	MASS PRODUCTION
1. Workers are very skilled.	1. Workers need limited skill.
2. Workers make a product from start to finish by themselves.	2. Workers work on only one part of the product.
3. Work is varied and interesting.	3. Work is routine and often dull.
4. Craftspeople get satisfaction by seeing the finished product (like a completed chair).	4. Factory workers see only the one part that they produce (like the chair leg).
5. Each part is hand crafted so no two are exactly alike.	5. Parts are machine made and are interchangeable.
6. Only one item is produced at a time.	6. Many items are produced during the production run.
7. It takes a long time to produce each item.	7. The average time it takes to produce each item is reduced.
8. The cost of each item is high.	8. The cost of each item is lowered.
9. Quality depends mainly upon the skill of the craftsperson.	9. Quality depends mainly upon the accuracy of the machines and how well they have been set up by people.

THE ENTREPRENEURS

An **entrepreneur** is a person who comes up with a good idea and uses that idea to make money. To compete with others who are producing similar products, entrepreneurs make improvements in their products. Or, they improve the way they make their products. Sometimes, they come up with ideas for new products.

Some entrepreneurs are **inventors.** An inventor comes up with a totally new idea. Examples of inventions are the safety razor, the laser, and the contact lens. Inventions can be patented. Once an individual receives a patent for an invention, that device cannot be produced by anyone else for seventeen years. Many inventions have become commonplace items. Some, like air-conditioned suits and hats with fans, were never accepted by the public.

Some entrepreneurs are **innovators.** An innovation improves an invention and leads to other uses. Innovations can start new industries or change existing ones. The

Fire Escape Parachute
Patent No. 221,855
March 26, 1879

In March 1879, Benjamin Oppenheimer patented a fire escape parachute, complete with headpiece and sponge-bottom shoes.

Entrepreneurship and McDonald's Restaurants

One famous entrepreneur is Ray Kroc. Ray was a salesman who sold electric mixers, called multimixers, to restaurants. One of the restaurants he serviced was owned by two brothers. In 1954, Ray made an agreement with the two brothers to franchise the concept of the restaurant they were operating. The menu included 15¢ hamburgers, 10¢ french fries, and 20¢ shakes. The brothers were Mac and Dick McDonald. The restaurant was McDonald's. There are now close to 10,000 McDonald's restaurants worldwide.

Ray A. Kroc (Courtesy of McDonald's)

electric guitar is an example of an innovation. An entirely new music industry sprung up around the rock music played on this instrument. Other examples of innovations are the diesel engine, power steering, and the integrated circuit.

RESOURCES FOR MANUFACTURING SYSTEMS

Manufacturing systems make use of the seven types of technological resources.

Manufacturing systems combine the seven technological resources to provide a finished product. The process adds value to the resources. Therefore, the finished product is worth more than the sum of the individual resources themselves. People, materials, energy, and other resources must be combined correctly for the manufactured product to serve its purpose and be a profitable item. Because manufacturing is a system, it depends upon all of the resources being combined at the right time and in the right way. The outputs of the manufacturing system must be monitored carefully. Any unexpected outputs should be carefully assessed. The manufacturing system should not cause undesired outputs (such as pollution). It should fit in with the needs of human beings and the environment.

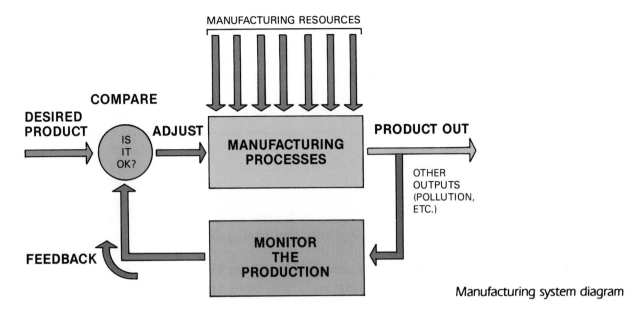

MANUFACTURING RESOURCES

DESIRED PRODUCT

COMPARE

IS IT OK?

ADJUST

MANUFACTURING PROCESSES

PRODUCT OUT

OTHER OUTPUTS (POLLUTION, ETC.)

FEEDBACK

MONITOR THE PRODUCTION

Manufacturing system diagram

People

People design the products. They also engineer the technical methods of production. People choose the materials, select the best tools and machines, and organize the production lines. People arrange for financing the company. People advertise, sell, and distribute the finished products. In past years, people provided all the labor in manufacturing factories. In fact, in the 1950s, about 30 percent of the work force in the United States was involved in manufacturing jobs. Today, machines have taken over many of the jobs people used to do in the factories. About 16 percent of the work force now works in manufacturing industries.

People have begun to play new roles in the manufacturing system. As workers have become more educated, they have expressed the desire to become more involved in making decisions that affect the company. **Quality circles** have been formed. These circles provide an opportunity for workers to meet with each other and with management during the work day. Within these quality circles, workers discuss problems that they feel should be overcome. They also suggest ways the company might improve the production process. Quality circles also give company management a way to point out what problems the company has with costs, profits, and competition. Such circles provide an opportunity for workers and managers to discuss these issues openly. The work environment is therefore made more pleasant, and productivity can go up.

Employees at Pittsburgh Plate Glass developed a better method for attaching the "button" that secures the rearview mirror to an automobile windshield. (Courtesy of PPG Industries)

As salaries continue to rise in the United States, competition from foreign countries has prompted companies to develop more efficient production methods. Some companies that have been unable to do so have been forced out of business.

American industries cannot afford to compete with foreign companies whose workers are paid much less. For example, U.S. steelworkers are paid six times as much as steelworkers in Brazil and about eight times as much as steelworkers in Korea. Because of foreign competition, some U.S. companies have reduced production or closed down entirely. In the steel industry alone, between 1975 and 1985 about 250,000 workers lost their jobs. Many people who lost high-paying jobs in manufacturing industries found that the only work available was in the service industries, at much lower salaries. It has been very difficult for people to readjust to a life-style that means a lower standard of living.

In the future, many of the career opportunities in manufacturing in the United States will center around the design and engineering of products. Most of the actual assembly-line work will be done by automated machines or by people overseas. People who lose jobs in manufacturing industries will have to be retrained to do other kinds of work.

Information

Manufacturing plants are often built in areas where knowledgeable workers like engineers and computer programmers are available. Sometimes, companies build their plants near universities. Professors can then provide technical assistance. Companies can also use university research facilities to obtain up-to-date technical information about new materials and production methods.

Companies also need information about product demand. Market research is done to collect information about what people will be buying and how tastes are changing. Data is collected about costs of materials. This data is processed into information. The information then helps managers make decisions about what materials to buy and when to substitute one material for another. The feedback obtained by monitoring the outputs of a production system is also a form of information. It is used to adjust the manufacturing process.

Here, information is being collected about how accurately the machine tool is performing. (Courtesy of Deere & Co.)

Materials

Raw materials are processed into basic industrial materials. Basic industrial materials are processed by manufacturing systems into finished products. The steel industry takes coal, limestone, and iron ore and processes these raw materials into basic iron ingots. The iron ingots are then processed into steel slabs and sheets and made into finished products, like automobile frames and bodies. Materials are processed through the forming, separating, combining, and conditioning techniques discussed in Chapter 4.

The cost of raw materials is very important to manufacturers. Sometimes, high costs cause them to seek alternatives. Costs include the actual per-pound cost of the material itself and other costs like mining and shipping. If a steel company in Indiana needs iron ore, should it buy that iron ore from U.S. sources on Lake Superior in Wisconsin, where the ore is only 20 percent pure? Or should it buy the ore from Brazil, where it is 65 percent pure, and pay the extra shipping costs? These are the kinds of cost-benefit trade-offs manufacturers have to make in choosing materials.

Tools and Machines

The tools used in modern manufacturing systems are very advanced and very precise. Generally, the machines are automatic. Computer programs or sensors provide feedback and guide the operation of the machine. Some machines, called **numerical control machines,** are controlled by punched tapes. The tapes consist of sections of punched holes. The holes represent a series of instruction codes for the machines. A **control unit** receives and stores all the coded data.

Many modern machine tools are controlled by computers. Computer-aided manufacturing will be discussed in detail later on in this chapter.

Manufacturing systems sometimes require specialized tools. This mustard dispenser used by McDonald's employees puts the same amount of mustard on each hamburger.
(Courtesy of McDonald's)

Energy

Approximately 40 percent of the energy consumed in the United States is used by manufacturing industries. Among the primary users are the industries that produce metals, chemicals, ceramics, paper, food, and equipment. Most manufacturing industries rely on fossil fuels (oil, coal, and natural gas). However, hydroelectric power and nuclear energy are used by some industries to generate electric power.

Plants are often built in areas where energy costs are low. The glass industry grew up in West Virginia because of an inexpensive supply of natural gas. The steel industry builds minimills (small steel mills that use electric furnaces to melt the metal) in places where electric power is cheap. The cost of energy has even caused some firms to relocate to places where energy costs are lower.

Some industries, like the paper-making industry, make good use of the heat energy generated by production processes. This energy is used to heat water and make steam. The steam then turns steam turbines and produces electricity. This process is called **cogeneration** and makes optimum use of the materials being processed.

Capital

Companies must secure capital to finance their operations. Capital is needed to buy land, build the factory, purchase equipment, pay the workers, maintain machines, and advertise the product. Capital is often obtained by selling shares of stock to the public. If the public believes that the corporation will make a profit, they become partners in the corporation by investing money in company stock. If the company does make a profit, the value of the stock may go up. The stockholders will then make a profit on their investment.

Private companies can raise money from private investors who contribute **venture capital.** A venture (like an adventure) is a trip into the unknown. Venture capital is money that is used to finance the costs of starting a new company. The investors take a risk. They expect to make more money once the company starts production. They actually own part of the company and receive a percentage of the profits. Often new companies started by entrepreneurs are financed through venture capital. Investors do not have to know very much about the technical side of the business. They supply the venture capital because they believe that the company is well managed, has a good idea that will have wide appeal to consumers, and will therefore make a profit.

Time

In manufacturing, time is money. The more products that are produced in a given time, the more profitable the company will be. Manufacturing industries strive for high productivity. **Productivity** means producing more in less time and at less cost. Since workers are paid by the hour, reducing the amount of production time reduces labor costs for the company.

A great deal of thought has been put into improving productivity. In 1910, Frederick W. Taylor developed an idea called **scientific management.** This idea involved observing workers very closely while they were working in the factory. Every movement that the worker made was analyzed to determine if there were wasted movements. The worker's routine was then reorganized to cut out wasted movements. This change provided increased output without making people do any more work.

Companies now try to ensure that products are constantly in motion, moving from one process to the next until the item is completed. The longer a product sits waiting for the next operation to occur, the more it costs to produce it.

HOW MANUFACTURING IS DONE

People need and want manufactured goods. However, before a company goes into production it does some **market research.** Market research helps the company find out exactly what kinds of features customers would prefer. Generally, companies survey a sample group of people. They try to pick a cross section of the total population. They ask this sample group what it thinks about a new product idea. The feedback they receive helps them decide whether or not to produce the product, or how the product should be modified.

Once the results of the market research are known, the product can be designed. Designers and product engineers prepare drawings, sometimes on computer-aided design equipment, and develop finished design ideas.

Research and Development (R & D) is done by product designers and engineers. R & D is used to develop

Expert craftspeople make a clay automobile model. (Courtesy of Ford Motor Company)

OBTAIN MATERIALS

HEAT AND FORGE BLADE

ANNEAL (SOFTEN) BLADE BY HEATING TO CHERRY-RED COLOR AND COOLING SLOWLY

FILE TWO FACES OF BLADE

GRIND TIP TO PROPER ANGLE

REMOVE ALL SCRATCHES WITH EMERY CLOTH

HARDEN BLADE BY HEATING TO CHERRY-RED COLOR AND COOLING QUICKLY

CLEAN BLADE WITH EMERY CLOTH

TEMPER BLADE BY HEATING TO A STRAW COLOR AND COOLING QUICKLY

CLEAN BLADE WITH FINE EMERY CLOTH

BUFF BLADE

INJECTION-MOLD PLASTIC HANDLE

This flowchart shows the steps in making a metal screwdriver with a plastic handle.

ideas for entirely new products and to improve methods for making existing products. R & D often leads to new inventions or product innovations.

Once designs have been developed, they are reviewed by company management. The company needs to determine how cost effective the product is and whether it can be sold at a price that will still bring in a profit. If the product appears to be too expensive to produce, trade-offs must be made. The product could be redesigned, perhaps using lower quality, less expensive materials. Sometimes an innovative production idea can reduce manufacturing costs. Occasionally a different material might be substituted that is less expensive but functions just as well.

Once the engineers and businessspeople agree on a design, a model called a **prototype** is made. Modern production systems need craftspeople who work closely with engineers and business managers to design and build these prototypes. Prototypes help manufacturers work out design and engineering problems. Prototypes are generally tested for a period of time before expensive machinery is purchased for large-scale production.

The final step involves setting up the production line. Proper forming, separating, combining, and conditioning tools must be selected. In addition, the operations must be organized in the best sequence to manufacture the product in quantity. The organization of a production line can be diagramed as a sequence of steps and drawn as a **flowchart.**

Companies advertise their products to sell them to more people. People who work in advertising take photos of the product, make radio and television commercials, and write catchy phrases that will make the product appeal to consumers.

Salespeople sell and distribute the finished product. Although these people are not directly involved in the production of the item, they are an important part of the business side of manufacturing.

Usually, the company's responsibilities do not end once the product has been sold. Sometimes, products need repair, or customers need technical help in using the product properly. Customer service departments provide after-sale support for purchasers. This includes advising them about improvements and servicing and repairing defective merchandise.

There are two subsystems within the manufacturing system: the material processing system and the business and management system.

ENSURING QUALITY IN MANUFACTURING

Companies try to build quality into the products they manufacture. They want each finished product to be usable. Making sure that quality remains high during the production run is called **quality control.** Quality control provides feedback about the output of the production line. It permits employees to adjust the manufacturing processes. Manufacturers try to ensure that the dimensions of the products do not vary from one item to the next. **Uniformity** in production is an important concern.

Without proper quality control, time and materials would be wasted on products that would have to be discarded. Good quality products keep customers satisfied. This means less returned merchandise and fewer repairs for the company to deal with.

In some factories, people inspect the parts by eye at various stages of the manufacturing line.
(Courtesy of Sperry Corporation)

AUTOMATED MANUFACTURING

Automation is the process of controlling machines automatically. Machines can be programmed to perform a series of operations, one after another. A robot, for example, can be programmed to pick up a part, move it a certain distance, and drop it into a bin. This kind of control is called **program control.** More accuracy can be provided if the automated machine also uses **feedback control.**

Feedback is used to adjust the operation of the machine. For example, feedback can be used to make an automated

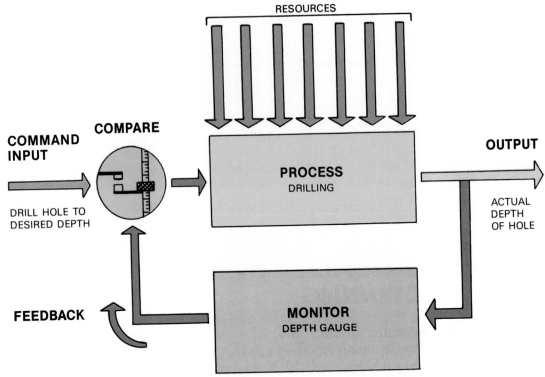

System diagram for an automatically controlled drill press. A switch activated by a depth gauge controls the drill press.

drilling machine stop when a hole has been drilled to the proper depth. A **sensor** "senses" when the drill has drilled deep enough. The sensor sends a feedback signal that switches the drill off automatically. This eliminates the need for a machine operator to turn the drill press on or off.

In order for U.S. industries to compete with lower-cost foreign products, some human workers have been replaced by automated machinery. By automating manufacturing, companies can save money on labor costs and remain competitive.

In modern automated factories, quality control is provided by sensors on machines. These sensors regulate the machines and make sure that they are performing to specifications. Accurate instruments are then used to inspect and test the product.

Robotics

Robots are advanced automated machines. They are "intelligent" machines and can be controlled by computers. They can be equipped with sensors so that they can do different jobs. Some robots can "feel." These robots can be used to hold fragile parts with just the right amount of

Automation has caused some jobs to be lost, but it has also created new jobs.

Automated machines attach heavy wheels to tractors. In the past it took three people to do this work. (Courtesy of Deere & Co.)

pressure. Some robots can "see" with television camera eyes. These kinds of robots can distinguish among parts with different shapes. Robots are replacing people in many modern factories. In Japan, there is already a "humanless" factory. There, the only function of the human being is to make sure that everything is operating properly.

Some people believe that robots are responsible for workers losing jobs. It is true that robots have replaced some people on assembly lines. However, robots have permitted manufacturers to make products of higher quality at a reduced cost. Robots do not take coffee breaks and don't mind uncomfortable work. They will work twenty-four hours a day without complaining and will never be late to work.

In manufacturing, robots are used primarily for welding, spray painting, picking up parts and placing them into machines, and loading objects onto platforms and conveyors. Most of the jobs robots do relieve people from dangerous, boring, heavy, or unpleasant work. A great advantage of robots over other machine tools is that robots can be **reprogrammed.** They can be programmed to weld a fender on one type of car. Then, they can be programmed differently to weld a fender on a different model. People are needed to install, maintain, and program robots.

Robots welding automobile parts (Courtesy of Ford Motor Company)

The Robot Revolution

Often we hear of the "robot revolution." Why is the use of robots a "revolution"? The Industrial Revolution replaced human energy (muscle power) with mechanical energy (machine power). Despite the new technology, human beings were still needed to operate the machines. Robots can actually replace human beings in a manufacturing plant. Although robots don't look human, they can act very much like people. They can perform movements just like a human arm, wrist, and hand. They can use sensors to "see" and "feel." Robots used in industry are called **industrial robots.**

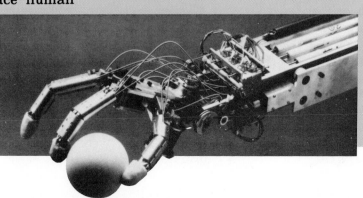

This robotic hand is light and compact. It has three human-like fingers and fourteen joints. (Courtesy of Hitachi, Ltd.)

In a CAD/CAM operation, a designer stores a drawing in a computer and sends it to a computer-controlled machine. There a prototype is produced. (Courtesy of Grumman Aerospace)

CAD/CAM

CAD (**computer-aided design**) is a modeling technique. Designers, drafters, and engineers use CAD to do drawings and designs on a computer. The drawings can then be stored in the computer memory. (See Chapter 5.)

CAM refers to **computer-aided manufacturing.** It is the process of using computers to control machines on the factory floor.

CAD/CAM is a new technology that joins CAD with CAM. CAD/CAM permits an individual to design a part on a computer screen. Then, all the necessary design information (like the size and shape of the part) is communicated directly to a machine tool. The machine tool produces the actual part. A CAD/CAM system links engineering and design with production. An engineer can design a part like a gear and run it through a series of tests on a computer screen without actually having to produce the part. This technique is called **simulation.** Once the computer-generated model is complete, the computer stores the finished design as numerical data. The numerical data is easily understood by a computer-controlled machine, which produces the actual part.

Managers involved in purchasing, shipping, accounting, and manufacturing can get a total picture of all the factory conditions by looking at the computer screen.

(Courtesy of International Business Machines Corp.)

Computer-Integrated Manufacturing (CIM)

CIM, or **computer-integrated manufacturing,** is a happy marriage between the manufacturing and business systems. Not only do the computers store drawings and designs and control machinery, but they also store information about raw materials and parts. In addition, computers

CIM combines the manufacturing, design, and business functions of a company under the control of a computer system.

schedule the purchase and delivery of materials, provide inventory reports on finished goods, and do the billing and accounting.

Flexible Manufacturing

One of the most promising new techniques of manufacturing is called **flexible manufacturing.** Today, many of the products produced by industry are manufactured in small quantities—batches of a hundred or a thousand, rather than millions of items. This is because customers have special requirements that the manufacturers try to fulfill. The General Electric Company in New Hampshire makes 2,000 versions of its basic electric meter. A John Deere factory in Iowa can make 5,000 different tractor variations for farmers with custom needs. Flexible manufacturing makes the production of small batches of custom-tailored products possible. The same production line is used, but the machines are programmed to do different operations as the specifications of the parts change.

Flexible Manufacturing at Deere & Company

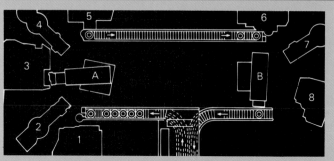

This plant makes construction equipment called backhoe loaders. Once a part enters the manufacturing system, it is tracked and controlled by a computer. It enters one end of the system and comes out the other end completely finished in as little as five minutes. The machines within the system are all computer-controlled. Each machine can be programmed to do many different kinds of machining jobs. Thus, the machines can turn out parts with different dimensions. An employee monitors the entire process.

The manufacturing system is diagrammed here to show part flow and system components. Parts progress from the number 1 machining center to the last machining stop at number 8. Parts move on automated conveyors until picked up by one of two robots (A and B). These robots feed parts to appropriate cutting machines (1, 3, 5, 6, or 8) at appropriate times. Stations 2, 4, and 7 are parts storage areas and automated tool carousels.

The entire manufacturing process is computer controlled.

A finished backhoe loader doing its job
(All photos courtesy of Deere & Co.)

Just-in-Time Manufacturing

Raw materials are delivered to factories by trucks, trains, or ships. They are usually delivered in large quantities. Large quantities of materials require large amounts of storage space. Companies who need storage space must pay rent, heating, and lighting costs for a warehouse. They must also employ people to transport the materials and deliver them to the production line.

Just-in-time (JIT) manufacturing is a method of making sure that raw materials and purchased parts arrive at the factory just in time to be used on the production line. When the product is completed, it is immediately shipped on its way to the customer. This again eliminates the need for storage space and personnel.

Automation has greatly increased productivity in manufacturing.

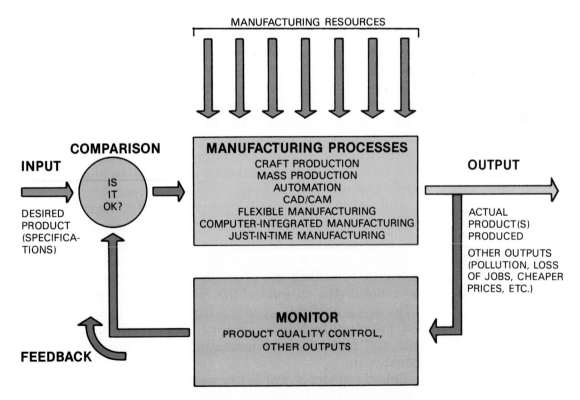

Well-designed manufacturing systems use modern processes to make products of good quality. The products must meet the desired specifications without harmful effects on people or the environment.

SUMMARY

Production systems include manufacturing and construction. Manufacturing refers to producing goods in a workshop or factory. Construction involves producing a structure on a construction site.

For thousands of years, manufacturing involved making things by hand. Craftspeople produced items for their own use and for the use of others. As the nation grew, people demanded more goods. Factories began to mass produce goods using assembly-line techniques.

Assembly lines divide the production process into specialized tasks. Each worker does one job. The assembly line produces standardized parts that are interchangeable.

The factory system improved the standard of living. It brought prices down and permitted the average person to own things that only the rich had been able to afford. The factory system made time a more important consideration since manufacturing operations were synchronized.

Manufacturing has a technical side and a business side. Entrepreneurs start companies, invent new products, and innovate existing ones. Many of today's successful businesses are small companies started by entrepreneurs.

Manufacturing, like all technological systems, requires the use of the seven technological resources: people, information, materials, tools and machines, energy, capital, and time. These resources should be combined to produce a quality product at the lowest possible cost. The outputs of the system should be carefully matched to the needs of people and the environment.

Automation has made manufacturing systems much more efficient. More goods can be produced in a shorter time with less labor cost. Automation can involve programming a machine to do a series of steps in sequence. Automated control often makes use of feedback from sensors. The sensors monitor the outputs of the manufacturing processes.

Ensuring uniform, good quality products is an important goal for manufacturers. Quality control techniques reduce costs and mean fewer rejects and repairs for the company to deal with.

Robotic systems are being used in modern manufacturing plants. These systems do jobs that are unpleasant or dangerous for humans. In some automated factories, robots are replacing people.

Computer-aided design and computer-aided manufacturing (CAD/CAM) links the engineering and design departments with the factory floor. Parts can be designed, modified, and stored as data in a computer. The data can then be sent directly to a machine, which produces the part.

Computer-integrated manufacturing, flexible manufacturing, and just-in-time manufacturing are new techniques for managing the production process. They depend upon computers to provide a total connection between the purchase, storage, and inventory of materials, the control of the manufacturing machinery, and the sales and distribution of the finished products.

Manufacturing technology has come a long way since craftspeople made clay pots 30,000 years ago.

AN ENTREPRENEURIAL COMPANY

Setting the Stage

Entrepreneurs are people who come up with good ideas for new or improved products or services. They then use those ideas to make a profit. The ideas usually start with defining a need that someone has, then figuring out a way to fill that need.

For example, you and your classmates probably have been assigned lockers. It's not easy, however, to keep lockers neat. They tend to become cluttered and messy. Wouldn't it be a great idea if someone designed a locker organizer that would help keep items stored in an organized fashion?

Your Challenge

Form your class into a company to design a locker organizer that would be popular at your school. The organizer must provide a place for pencils, pens, and notebooks. You may also want to provide a way to organize other items you keep in your locker.

Procedure

1. Decide on a name for your company. Then, divide into R & D quality circles, four or five persons to each circle. Each circle will work independently throughout the design process.
2. Do market research to find out the preferred features for a locker organizer. Measure several different pencils, pens, and notebooks to determine size needs.
3. Make several rough sketches of your ideas. Consider materials best suited for the proposed designs.
4. Combine several ideas or select a good idea for your design.
5. Make a drawing of your design. Add dimensions.
6. Using cardboard as a substitute for the actual material you plan to use, make a prototype of your design. Color the prototype as you wish the actual product to appear.
7. Make a list of materials you would need to make each organizer. Determine the cost of the materials needed to produce your design. Use catalogs to compare different suppliers.

Suggested Resources

Sketch pad and soft lead pencils or felt pens
Ruler
Scissors
Transparent tape
Tagboard or thin cardboard
Tempera colors

8. Each circle will present its design to the class. Show the drawings and prototype, and discuss the features of the design that would make it popular. Tell how much it would cost to make.
9. Take a class vote on which organizer the company should make.

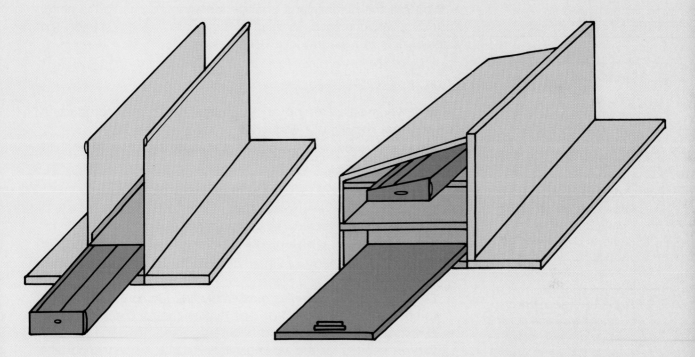

Technology Connections

1. Quality circles allow workers to become directly involved in decisions about the production process. What effect has this had on the workers and the companies?
2. Many of the products you use every day are manufactured by machines and automated equipment. How would your life change if everything you use had to be made by hand?
3. New production systems make increasing use of technologies such as robotics, CAD/CAM, and CIM, just to name a few. How will this affect you as a future worker and consumer?

Science and Math Concepts

▶ Control of modern production machines depends upon numerical data which counts in thousandths of an inch or less.
▶ Many modern production machines are combinations of electrical, hydraulic, and pneumatic systems.

279

T-SQUARE ASSEMBLY LINE

Setting the Stage

The newspaper clipping read: "Engineering firm needs designer/drafter helper for summer work: no experience necessary." You have always liked to draw, especially since you learned how to use drawing tools and a CAD system in your 7th and 8th grade Technology Education class.

You call the number and arrange for an interview. You're hired! Your first assignment? To help design, plan, and establish an assembly line to mass produce T-squares.

Your Challenge

With your class, set up an assembly line to mass produce a drawing tool known as a T-square.

Jigs and fixtures, which speed up the assembly line, can be built by the entire class.

Procedure

1. Be sure to wear safety glasses and a lab coat.
2. The teacher should precut the ½" × 1½" strips and ¼" × 1½" strips of clear cherry or other hardwood using the table saw, band saw, jointer, and planer.
3. Develop a flow chart of the assembly line. Be sure to include a foreperson, safety supervisor, and quality control check points.
4. Everyone in the class should help design and make the jigs and fixtures needed for the assembly line. Accuracy is very important at this stage.
5. Elect a foreperson to assign positions on the assembly line.
6. Rotate assembly line positions periodically.
7. When people switch jobs in the assembly line, they may have to be retrained for their new positions.
8. The miter boxes should be set up with stops to cut the head and blade pieces of the T-square to their proper length. The head is an 8" long piece of the ½" × 1½" strip. The blade is a 19½" long piece of the ¼" × 1½" strip.
9. In order to align the three pilot holes drilled in the T-square blade, use a student-made fixture clamped to the drill press. This technique is also used to drill the ⅜" hang-up hole in the other end of the blade. Two or more drill presses help prevent a log-jam at this point in the assembly line.
10. Accurate alignment jigs are needed to hold the pieces in place during the final assembly.

Suggested Resources

Safety glasses and lab aprons
Computer with software for computer aided drawing (CAD)
Dot matrix printer and/or laser printer
Pieces of clear cherry or other hardwood (for the head of the T-square)—½" × 1½"
Pieces of clear cherry or other hardwood (for the blade of the T-square)—¼" × 1½"
Miter boxes
Drill press
Power screwdrivers and/or variable speed electric drills with driver bits
Twist drills for the shank holes—⅛"
Twist drills for the pilot holes—1/16"
Round head wood screws—⅝" #4
Plywood (for the assembly jigs and fixtures)—⅜"
Heavy gauge aluminum strips (for quality control templates)—1½" × 19½"
Tri squares (for quality control checks)

11. Pilot holes of ¹⁄₁₆″ diameter should be drilled in the head to facilitate the assembly process.
12. Use power screwdrivers and/or variable speed electric drills with driver bits to assemble the T-square. Do NOT overtighten the screws.

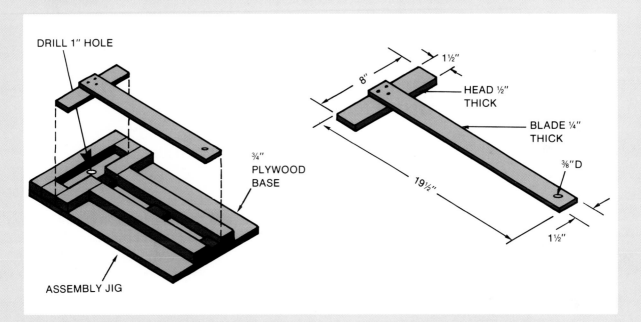

DRILL 1″ HOLE

¾″ PLYWOOD BASE

ASSEMBLY JIG

8″

1½″

HEAD ½″ THICK

BLADE ¼″ THICK

19½″

⅜″D

1½″

Technology Connections

1. What are some advantages of CAD?
2. Why is accuracy so important in designing and making the jigs and fixtures?
3. Why are quality control checkpoints important in an assembly line? Interchangeable parts?
4. There are two basic subsystems of the manufacturing system: (1) material processing and (2) business and management. Which subsystem does developing an assembly line flowchart fall into? Marketing? Final assembly?
5. Why is it important to use *standardized* measuring systems in manufacturing?

Science and Math Concepts

▶ The *metric system,* which is based on decimals, is a standardized measuring system used in science and technology all over the world.

▶ Threaded fasteners, such as screws, nuts, and bolts, use the principle of the *inclined plane.* This inclined plane, instead of going in a straight line, continually circles around the fastener in a spiral shape.

ROBOTIC RETRIEVER

Setting the Stage

The first thing you notice as you enter the nuclear research building at Schleemer Enterprises, Inc. are the sirens and floodlights. Some men in white coats are huddled around a floorplan of the building. Fear is written on their faces. "Am I glad to see you!" says one of them as they notice your arrival. "Two units of Radium-X66 have melted through their canisters. There's not much time!"

Your Challenge

Using a teach pendant, program a robot to pick up and remove two Radium-X66 canisters from a designated area.

Procedure

1. Using the tape rule, T-square, and felt tip marker, measure and draw a grid made of 2" squares on the sheet of plywood. If done correctly, the plywood should have 24 rows of squares going each way.
2. The plywood will be the floor of the building. Make the walls using the lath material as shown.
3. Fasten the magnet to the bottom of the robot in a convenient spot. Epoxy cement may be used to make this more permanent.
4. Study the Model 603 Robot Manual and carefully follow the operating instructions.
5. Experiment with these commands by *measuring with the tape rule* how far the robot travels with each GO FORWARD command. Also, determine how far the robot turns with each TURN command.
6. Using the "building" you made earlier, place the robot in one of the corners opposite the door. Place one of the "canisters" in any one of the squares in the room, *except the squares along the edges of the wall.*
7. Using your graph paper, make a map of the room showing the location of the robot, the canister, and the door. Sketch on the graph paper the path the robot should follow to magnetically pick up the canister and carry it out the door. (*Hint:* Use only 90° turns at this time.)
8. On your sketch, write in the commands you must program into the robot's memory. Be very careful and take your time doing this.
9. Enter your program into the robot using the teach pendant. Replay the program and compare the robot's actual path with the path you sketched.

Suggested Resources

Plywood or masonite (any thickness)—4' × 4'
Lath strips (or similar material)—16' long
Model 603 programmable robot (Graymark International, Inc.)
Two steel ¼" washers (Radium-X66 canisters)
Magnet (approximately 1" square × ⅛" thick)
Graph paper (with ¼" squares)
Tape rule
T-square
Pencil
Felt tip marker
Glue
Epoxy cement (optional)
Saw
Hammer
Nails

282

10. Debug your program so that it works properly.

11. Try using the robot to pick up two canisters from different squares. You may want to challenge other students in a canister-fetching competition.

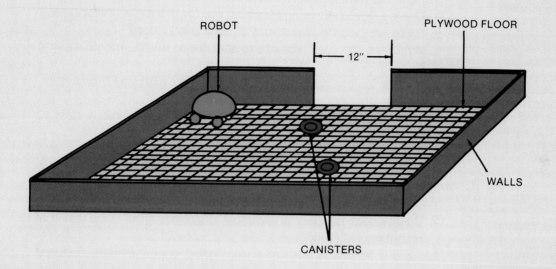

ROBOT

PLYWOOD FLOOR

12"

WALLS

CANISTERS

Technology Connections

1. Robots became practical only after the invention of the computer. Can you name what tasks are performed by the computer inside the 603 robot?

2. More advanced robots are able to sense certain things in their environment. What kinds of senses would make your robot more useful?

Science and Math Concepts

▶ How many TURN LEFT commands are necessary to make the 603 robot turn in a complete circle?

▶ With 360° in a circle, how many degrees does each TURN LEFT command use?

REVIEW QUESTIONS

1. Give five examples of how manufacturing technology has helped satisfy people's needs and wants.
2. Explain how craft production differs from mass production.
3. How did mass production improve people's standard of living?
4. Draw a flowchart of an assembly line for making a greeting card.
5. What are two disadvantages of mass production?
6. What are the two subsystems that make up the manufacturing system?
7. Suggest an invention that would help you with your homework.
8. How might you innovate a soda can?
9. What would be a suitable product for your technology class to manufacture?
10. If you could form a quality circle with your classmates and teacher, what improvements would you suggest be made in the technology laboratory?
11. Why do manufacturers want their products to be uniform?
12. What could be two undesirable outcomes from a system that manufactures computers?
13. How is feedback used to ensure good quality in manufactured products?
14. Do you think that robots should be used instead of people on assembly lines? Explain why or why not.
15. How have computers affected the manufacturing industry?
16. Draw a system diagram for a system that manufactures chewing gum. Label the input command, resources, process, output, monitor, and comparison.

KEY WORDS

Assembly line	Factory	Just-in-time manu-	Prototype
Automation	Feedback	facturing	Quality control
CAD/CAM	Flexible manufac-	Manufacturing	Robot
CIM	turing	Mass production	Uniformity
Custom-made	Interchangeable	Numerical control	Union
Entrepreneur	Inventor	Production	

SEE YOUR TEACHER FOR THE CROSSTECH PUZZLE

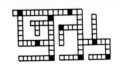

CHAPTER 10

CONSTRUCTION

MAJOR CONCEPTS

After reading this chapter, you will know that:

- Construction refers to producing a structure on a site.
- A construction system combines resources to provide a structure as an output.
- Three subsystems within the construction system are designing, managing, and building.
- Construction sites must be chosen to fit in with the needs of people and the environment.
- A foundation is built to support a structure.
- The usable part of a structure is called the superstructure.
- Structures include bridges, buildings, dams, harbors, roads, towers, and tunnels.

CONSTRUCTION SYSTEMS

There once was a time when people had no houses to live in. They may have lived in caves or under a bush. They didn't have the technology to build structures that would shelter them from the weather or from dangerous animals.

As people began to roam away from their own territory to hunt for food, they developed the need to build shelters. The first type of building construction was probably a simple shelter to protect people from rain and wind. Teepees made from animal hides stretched over a wooden frame were in use as long ago as 20,000 B.C. These homes were movable, and people could take them along as they searched for new sources of food. As people began to settle in villages, they needed more permanent houses. Natural materials like wood, stone, and mud were used. Later, bricks made from straw and mud were used to construct houses.

The early Egyptians, Greeks, and Romans were very good at building structures. The great pyramids in Eygpt were built from blocks of limestone as early as 3,000 B.C. The largest pyramid was the Great Pyramid of King Khufu. It is almost 500 feet tall and is made from over two million blocks of limestone, each weighing over two thousand pounds.

The Romans were known as great engineers. They constructed cities, built roads for their armies, and erected bridges. About 300 B.C. they developed systems (called aqueducts) for supplying water to cities. These ancient pipelines were built from stone and cement. They brought water from nearby rivers to the cities.

Even today, people in some parts of the world use natural materials and anything else they can find to construct housing. (Photo by Michael Hacker)

The Romans used the arch to support water pipelines made of stone. These were called aqueducts. (Courtesy of Harvey Binder)

286

The Romans were very concerned about health. They built sewer systems to carry waste from the city of Rome into the Tiber river. People were advised not to drink the water because it was polluted. Other civilizations did not practice such good health habits. Sicknesses occurred where people drank water polluted by sewers.

One of the greatest Roman contributions to construction technology was the use of the **arch** to support structures. Although the arch had been invented earlier, the Romans improved upon it and used it to build many of their structures.

Construction refers to building a structure right on the location where it will be used. The location is called a **construction site.** Today, about six million people work in the construction industry in the United States. They are employed in constructing buildings, roads, bridges, tunnels, dams, and towers, and in managing the business side of the construction. In total, the construction industry produces over a hundred billion dollars' worth of goods and services for the U.S. economy every year.

Construction refers to producing a structure on a site.

An example of modern construction technology
(Courtesy of NY Convention & Visitors Bureau)

RESOURCES FOR CONSTRUCTION SYSTEMS

A construction system processes the seven technological resources to provide a structure. If the resources are combined correctly, the result is a usable structure, such as a building, bridge, dam, road, or tunnel. Each of these structures requires specific construction materials and equipment. Special techniques and carefully trained people to perform them are needed. The construction system depends upon all of the resources being combined in the right way

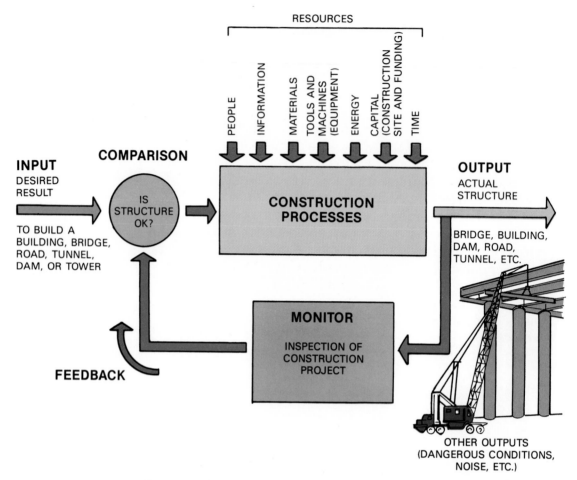

A construction system combines the technological resources to produce a structure on a site.

A construction system combines resources to provide a structure as an output.

at the right time. The failure of any one resource to do its job correctly could result in poor work, dangerous conditions, or total failure.

People

People perform many different kinds of jobs in the construction industry. Some of the jobs involve the design and engineering of the construction project. Some involve managing the construction system. Other jobs involve the actual "hands-on" construction work.

Three subsystems within the construction system are designing, managing, and building.

The **land owner** determines that there is a need to build a particular structure. The owner may be a person who wants to build a private home. The owner may also be a commercial developer who decides to build a housing development or shopping center.

Architects and engineers do the design and planning. **Architects** play the key role in designing buildings. They develop the overall site and building plans for offices, schools, shopping centers, and hospitals. Architects must attend college and study mathematics, engineering sciences, technical and architectural drawing, and art.

Civil engineers are responsible for the technical parts of the project. Architects work with them when designing buildings. Civil engineers help the architects decide if the building can be built exactly as desired. Civil engineers prepare exact drawings and plans for the building framework and foundation. The drawings and plans include information about the size and strength of the construction materials. The plans also give details about **utilities** (plumbing, heating, air conditioning, and electrical wiring).

Civil engineers generally develop the plans for bridges, roads, tunnels, dams, and towers. Engineers must attend college and study subjects such as mathematics, science, technology of materials, and technical drawing. The structural design of the construction project is the responsibility of a **structural engineer.** He or she designs the structure to be strong enough to support the load it is intended to carry.

General contractors own their own construction companies. They hire workers and are responsible for building part or all of the construction project. General contractors work closely with all parties involved in the project, including the owner, the architects, engineers, and craftspeople.

Estimators work for the contractor. They prepare an estimate of the cost of the construction project. Once they figure out what the project will cost, they prepare a cost proposal called a **bid.** If the customer accepts the bid, it means an agreement has been reached about costs and specifications.

Project managers ensure that the construction is planned properly and meets local building codes. They hire and supervise the workers and make sure that the cost of the project stays within their budget. Some colleges and universities offer programs in construction education to train project managers. Some of the subjects covered in these pro-

Project managers must review construction plans with the architects and engineers before actual construction is begun.
(Courtesy of The Turner Corporation, photograph by Michael Sporzarsky)

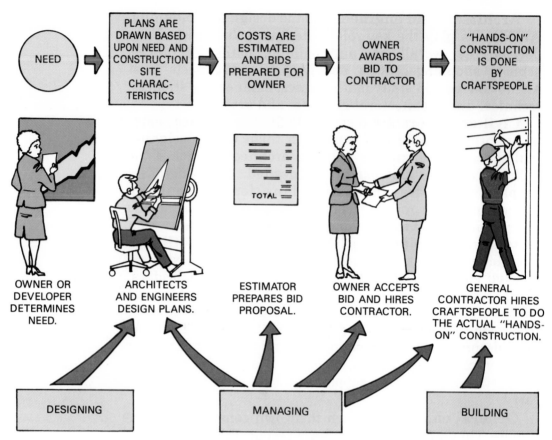

THE THREE SUBSYSTEMS OF CONSTRUCTION (DESIGNING, MANAGING, AND BUILDING) INVOLVE CONTINUOUS COOPERATION AMONG THE OWNER, THE ARCHITECTS AND ENGINEERS, THE CONTRACTOR, AND THE CRAFTSPEOPLE.

People in the construction industry

grams include materials and techniques, planning and scheduling, engineering, and accounting.

Craftspeople work on projects from small housing developments to huge skyscrapers. They may construct small swimming pools or gigantic dams. A craftsperson must be expert in the use of the tools and materials of the trade. Some craftspeople work as carpenters. Others are electricians, plumbers, or cement masons. Some operate heavy construction equipment like bulldozers.

Craftspeople may study the construction trade in high school or at a technical or vocational school. They can also learn by working alongside more experienced workers "on the job" as apprentices.

Information

In the construction industry, information is provided by the people who want the construction done. This information is provided by means of plans and specifications. The plans and specifications are the command inputs to the construction system. They are sometimes called the design criteria. These are the customer's desired results. The specifications are written by architects and engineers. Once the contractors know the specifications. they can provide the customer with a cost estimate (a bid). Specifications include data like the materials to be used, type of foundation, and even the kind of trees and bushes to be planted around a residence.

An ability to prepare and to read mechanical and architectural drawings and blueprints is necessary for people in construction. Information is communicated by engineers and architects to the builder through drawings. In construction technology, a great deal of the planning and designing is done through technical drawings and sketches.

Information about construction techniques is necessary for the builders and craftspeople. These techniques have been learned through the accumulation of centuries of experience.

Engineers must have a good mathematics background. They have to calculate all the stresses (forces) on the structure. Engineers select the proper construction materials to withstand the stresses. If they make a mistake, the structure may fail.

Materials

Materials used in construction are called **building materials.** Among the most popular building materials are concrete, wood (called lumber), steel, glass, and brick.

Concrete is made from stone, sand, water, and cement (cement is made from limestone and clay). When concrete is wet, it looks and feels like mud. After it has been mixed, it sets (gets hard). While it is wet, the concrete can be poured into molds to form desired shapes. These molds are called **forms** and are often made out of wood.

Concrete is brittle. To give it more strength, round steel rods called reinforcing rods are fixed in the forms before concrete is poured. Sometimes wire screens are placed in the concrete. This kind of reinforcement is called reinforcing mesh. Concrete is used to build dams, roadways, tunnels, and buildings.

Wooden lumber is a building material used to provide the framework for private homes. Wood is easy to work with and relatively inexpensive. Some composite materials used in construction (like plywood and particle board) are made from wood. These provide great strength at reasonable cost.

This six-building complex uses energy-efficient glass as a primary building material. (Courtesy of PPG Industries)

Steel is used to provide the framework for skyscrapers, bridges, and towers. Steel is very strong and can be made into cables, beams, and columns.

Glass not only provides daylight, but can add great beauty to a construction project. Glass can be installed as single, double, or triple panes. Thermal pane glass (two or three thicknesses) is used to conserve energy.

Brick is made from clay. The clay is fired in an oven called a kiln and then becomes hard. Brick houses are expensive. Labor costs are high because each brick has to be set in place individually. Brick houses generally last longer and are much more fire-resistant than houses made from wood.

Tools and Machines

The tools of the construction industry range from hand tools used by carpenters, plumbers, and masons to huge pieces of equipment like cranes and bulldozers. The largest pieces of equipment are used in earth-moving operations. Special equipment is used for excavating (digging), lifting heavy materials, and building bridges, dams, roads, pipelines, and tunnels.

Construction equipment has come a long way since this steam shovel from the early 1900s. (Courtesy of Perini Corporation)

A wide variety of power tools and equipment is used to do construction projects. (Courtesy of The Stanley Works)

Robots are being used more and more frequently by the construction industry. Construction is a good place to use robots because of the danger involved in working high in the air, below ground, and deep underwater.

Energy

Construction systems use energy to operate huge earth-moving equipment and other machines. However, more energy is used by the industries that support the construction industry than by the construction industry itself. Vast amounts of energy are used by the manufacturing industries, which produce building materials (like concrete, bricks, and steel) for the construction industry. Trucks, railroads, and ships consume energy as part of the transportation system, which brings building materials from one location to another.

Capital

Construction projects are very expensive. Equipment costs a lot to purchase or rent, labor costs are high, and building materials are costly. People wishing to construct a large structure must obtain financing for their project. Before a bank lends money for construction, it must be convinced that the construction will be profitable. The bank wants to be sure that the person applying for a loan will be able to pay it back. A loan for a home is called a **mortgage.** Big construction projects often require financial packages combining private loans and governmental aid.

A loan is paid back over a period of years with interest added to each payment. **Interest** is the fee the bank charges for loaning a customer money. Mortgages for private homes are generally paid back within fifteen to thirty years. If a person borrows $50,000 to construct a house, the amount paid back is much more. However, the payments are spread out over many years so that the borrower can afford each monthly payment.

Land purchase is often one of the highest construction costs. In some parts of the country, land is extremely expensive. Can you imagine how much it would cost to buy even a quarter acre of land in the downtown area of a major city? In the middle of a city like New York or San Francisco, land costs could reach $1,000 per square foot. That comes to about $40 million per acre. People who construct buildings in the cities must charge high rents for offices and apartments because the cost of the land is so high. Skyscrapers are often built in downtown areas to make the best use of the expensive land. People who own expensive land generally have little trouble arranging for a construction loan from a bank. If the loan is not repaid, the bank takes over the ownership of the land.

Time

Construction projects generally involve long periods of time. Years can be involved in constructing a major structure like a bridge or tunnel. The use of modern tools and equipment can help contractors save time. For example, powerful air-driven staple guns are replacing the hammer-and-nail system of fastening shingles to roofs.

Residential housing can be completed in a period of several months. New techniques of construction reduce the time it takes to build structures.

Instead of building a structure entirely on site, parts of the structure can be **prefabricated** in a factory. For example, entire walls may be built in a factory. The walls are then transported to the construction site, where they are assembled. Time is saved because mass-production techniques can be employed in prefabricating the parts.

A type of construction that is becoming more popular is **modular construction.** A module is a basic unit like a room. Modules can be combined to form structures with different dimensions, depending upon the number of modules used.

SELECTING THE CONSTRUCTION SITE

One of the subsystems of the construction system is the management system. Management involves all the design, engineering, and planning that comes before the actual "hands-on" construction. The selection of a construction site is one of the most important management decisions. Land costs and yearly taxes can vary greatly from one location to another. Often, a few miles can make the difference between high and low land costs. The actual conditions of the ground must be suitable for construction. If the location is very hilly, too much time and money will have to be spent **grading** (leveling) the soil. If the ground is too rocky, it will be very difficult to dig a foundation.

There are many things to consider when choosing a construction site. If the construction project is a factory, management must think about how manufacturing materials and equipment will be delivered and shipped. A neighborhood school should be located within a short distance of a residential area. A bank should be located close to the business areas of a city. If the project is a housing development, it should be located near schools and shopping. If the structure is a bridge, it should be located at a point where conditions are suitable for the foundations.

It is important to think about what the construction will do to the surrounding environment. The scenic beauty must be preserved. The community needs must be considered. For example, it would not be a good idea to select a site for an airport right in the middle of a quiet residential community. If the site is in an historical area, the construction project should fit in with the historical nature of the surrounding buildings. In some cases, roads or tunnels must be routed around or underneath existing historical landmarks in order to preserve the landmarks.

A good construction site must be near roads, railroads, or ports so that building materials and heavy equipment can be delivered. Utilities like water and electricity must be available. Ease of waste disposal must also be considered. In addition, construction workers should be able to reach the site easily. The best site for construction is the one that meets all the specifications at the lowest cost.

Construction sites must be chosen to fit in with the needs of people and the environment.

PREPARING THE CONSTRUCTION SITE

The ground where a structure will be built must be prepared. Heavy construction equipment clears the ground. If there are unwanted buildings on the site, they are removed. Wrecking balls or explosives are used to demolish buildings. Bulldozers clear timber or brush from an area. Unwanted materials are hauled away by dump trucks.

Giant shovels remove earth to prepare a site for construction. (Courtesy of RCA)

The immense construction site for the Miho Dam dwarfs the huge pieces of machinery and equipment below.
(Courtesy of Kajima Corporation)

Before any construction can begin, the structure must be laid out on the site. A **surveyor** is a person who shows the construction workers exactly where the structure should be placed. Surveyors use a **transit** to measure and lay out angles. They use an **engineer's level** to set the elevation (height above ground) of different points.

BUILDING THE FOUNDATION

A building **foundation** supports the weight of a structure. If you were standing on soft earth, you might sink down. If you stepped on a wide board, your weight would be spread out over a larger area. You wouldn't sink down as far. A foundation spreads the weight of a structure out over a larger area of ground so that the structure is well supported. In cold climates, foundations also prevent frost damage to the structure. Foundations are sometimes called **substructures** because they are underneath the main structure.

A foundation is built to support a structure.

Foundations include the earth (on which the entire structure will rest), the footing (which transfers the structure's weight to the earth), and the vertical supports (which rest on the footing).

If the structure will be built on hard soil, a **spread footing** is used. It spreads out the weight of the structure, just like your body weight is spread out over an area as wide as your foot. If the structure will be built on soft earth, on marsh land, or underwater, **piles** are used. Piles are like stilts. They are driven down into the earth until they reach firm, hard ground or rock.

Foundations are usually made from concrete. The wet concrete is poured into metal or wooden forms, which act like molds. When the concrete sets, the forms are removed. Often, reinforcing rods are buried in the concrete to make the footings stronger. Sometimes concrete blocks are used to form the walls of a foundation. The blocks are stacked on top of each other and cemented together by mortar.

Piles support the Columbia River Bridge in Portland, Oregon. *(Courtesy of Sverdrup Corporation)*

BUILDING THE SUPERSTRUCTURE

The usable part of a structure is called the superstructure.

The **superstructure** is usually the part of the structure that is visible above the ground (unless the structure is a tunnel or a pipeline).

Mass superstructures are made from large masses of materials. They have little or no space inside them. Dams

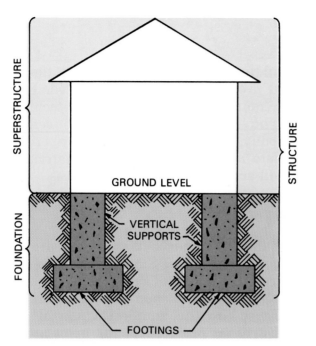

SUPERSTRUCTURE

FOUNDATION

STRUCTURE

GROUND LEVEL

VERTICAL SUPPORTS

FOOTINGS

A structure includes a foundation and a superstructure.

and monuments are mass superstructures. They are built from brick, concrete, earth, or stone.

Bearing wall superstructures are built from brick, concrete, or stone. They enclose a space with walls. Examples of bearing wall superstructures are the castles and chateaus built in the middle ages. Often the walls of these structures were 20 feet thick or more at the base.

Framed superstructures use a skeleton for support of the structure. A tower is an example of a framed superstructure. Today, most buildings use framed superstructures. The framing is often done with wood in residential construction. Reinforced concrete or steel are often used to frame commercial buildings. Reinforced concrete and steel framing last a long time and make a building more fire-resistant than does wooden framing.

This castle is an example of a bearing wall superstructure. (Photo by Michael Hacker)

This airport terminal in Riyadh, Saudi Arabia has a framed superstructure. (Courtesy of The Turner Corporation)

A dam is an example of a mass superstructure. (Courtesy of U.S. Department of the Interior, Bureau of Reclamation)

TYPES OF STRUCTURES

Bridges, buildings, dams, harbors, roads, towers, and tunnels are all projects that are constructed. Each of these structures requires special construction techniques. These techniques have been learned through centuries of experience. Today, we combine that past experience with engineering data about the properties of materials and about how they will act under conditions of stress.

Bridges

Bridge building stands out as one of people's greatest engineering achievements. Since prehistoric times, people have constructed bridges of one kind or another. Bridges that use a single beam to span a distance are called **beam bridges.** These are the simplest kinds of bridges. The strength of a beam bridge depends upon how strong the beam is.

The arch is a strong shape that was used a great deal by the Romans to support viaducts (bridges) and aqueducts (elevated trenches used for carrying water from one location to another). **Arch bridges** were originally made from stone. They are now made from concrete because of its high strength under compression.

A **cantilever bridge** works like two diving boards facing each other. Each section is firmly attached at the ends. If the tips do not meet, a short section may be added to link the two sections together.

An early beam bridge

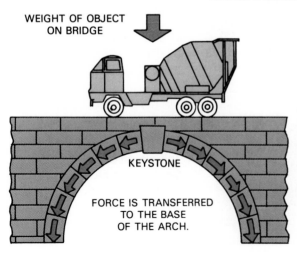

WEIGHT OF OBJECT ON BRIDGE

KEYSTONE

FORCE IS TRANSFERRED TO THE BASE OF THE ARCH.

A stone arch bridge. The arch transfers weight to the supports at the base of the arch.

Cantilever bridges extend outwards and are secured at the ends.

A cantilever bridge requires a huge force to support each end of the structure. Engineers have designed **double cantilever bridges** to overcome this problem.

Suspension bridges are used to bridge wide spans. The deck of a suspension bridge is suspended from steel cables. These cables are attached to towers firmly implanted on a strong foundation. Some of the most famous bridges in the world are suspension bridges. The Golden Gate Bridge in San Francisco is 4,200 feet long. The George Washington Bridge, which links New Jersey and New York, is 3,500 feet long.

The Severn Bridge in Avon, England is a suspension bridge that took twenty-one years to build. It was completed in 1966 after years of planning and construction. The deck was specially designed not to sway very much in the wind. The sections of the deck were built in a factory and were then transported to the construction site. This method of construction kept the cost down, since it is easier to build in a factory than a hundred feet up in the air. The bridge towers rise about 300 feet above the water.

The Severn Bridge spans 3,250 feet. (Courtesy of British Tourist Authority)

Buildings

Building construction can be divided into four categories:

1. residential (hotels and housing);
2. commercial (banks, stores, and offices);
3. institutional (schools and hospitals); and
4. industrial (factories).

In the first three cases, the architectural design and the beauty of the project are of primary importance. In the fourth case (industrial construction), the design is not generally based on good looks. Rather, the design of a factory is based on how well heavy machinery, equipment, and material transport systems (like conveyors and big moving cranes) will fit in.

Land costs for residential, commercial, and institutional construction sites are generally higher than for industrial construction sites. Houses, offices, schools, and hospitals must be located near the center of a residential area. Factories, on the other hand, can be located in the countryside. In fact, it is usually desirable to build factories far away from heavily populated areas. Then noise and pollution will not interfere with people's lives.

Residential construction usually involves framed super-structures. The skeleton of a private house is most often framed with wood. The lumber used is called **dimensional**

Electrical wiring and utilities are installed before the walls are completed.
(Photo by Michael Hacker)

lumber because it comes in standard thicknesses and lengths. Most walls in residences are framed with 8-foot lengths of 2″ × 4″ lumber. These lengths are spaced so that the center of each one is 16″ from the center of the next. Insulation, electrical wiring, and pipes for plumbing are installed within the structural framework before the walls are attached. The inside walls themselves are usually made from panels of **plasterboard** (made from a plaster core covered with heavy paper or cardboard). The plasterboard panels are nailed or glued to the framing.

Outside walls are made from composite or plywood panels. They are often covered with brick veneer or **siding** made from aluminum, vinyl, or wood. Siding provides an attractive appearance and protects the structure from the weather. Wood siding needs to be painted or stained every few years. Aluminum or vinyl siding needs less maintenance. **Insulation** made from fiberglass or a plastic foam material is installed between the inside and outside walls to conserve energy. Insulation helps maintain the temperature inside the building and saves heating and cooling costs.

For conventional houses, all the construction is done at the site. Sometimes, all the building materials are precut to size in a factory. The materials are then assembled on site.

A recent trend in residential construction is to build entire houses in sections in a factory. This is called **panelized construction.** The walls and roof trusses (frame for the roof) are prefabricated. Complete walls are built in panels and are mass produced in standard sizes in the factory. The walls may include insulation, wiring, windows, and doors. The labor cost of the actual construction is therefore reduced.

Apartment houses and commercial and institutional buildings are framed with beams and columns made from steel or reinforced concrete. The beams are the horizontal pieces. They carry the weight of the floors and the walls. The columns are the vertical members. Columns transfer weight from the beams directly down to the foundation. A steel framework is often prefabricated and transported to the construction site. There it is lifted into place by large cranes. Reinforced concrete beams and columns are cast in place at the site. Floors may be made of concrete or steel. The outside walls are usually made of brick or panels of concrete, glass, metal, or plastic.

This building has a steel framework. (Photo by Michael Hacker)

The Tacoma Dome

An example of commercial construction is the Tacoma Dome in Tacoma, Washington. The dome will be part of a $44 million sports and convention facility. The 100,000 square-foot arena seats 26,000 people. It is topped by a 530-foot diameter wooden dome, the largest in the world.

A dome is a lightweight structure that is very strong. It works on the principle of the arch. Weight is transmitted from the very top outward and downward to the base. A dome is really a combination of a roof and walls. Notice that the dome is composed of triangular sections. A triangle is a very rigid shape that provides great strength.

(Courtesy of Sverdrup Corporation)

Tunnels

Perhaps it was from watching animals burrow underground that people got the idea for tunnels. Tunnels under water and through mountains have shortened travel routes for people since ancient times. The earliest tunnels were dug by the Babylonians about 2,000 B.C.

The construction of tunnels presents a series of complex problems to engineers. Early tunnels were dug by hand because there was very little machinery available. The earth that was removed was hauled away by carts and ponies. Workers were always in fear of the tunnel caving in around them. In 1818, the **tunneling shield** was invented. This device supported the earth while the tunnel was being dug. The shield was pushed along as the work progressed.

Tunnels constructed in rock formations are drilled and blasted with explosives. The resulting opening is supported by steel arches. A new technique involves the spraying of a concrete slurry (mixture) called **shotcrete** onto the walls. Shotcrete prevents water from seeping through the rock.

In the early 1900s, tunnels were dug by hand. Workers erected wooden arches and supports as the construction progressed. (Courtesy of Perini Corporation)

Tunnels drilled in soft ground are now being dug with tunneling machines. These machines have huge rotating cutters that may be as large as 15 feet in diameter. The cutters rotate at about 5 rpm and push against the earth with a force of close to a million pounds. As the earth is cut, precast concrete or steel rings are pushed into place to provide support.

One of the longest automobile tunnels in the world is the tunnel that connects France and Italy under Mont Blanc in the Alps. The tunnel is about eight miles long and saves around a hundred miles of driving. One French team and one Italian team started digging from opposite sides of the mountain. They met after over two years of construction in 1962.

The Fort McHenry Tunnel

Fort McHenry in Baltimore Harbor is an historic landmark. During the war of 1812, the British bombarded the fort with over 1,500 shells. U.S. soldiers were able to hold on and turned back the British force of fifty ships. From an American ship in the harbor, a young attorney, Francis Scott Key, could see that the American flag still waved. The waving flag inspired him to write "The Star Spangled Banner."

Recently, a tunnel was built near Fort McHenry in order to link parts of the interstate highway system. A bridge was considered first, but it would have ruined the historical quality of the surroundings. The Fort McHenry tunnel was built entirely out of sight and earshot of the fort. The completed tunnel carries eight lanes of traffic and 66,000 cars and trucks a day. The $750 million project took seven years to complete.

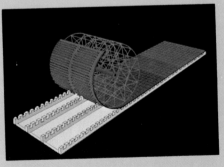

Tube making for the Fort McHenry Tunnel begins with steel panels. The panels are welded together to form a shell plate. Stiffeners are added for strength.

Modules are shaped by wrapping the shell plate around a specially designed reel. More structural pieces and various form plates are added.

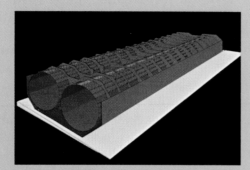

Sixteen modules (eight for each tube) are joined. They form one section of the double-barreled tube. Each tube holds two lanes of roadway.

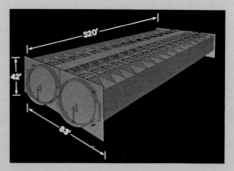

Dam plates seal each end of the tube. Keel concrete is added for strength and rigidity. The section is then launched for a 12-hour tow to the Fort McHenry Tunnel site.

A heavy, plow-like beam from a screed barge dredges along the harbor bottom. This forms the trench that will hold the tunnel.

From a lay barge, the tube section is lowered into the trench. It is then attached to another tunnel segment. Thirty-two tube sections were carefully lowered and connected to form the Fort McHenry Tunnel.

The Fort McHenry Tunnel site

(All photos courtesy of Sverdrup Corporation)

Roads

Before the time of ancient Rome, roads were only narrow paths used by two-wheeled carts. Roman engineers built over 40,000 miles of roads in order to transport the Roman Legions (the Roman armies). Tunnels and bridges allowed the roads to be built in a straight line from one place to another. The Roman roads were even paved with stones.

Modern road-building techniques began with ideas proposed in the late 1700s by a Scotsman named John Loudon McAdam. McAdam developed procedures for making roads last longer by improving their drainage. He designed roads with a base of hard soil, a stone layer for drainage, and a covering of tar. He also made roads a little higher in the center than at the edges. This made water flow away from the roadbed. McAdam's name was given to the material that is now a popular surface for roads. It is called **macadam.** Since the early 1900s, most roads have been built of concrete or macadam.

The coming of the automobile created a need for new and improved roads. Prior to the early twentieth century, few roads were paved. Many roads were made of a crushed stone surface. The crushed stone worked well for horses and buggies, which provided the major means of land transportation. However, rubber automobile tires kicked out the stones and ruined the roads. A new type of surface was needed.

Today, roads must be designed to support heavy loads carried by vehicles moving at high speeds. Roads must be safe and reliable. Techniques of road construction have become very well refined.

Now that good roads are available, people can live miles away from their work and commute daily. Suburbs have sprung up around cities. Roads have also improved access to all areas of the country. The interstate highway system has linked our great cities and opened up new markets for many businesses.

Road construction begins with choosing the exact location of the road. When possible, the proposed route will avoid interfering with existing construction. Sometimes, buildings that are in the way must be destroyed because it would cost too much to build the road around them.

The ground must be prepared by bulldozers and a smooth surface made. Modern highways are built on soil pressed down by heavy rollers. The soil is covered with a layer of stone, which spreads out the load and provides drainage. The pavement is made of concrete about one foot thick or from blacktop materials like macadam or asphalt.

Road construction technology also involves design of center barriers and proper lighting. Highway control systems, such as traffic lights, are included as a part of the road-building system.

Today's highways are safe for high-speed long distance travel. (Courtesy of New York State Department of Transportation— Clough, Harbour, & Associates)

Other Structures

Airports, canals, dams, harbors, pipelines, and towers are other constructed projects. Large construction companies often have divisions that are responsible for each type of structure. For example, a large construction company may have a tunnel division, a building division, a pipeline division, and so on. Smaller companies usually specialize in only one or two types of construction.

Structures include bridges, buildings, dams, harbors, roads, towers, and tunnels.

(Courtesy of Perini Corporation)

(Courtesy of Mardian Construction Company)

(Courtesy of Perini Corporation)

(Courtesy of Perini Corporation)

Construction technology provides us with many benefits.

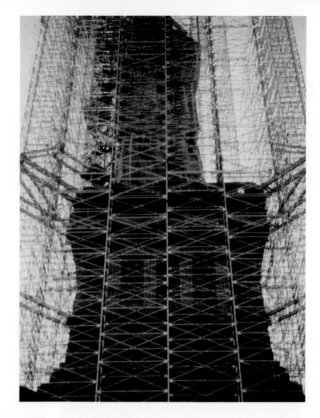

The Statue of Liberty was recently renovated. After 100 years, salt water and air pollution had worn parts of the statue away. The renovated statue was rededicated on July 4, 1986. (Courtesy of NASA)

RENOVATION

Renovation is the process of rebuilding an existing building. Sometimes it is done to change the style of a building. Many times structures are renovated because they need repair. As a structure ages, dust, wind, and moisture slowly eat away the materials from which the structure is made. When a structure is renovated, old materials are replaced or renewed. Renovation is a good alternative to demolishing the structure and starting to build all over again.

SUMMARY

Construction refers to the production of a structure on a site. Construction systems combine technological resources to provide a structure as an output. People engineer, design, manage, and build structures. They use information in the form of engineering science, building techniques, drawings, and mathematical data. Building materials like concrete, steel, and wood are used in large quantities during the construction process. Special tools and equipment are needed to do the heavy construction work. The heavy equipment used in earth-moving operations consumes a great deal of energy. Capital costs are generally quite high in construction projects because of the high price of land, material, equipment, and labor. Construction projects generally extend over a rather long period of time. Modern techniques like prefabrication and modular construction reduce construction time.

The construction process involves choosing and preparing the site, building the foundation, building the superstructure, installing utilities, and finishing and enclosing the outside surfaces.

Site selection depends upon the intended use of the structure. Commercial buildings should be located near the business district of a town. Schools should be located in residential areas. The effects of the construction project on the environment and on the community should be considered.

Before construction can begin, the site must be cleared. Surveyors lay out the placement of the structure exactly on the actual site. The structure must be supported by a foundation. The foundation spreads the weight of the structure over a larger area of ground. The superstructure is the usable part of the structure. Mass, bearing wall, and framed superstructures are common types.

Techniques of building structures have been learned through many years of practical experience. Today this experience is combined with a knowledge of the properties and limitations of materials to build airports, bridges, buildings, canals, dams, harbors, pipelines, roads, towers, and tunnels.

(Courtesy of British Rail)

DOME CONSTRUCTION

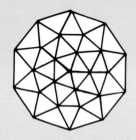

Setting the Stage

What sort of structure can be built with no internal supports at all? A hint . . . Eskimos have built them for years. The structure we are talking about is, of course, the **dome.** Domed structures are often used for sports stadiums like the Astrodome in Houston, Texas, or for World's Fairs, Expos, or amusement parks.

The weight of a dome is transmitted uniformly from its top outward and downward to its base. As a result, no internal weight-bearing supports are needed. Domes can be made from standardized parts. These parts can be made in a factory and assembled later at the building site. Domes are also fairly light in weight. This means that the foundation for this type of structure can be smaller than conventional foundations.

Your Challenge

Build a small, dome-shaped structure that can be used as a greenhouse.

Suggested Resources

Safety glasses
Galvanized steel—22-24
    gauge
Round head machine
    screws—½", 4–40 NC
    with nuts
Bar folder
Box and pan brake
Notcher
Sheetmetal hole punch—⅛"
Squaring shear
Assorted sheetmetal layout
    and hand tools

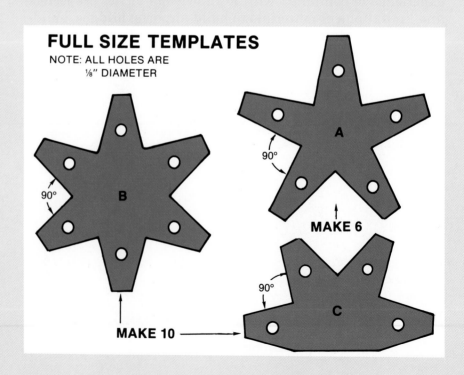

FULL SIZE TEMPLATES

NOTE: ALL HOLES ARE ⅛" DIAMETER

90° B

MAKE 10

90° A

MAKE 6

90° C

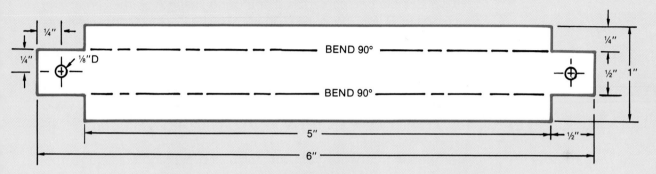

MAKE 35

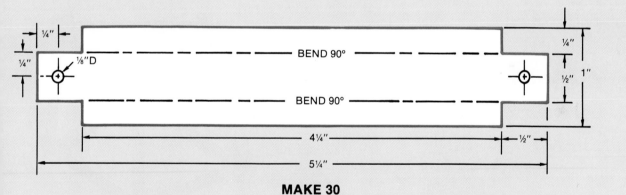

MAKE 30

Procedure

1. Be sure to wear safety glasses.
2. If you have not been told how to use any tool or machine that you need to use, check with your teacher BEFORE going any further.
3. Obtain a copy of the design information from your teacher.
4. Duplicate the full-size shapes A, B, and C and use them as templates.
5. Use a notcher to cut out 6 of shape A, 10 of shape B, and 10 of shape C.
6. Punch ⅛" diameter holes as marked.
7. Use the squaring shear and notcher to cut out the 6" and 5¼" connecting elements of the dome. Make 35 of the 6" length and 30 of the 5¼" length.

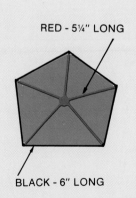

RED - 5¼" LONG

BLACK - 6" LONG

MAKE 5 PENTAGONS

309

8. Punch two ⅛" diameter holes as marked on the ends of these elements.
9. Bend ¼" on the long sides of the connecting elements to 90 degree angles. Also place a slight bend (approximately 10 degrees) on the ends of these pieces. See detail #1.
10. Assemble one pentagon shape using shape A as the center and shape B for the outer corners. The 5¼" connecting elements always radiate from the center of the pentagon. The 6" connecting elements are used on the perimeter. This will later become the top center of the dome.
11. Fasten the connecting elements to the corner shapes using ½" 4-40 round head machine screws and nuts. DO NOT COMPLETELY TIGHTEN THE SCREWS UNTIL THE ENTIRE STRUCTURE IS ASSEMBLED.
12. Working outward from the completed pentagon, assemble the rest of the structure. Follow detail drawing #2 and #3 carefully.
13. The dome can now be covered with plastic and used as a greenhouse.

SLIGHT BEND

DETAIL #1

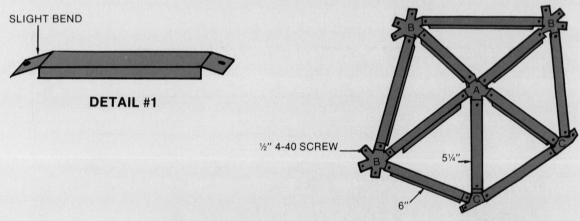

½" 4-40 SCREW

5¼"

6"

DETAIL #2

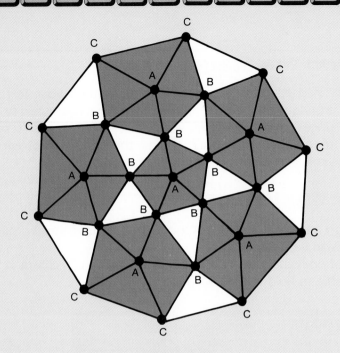

DETAIL #3

Technology Connections

1. A simple dome is not hard to build. It needs no columns or supports to hold it up. Why not?
2. How is the heat from the sun concentrated by a greenhouse? What is the *greenhouse effect?*
3. Why is it possible to use smaller foundations with dome structures?
4. Which two geometric shapes are usually part of the design of a dome? Are there any more?
5. What materials can be used to build a dome? Is an igloo a dome?
6. Notice how the dome is constructed of triangular sections. A triangle is a very rigid shape that provides great strength.

Science and Math Concepts

▶ A *pentagon* has five sides. A hexagon has six sides.
▶ *Infrared light* or *thermal radiation* has a wavelength longer than that of red light. It is below the visible light spectrum.

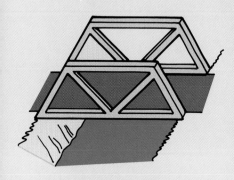

Setting the Stage

If you had to cross a stream without getting your feet wet, building a bridge might be the answer. Logs have been used to cross streams since prehistoric times. But what if the log isn't strong enough to hold your weight? This is the sort of problem that engineers still face when building bridges today.

Your Challenge

Design and build a model of a bridge. If desired, small groups of students can work together.

The bridge must be 4" tall and able to span an 8" chasm. It must also be able to hold at least 40 pounds.

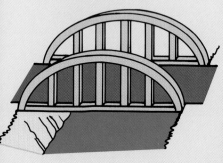

Suggested Resources

Safety glasses and
  lab apron
3 pieces of pine—
  ¼" x ¼" x 3'
1 piece of heavy
  cardboard—4" x 9".
  Matting board is good.
1 piece of cardstock—
  8½" x 11". A manila
  file folder works fine.
White glue
Wax paper
Assorted tools and
  machines

Procedure

1. Be sure to wear safety glasses and a lab coat.
2. Make a full-scale drawing of your bridge on graph paper. Do both the front and side views. When you are satisfied with your bridge design, show it to your teacher.
3. Cut the strips of pine to the exact lengths and angles needed for your design. Lay the cut pieces on top of your drawing to check for accuracy.
4. Before gluing the wood pieces in place, cover your drawing with a sheet of wax paper. You can glue the pieces using your drawing as a guide. The wax paper will prevent your wood from sticking to the drawing.
5. Cut small triangles from the cardstock. Use these triangles to reinforce the corners of the bridge. They are known as *gussets*.
6. Give the glued wooden pieces a chance to dry before you start gluing on the gussets.
7. Use the heavy cardboard to make the floor of your bridge. Glue it in place.
8. After it is dry, measure your bridge for the minimum height (4") and ability to cross an 8" chasm. Place 40 pounds *on top* of the bridge to test its strength.

Technology Connections

1. *Construction* refers to building a structure on a site. What factors must be considered when choosing a site for a bridge?
2. What type of bridge has its deck supported from above by steel cables?
3. What construction material is made from a mixture of cement, sand, stone, and water?
4. Engineers must calculate all of the stresses (forces) on bridge structures they design. Can you pick out some of the major stress points on your bridge?
5. What can happen to a bridge if it is not engineered properly? Has this ever happened?

Science and Math Concepts

▶ A *chasm* is a deep cleft or gorge in the earth.
▶ Concrete has a high *compressive* strength. It resists being compressed or crushed.
▶ Steel cable has a high *tensile* strength. It is very strong under tension and resists being stretched.

313

REVIEW QUESTIONS

1. What is the major difference between manufacturing and construction systems?
2. Draw a labeled systems diagram of the construction system.
3. Describe four career opportunities provided by the construction industry.
4. How could you finance the construction of a private home?
5. If you were to choose a site for a movie theater, what are five things you would have to consider?
6. What kinds of superstructures do the following structures have?
   a. the Washington Monument
   b. a tower that supports electrical wires
   c. a skyscraper
   d. a large dam
   e. a castle from the Middle Ages
7. Make a sketch of a suspension bridge.
8. Explain why an arch can support a great amount of weight.
9. Design a tower, using rolls of newspaper as your building material.
10. Name five different types of structures.

KEY WORDS

Arch	Engineer	Macadam	Structure
Architect	Estimator	Mortgage	Superstructure
Cement	Forms	Prefabricate	Surveyor
Concrete	Foundation	Specifications	
Construction	General contractor	Steel	

SEE YOUR TEACHER FOR
THE CROSSTECH PUZZLE

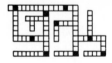

CHAPTER 11

CONSTRUCTING THE BUILDING

MAJOR CONCEPTS

After reading this chapter, you will know that:

- Buildings are constructed in steps. The footing, foundation, floors, walls, and roof are built. The utilities and insulation are installed, and the structure is finished.
- The footing acts as the base of the foundation. It distributes the weight of the structure over a wider area of ground.
- The foundation wall supports the entire weight of the structure and transmits it to the footing.
- Walls transmit the load from above to the foundation. They also serve as partitions between rooms.
- The roof protects the house against weather, prevents heat from escaping, and beautifies it.
- Insulation helps maintain a constant temperature in the house.
- Utilities include plumbing, heating, and electrical systems.
- Manufactured houses are built in a factory and delivered to the construction site.
- Wind effects influence the design and construction of skyscrapers.

INTRODUCTION

Through many years of practical experience, people have learned a great deal about how to build structures. Practical knowledge, engineering principles, and management skills are used in modern construction technology.

Residential and commercial buildings are the most common structures. This chapter will illustrate how these structures are built.

HOUSEBUILDING

So you want to build a house! Sounds like a pretty difficult thing to do. But building a house can be simplified to a series of steps.

Building even the most beautiful homes involves a series of well-defined steps.
(Courtesy of Western Wood Products Association)

PRECONSTRUCTION

A few important steps must be taken before the actual construction. These are done during the **preconstruction** phase. One step involves picking the right location for the house.

Picking the Location

Who decides where a house will be built? The builder usually picks the location. When a house is custom built, the builder, the architect, and the owner confer and agree on the best location.

How does a builder decide where to build a house? The builder looks at the cost of the land and how easy it will be to bring utilities to the home. Utilities include water, electricity, and sewers. The cost of the land will depend on how desirable the location is. If the house will be built near a famous ski resort or in an area with a beautiful view, the price of the land may be very high.

The builder also considers whether roads already serve the proposed location. If a new road will have to be constructed to reach the house, the expense could be very high.

Within the last twenty years, energy costs have risen. Builders, therefore, try to take advantage of the sun for heating houses. The sun rises in the east and sets in the west. Houses built in the northern hemisphere should face south to take advantage of solar energy. The builder generally tries to take advantage of deciduous trees on the building site. Deciduous trees have broad leaves. In the summer, the leaves provide shade. In the winter, after the trees have shed their leaves, the sun can shine through.

Buildings are constructed in steps. The footing, foundation, floors, walls, and roof are built. The utilities and insulation are installed, and the structure is finished.

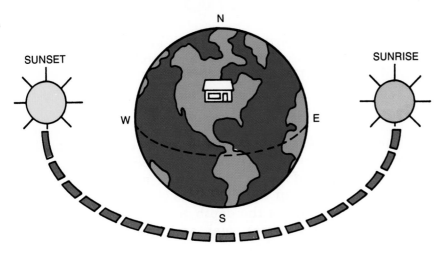

The sun is closest to the earth at the equator. A house built in the northern hemisphere should have large windows facing south to take advantage of the heat from the sun.

The building permit must be displayed during construction. (Photo by Michael Hacker)

Building Permits and Codes

Another step in the preconstruction process is obtaining the necessary **building permits**. A building permit authorizes the builder to begin construction. Before a permit is issued, complete construction plans must be approved by officials of the local building department.

A permit is issued only if the plans comply with local **building codes**. Each city, town, and village has its own building codes. These are a set of regulations intended to ensure that all buildings are constructed safely. Building codes give builders guidelines to follow when building a house. For example, a building code might specify how high windows must be above the floor, or the sizes of wood to be used in construction.

The building department generally sends **building inspectors** to monitor the construction project. If the structure meets all of the building codes after it is completed, the building department issues a **Certificate of Occupancy (CO)** to the builder. The CO means that the house has passed all of the inspections and is ready to be occupied.

CONSTRUCTING THE FOOTING AND FOUNDATION

All constructed objects have some kind of a foundation. House foundations have two parts: the **footing** and the **foundation wall**.

The Footing

The footing for most homes is made of concrete. It acts as a base for the foundation wall. The **spread footing** distributes the weight of the structure over a wider area of the ground. The depth of the footing is governed by the **frost line**. The frost line is the depth to which the ground will freeze. In ·colder climates, such as in northern Maine, the

The ground under the leaning Tower of Pisa settled unevenly. This uneven settling caused the tower to lean. The tower is 191 feet tall. It continues to tilt at the rate of one inch every eight years. The top is about 16 feet out of plumb (off exact vertical). (Courtesy of Italian Government Tourist Authority)

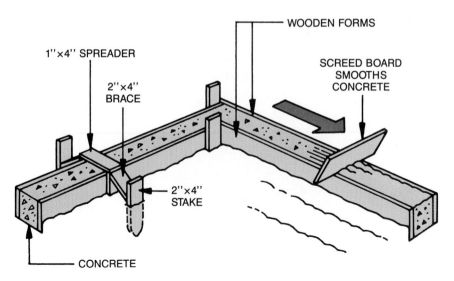

Wooden boards contain the concrete as it is placed.

The footing is dug to a depth below the frost line. (Courtesy of American Wood Preservers Bureau)

frost line could be as much as 7 feet below the surface (below grade). In warmer climates, such as in southern California, the frost line might be only an inch or two below grade.

The footing must be deeper than the frost line because when the water in the ground freezes, it expands. The movement of the earth could cause the footing to move or buckle. When the footing rests on earth below the frost line, the ground is stable.

The footing acts as a base for the foundation. It distributes the weight of the structure over a wider area of ground.

To build the footing, a trench is dug around the building site underneath where the foundation will be. The footing is usually about twice as wide as the foundation. Long boards made of 2 × 8 inch lumber are placed along the sides of the trench. These boards provide a form for the concrete that will be placed in the trench. Wooden braces keep the boards in position while the concrete is being placed. When the concrete cures (becomes rigid), the boards are removed.

The Foundation Wall

The foundation wall is built upon the footing. It supports the weight of the structure. The foundation wall is most often made from concrete blocks. The size of these blocks is about 8 × 8 × 16 inches. They have hollow cores to reduce the weight of the blocks. The wall is sometimes made from concrete. The foundation wall can also be made from lumber that is pressure-treated to resist rotting. These wooden foundations are gaining in popularity because they provide a dry basement. The basement can then become as comfortable a living space as any other room in the house.

The foundation wall supports the entire weight of the structure and transmits it to the footing.

A concrete block foundation (Courtesy of Michael Hacker)

Concrete block foundations are laid by a **mason**. A mason is a craftsperson skilled in the art of working with concrete. The concrete blocks are attached to the footing with **mortar**. Mortar is a mixture of cement, lime, sand, and water. The mortar acts like glue and bonds the concrete blocks to the footing, and to each other.

Steel **reinforcing rods** may be used to anchor the foundation walls to the footing. These circular steel rods are imbedded in the footing and cemented in place in the middle of the block.

It is important that the foundation be **level** (exactly horizontal), and **plumb** (exactly vertical). If the foundation is not perfectly level and plumb, the house will be crooked.

Concrete is strong under compression, but weak under tension. Steel reinforcing rods are used to give the foundation more tensile strength.

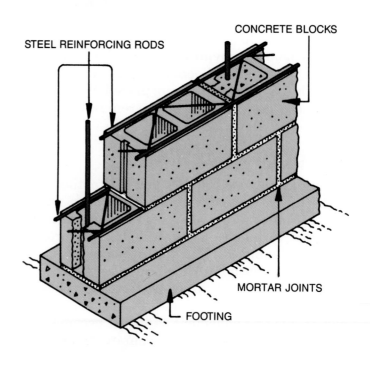

The foundation wall must be constructed so that it is perfectly level. (Courtesy of Portland Cement Association)

Mason checking plumbness with a level held vertically as he lays the first course of block on the footing. (Courtesy of Portland Cement Association)

If a house is to have a basement, the foundation walls are built 12 blocks high from the top of the footing. Since the blocks are each 8 inches high, an 8-foot high basement will be provided by the foundation wall. In houses without basements, the foundation walls can be shorter. A shorter foundation wall provides a **crawl space** instead of a basement. A crawl space gives only enough room to install and service heating ducts, electrical conduit, and plumbing pipe.

In some warmer climates (where the frost line is not far below grade) **slab foundations** are used. Slab foundations are also used where the soil is very loose. These foundations are made from a large slab of concrete. Slab foundations provide a very large area over which the weight of the structure may be spread. Homes erected on a slab foundation will have no basement.

BUILDING THE FLOOR

The floor is built on top of the foundation. It is constructed from long boards called **floor joists**. Floor joists are usually spaced 16 inches on center (apart). That means the center of one joist is 16 inches from the center of the next joist. Floor joists are made from lumber that is 2 inches thick and from 6 to 12 inches wide. The length of these boards varies from 10 to 16 feet. They are attached from one side of the foundation to the other.

Very often, the foundation is wider than 16 feet. The builder can buy longer boards, but lumber yards charge a

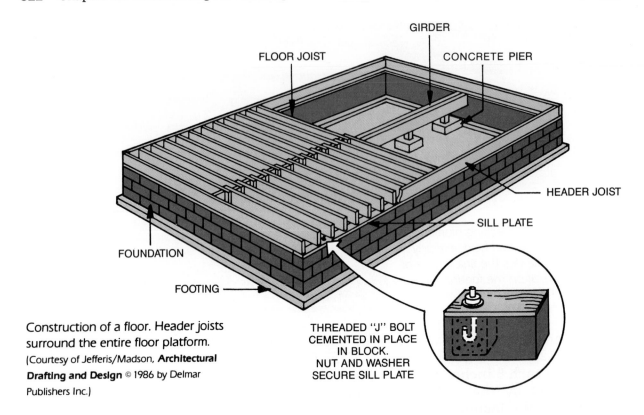

Construction of a floor. Header joists surround the entire floor platform.
(Courtesy of Jefferis/Madson, **Architectural Drafting and Design** © 1986 by Delmar Publishers Inc.)

THREADED "J" BOLT CEMENTED IN PLACE IN BLOCK. NUT AND WASHER SECURE SILL PLATE

much higher price for them. Normally, when the foundation is wide, a strong wooden or steel girder is installed. The girder serves as a midpoint support for the joists to rest upon. It also permits the builder to use shorter lumber for joists. The girder is supported by a pier or column.

Since the foundation is made from concrete, the wooden joists cannot be nailed directly to it. A wooden **sill plate** is installed on top of the foundation before the joists are attached. The sill plate is a 2 inch thick piece of lumber that is attached to bolts that have been cemented into the foundation. The joists are then nailed to the sill plate.

After the joists have been installed, the **subfloor** is nailed to them. The subfloor is often made from plywood. Plywood comes in panels that are 4 × 8 feet in size. For flooring, plywood panels that are 1/2 inch or 5/8 inch thick are nailed to the joists. The subfloor provides the surface to which the finished floor will be attached.

Plywood subfloor
(Courtesy of Michael Hacker)

FRAMING THE WALLS

Walls are normally constructed after the floor is built. Most houses are framed superstructures. **Framing** the walls refers to cutting and assembling their parts. Walls serve two main purposes. They carry the load from above, and they serve as partitions between rooms. A wall that has been designed to support weight from above is called a **load-bearing** wall.

The vertical members of a wall are called **studs**. These are normally made from 2 × 4 inch lumber and are about 8 feet long. The height of a typical wall is eight feet. **Jack studs** are short pieces of wood which are used above and below windows and above doors, see page 325.

Walls transmit the load from above to the foundation. They also serve as partitions between rooms.

Spacing Between Studs

Building codes specify how much space must exist between studs. The spacing between the studs is governed by the ability of the lumber to support a load from above. When 2 × 4 inch studs are used, they are typically installed 16 inches on center. When 2 × 6 inch studs are used, they are spaced 24 inches on center.

Since 2 × 6 inch studs are spaced further apart, fewer of them are needed in a wall section. In a 48-inch section of wall, only three 2 × 6s are needed. Four 2 × 4s would be needed to frame the same space.

The volume of the wood used is about the same, but there is a real advantage in using 2 × 6s. That advantage is due to the fact that in the winter, the wooden studs will conduct heat from the inside of the house to the outside. Three 2 × 6s would have less surface area in contact with the outside wall than would four 2 × 4s. Therefore, less heat would be lost. Also, there is more room for insulation when 2 × 6s are used. Insulation is often placed between the studs. The extra thickness permits the use of 6 inches of insulation.

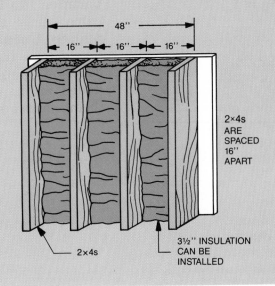

The wall is built horizontally.
(Courtesy of Mark Huth)

After completion, the wall is raised to the vertical position.
(Courtesy of Mark Huth)

Although walls are vertical, they are constructed horizontally. They are built on the surface of the subfloor. It is easier to construct the wall section when the wood can be supported by the floor. After the entire wall is constructed, it is raised into a vertical position and nailed in place.

The top and bottom of a wall are made from horizontal pieces of lumber called **top plates** and **bottom plates**. The top plate is made from a double layer of 2 × 4s. The bottom plate is made from a single 2 × 4. When the wall is being built, the studs are first cut to size. Then they are laid on edge on the subfloor between the top and bottom plates.

The studs are fastened to the plates with nails. If a partition (an inside wall) will connect to the outside wall, extra studs are installed to provide more surfaces for nailing. **Corner posts** are installed at the corners. These provide an inside and an outside surface to which wall covering material can be nailed. Bracing is sometimes added in the corners to prevent the frame from racking (being forced out of plumb).

Carpenter using an automatic nailer to frame walls (Courtesy of Paslode Corporation, an ITW Company)

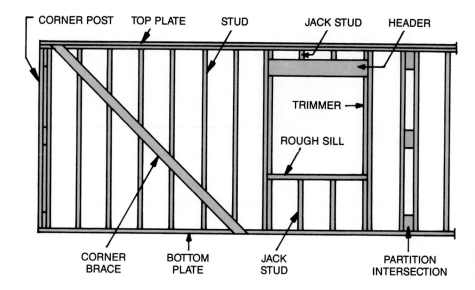

CORNER POST   TOP PLATE   STUD   JACK STUD   HEADER

TRIMMER

ROUGH SILL

CORNER BRACE   BOTTOM PLATE   JACK STUD   PARTITION INTERSECTION

Typical framing for an exterior wall (Courtesy of Lewis, *Carpentry*, © 1984 by Delmar Publishers Inc.)

When spaces are made for windows and doors, a strong piece of wood called a **header** is placed across the top of the space. The header carries the weight from above. It transmits the load to the studs. **Trimmers** are short studs that support the header. They are attached to the studs along the window opening.

SHEATHING

After the wall has been framed, it is covered with **wall sheathing**. Wall sheathing can be plywood, particle board, wooden planks, or rigid foam board. Most often, 4 × 8 foot sheets of sheathing are used.

Wall sheathing is used to make the frame of the house more rigid. It also provides a surface on which to attach exterior wall coverings such as wood or vinyl siding. Sheathing closes in the structure, and protects it from the weather.

Sometimes the long studs bow up toward the center. The carpenter sights along the length of the stud to see which edge bows up. This is called a **crown**. All of the crowned edges should face in the same direction, toward the outside of the house. In that way, the entire wall will have the same kind of bow, rather than having studs alternately bow in different directions. It is much easier to install sheathing and interior finish when the studs bow only in one direction.

Sheathing is nailed to the studs before the wall section is raised to the vertical position. It is generally installed over

A laser beam can provide a perfectly horizontal line from which measurements may be made. (Courtesy of Spectra-Physics)

This house uses a combination of plywood and rigid foam board sheathing. (Courtesy of Clara Littlejohn)

the entire wall section, including windows and door openings. After it has been installed, a carpenter saws out any sheathing that covers the openings. This may seem to be wasteful of materials. However, it is a fast method of construction which saves on the cost of labor. Pieces of sheathing that are cut from large openings can sometimes be used to cover other areas of the wall.

ROOF CONSTRUCTION

The roof protects the house against weather, prevents heat from escaping, and beautifies it.

The purpose of a roof is to protect the house against weather, and to prevent heat from escaping in cold weather. The roof also affects the way the house looks. In a single-story house, the roof provides the ceiling above the first floor. In multi-story houses, ceilings are built by laying joists across the top plate of the walls right above the studs. Ceiling construction is very similar to the construction of wood floors.

The most common type of roof is a gable roof. The gable roof slopes down from the top. The slope (or **pitch**) of the roof is determined by the climate. Where there is heavy snowfall, the pitch of the roof will be steep. In this way, snow will slide off the roof. It will not melt and seep into the house.

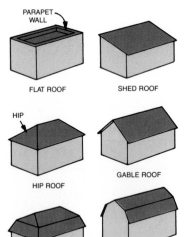

Principal types of roofs

Framing the Roof

In the past, most roofs were framed. To frame a roof, joists are installed across the top of the house. The joists are nailed to the top plates directly above the wall studs. Then roof **rafters** are installed. One end of the rafter is nailed to a joist. The other end is raised up and attached to a **ridge board**. The ridge board acts as a central support for the roof rafters. It runs along the entire length of the house. After the roof has been framed, plywood sheathing is installed. This is called **roof decking**. Finally, weatherproof material is used to cover the decking.

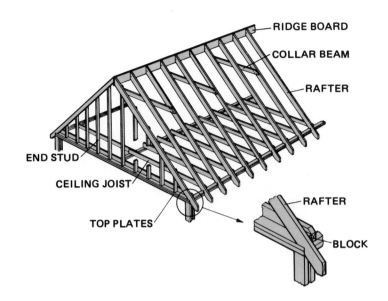

Conventional method of framing a roof (Courtesy of USDA Forest Service, Forest Products Laboratory)

Roof Trusses

Technology has made roof construction much easier. Most roofs today are made of prefabricated wooden parts. These parts are shaped like big triangles and are called **roof trusses**.

Roof trusses are made in a factory and delivered to the construction site by truck. The trusses are first suspended upside down (like a *V*). They are hung across the width of the house and rested on the top plates of the walls. They are then swung upright by a team of carpenters or lifted by a small crane and nailed in place.

Roof trusses are carefully designed and engineered. Because they are mathematically precise, trusses use thinner lumber than do framed roofs. The cost of manufactured trusses (including the labor to make them) is about the same as the cost of materials for a framed roof. However, the use of manufactured trusses reduces the time needed to construct the roof. Instead of taking 2–3 days, it now takes 2–3 hours.

Roof trusses have a triangular construction. The sides of the triangle are called **chords.** There are two top chords and one bottom chord. The supports in the middle of the truss are called **web members.** (Courtesy of Mark Huth)

Earthquakes and No-Fault Houses

At the center of the earth is a core of molten rock. When the molten core moves, cracks are created in the earth's crust. These cracks are called **faults**. They occur along areas known as **fault lines**. Structures built on a fault line might be subjected to sudden stress.

In 1964, an Alaskan earthquake measuring 8.5 on the Richter scale left death and destruction in its wake. Many homes collapsed due to the movement of the earth.

A team of construction specialists and engineers from the U.S. Forest Service went to Alaska to survey the damage. They studied the houses that were destroyed. The team found that the weak points in the houses were where parts of the frame were nailed to other parts. An effort was made to design a new, stronger type of construction.

Research was done for about ten years. The breakthrough came in the late 1970s. The **Truss-frame system (TFS)** was developed.

The TFS makes a single structural unit out of a roof truss, two wall studs, and a floor truss. Special rigid metal connectors with sharp teeth hold the wooden parts together. The TFS units are installed and covered with sheathing. The sheathing makes the structure stable. When sheathing is attached, these engineered parts make a building that is stronger and more durable than most conventional structures.

Truss-frame systems reduce the time needed to build a house by up to 50 percent. The selling price can be reduced by as much as 10–25 percent. The Truss-frame system also helps to conserve forest reserves. The house requires 30 percent less lumber since all the framing is done with 2×4s with 24-inch spacing on centers.

The principal designer of the TFS was a man named Roger Tuomi. He took out a public patent on his design. This means that anyone can use it. Within the last five years, buildings have been built in 35 states using the Truss-frame system. The TFS is a good example of how modern technology can make industrial processes more efficient.

Roofing

Once the roof is constructed and roof decking is installed, **roofing** material is applied. Roofing refers to covering the roof with weatherproof material. Roofing protects the decking from rotting. It is installed as soon as the decking is in place to prevent rain and snow from entering the rest of the house.

Roofing material is sold in **squares**. A square of roofing is the amount of material needed to cover 100 square feet of roof surface. Asphalt, fiberglass, or wooden shingles are most often used as roofing materials.

Before shingles are installed, the roof is covered with paper that has been soaked with asphalt. This material is called **asphalt felt**. It serves as an **underlayment** for the shingles. Asphalt is the same material that is used to pave roads. It is a tar-like substance that offers good protection against the weather. Asphalt felt is stapled to the decking in horizontal rows. Each row (called a **course**) overlaps the previous course by about 2 inches.

Asphalt Shingles

Asphalt shingles are applied on top of the underlayment. Asphalt shingles are a very effective roofing material. Their surface is covered with granules of slate, granite, or some other mineral. The granular material comes in different colors. It gives the roof an attractive appearance. It also

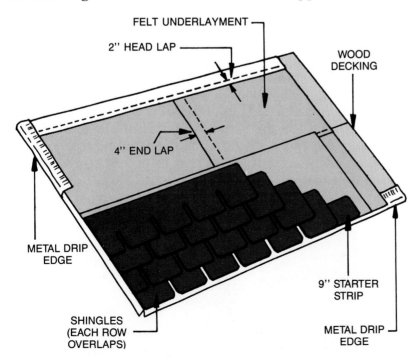

Asphalt shingles are installed in strips.

protects the shingles from damage. The granules reflect the sun's heat, which would otherwise soften the shingles. Black roofs absorb heat and raise the inside temperature of the house. White roofs reflect the sun's rays and keep the house cooler.

Asphalt shingles are nailed or stapled in place. One course of shingles is installed overlapping the previous course. The overlap ensures good protection against weather.

Wooden Shingles and Shakes

Wooden shingles have been used since Roman times. The Romans used oak to make their shingles. Wooden shingles were used throughout Europe until around A.D. 1600 when a shortage of wood developed.

Wooden shingles are sawed from a log. Wooden pieces split from a log are called **shakes**. Wooden shingles and shakes are very attractive. They are being used again as roofing material. Wooden shingles and shakes are made from Western red cedar, cypress, and redwood trees. These woods resist decay because of their natural oils. Western red cedar, cypress, and redwood trees grow slowly. Because of this, they have a very close grain. The close grain does not expand very much when the wood is soaked with water. The wood, therefore, resists cracking and splitting.

Wooden shingles and shakes are usually applied directly over the roof decking. No underlayment is used. Underlayment would trap water between itself and the wood. The shingles and shakes would then be more apt to get wet and to rot.

INSULATING THE HOUSE

Insulation is a material that does not conduct heat well. It is placed between the inside and the outside surfaces of the house. The insulation keeps the inside of the house isolated from the outside temperature. In this way, heat is prevented from escaping to the outside during cold winter days. Heat is also prevented from entering the house during hot summer days.

Hand-splitting shakes using a mallet and a froe (Courtesy of Red Cedar Shingle and Shake Bureau)

Roof with wooden shingles
(Courtesy of Red Cedar Shingle and Shake Bureau)

Certain materials are better insulators than others. Fiberglass and some kinds of plastic are commonly used to insulate structures. Fiberglass is generally formed into **blankets** that are 3 1/2 inches thick. The blankets are rolls of fiberglass that are 30 to 80 feet long. The blankets are often cut into **batts** that are 2, 4, or 8 feet long. The fiberglass blankets are covered with plastic sheet, waterproof paper, or aluminum foil on one side. Fiberglass insulation is also produced as loose fibers. That is blown into walls through drilled holes, after the walls have been installed.

Insulation is also made from plastics like polyurethane. A gas is mixed with the plastic while it is being manufactured. The gas causes the plastic to expand like shaving cream foam. The polyurethane foam becomes rigid. It is formed into boards 4 feet wide × 8 feet long and 1/2–2 inches thick. The insulation board is covered with metal foil on one or both sides. The boards are attached to the studs with special fasteners that have wide heads.

The plastic sheet, waterproof paper, or aluminum foil on the outside of the insulation acts as a **vapor barrier**. The vapor barrier is placed on the warm side of the house. It protects against **condensation**. To understand condensation, think about what happens to your bathroom mirror after you have taken a hot shower.

Warm air carries more moisture than cool air does. When warm, moist air cools, the moisture condenses. After a hot shower, the bathroom is filled with warm, moist air. This air touches the cooler mirror surface and droplets of moisture form (condense) on the mirror.

If moisture were allowed to condense inside a wall, the wood would get wet and it might rot. The vapor barrier prevents warm air from reaching the colder side of the house. Thus, it eliminates condensation.

The degree to which a material is a good insulator is indicated by its **R value**. The better the insulator, the higher its R value is. For example, fiberglass batts that are 3 1/2 inches thick have an R value of 11. Those that are 6 inches thick have an R value of 16.

Insulation helps maintain a constant temperature in the house.

Fiberglass insulation installed between wall studs and roof rafters in a garage (Courtesy of Clara Littlejohn)

FINISHING THE HOUSE

Once the framing, sheathing, and roofing are completed, the inside and outside of the house can be finished. **Exterior finishing** refers to applying material like siding, shingles, shakes, or paint to the outside of the house. **Interior finishing** refers to finishing the inside walls and ceilings.

Exterior Finishing

Exterior finishing makes the outside of a house more attractive. Wood siding is a commonly applied exterior finish. It is made from wooden boards that are nailed horizontally or vertically to the sheathing. A form of siding called **clapboard** is thicker on one edge than on the other. When it is installed, one piece of clapboard overlays the next.

Wooden siding will rot if it is not protected. Some sort of a coating must be applied over the wood. Sometimes, an oil-based stain is applied. More often, the siding is painted.

Shingles and shakes are also used as exterior siding. These are very similar to the materials used in roofing. Shingles and shakes used as siding are generally bought in panels. Several shingles or shakes are glued to a piece of plywood and nailed to the sheathing. These panels can be installed easier and faster.

Many homes today have vinyl siding. Vinyl siding looks like wood, but it is made from plastic. It never needs painting and will not rot.

Other kinds of exterior finishes include plywood, aluminum siding, brick veneer, and stucco.

Siding makes a house look attractive.
(Courtesy of Western Wood Products Association)

Wooden shingles are often used as siding.
(Courtesy of Red Cedar Shingle and Shake Bureau)

Interior Finishing

Interior finishing is often referred to as **drywall** construction. It involves finishing the inside walls and ceilings with gypsum wallboard (also called plasterboard or sheetrock). Interior finishing also involves spackling and taping the joints, and applying paint, paneling, or wallpaper.

Gypsum wallboard is made from plaster that is sandwiched between two layers of paper. It is purchased in 4 × 8 foot panels. The wallboard is attached to the studs with special drywall nails or screws. The fasteners are set beneath the surface of the wallboard, forming a small dimple. **Spackle** (a plaster-like substance) is then applied over the dimple to conceal the fasteners.

The joints between the pieces of wallboard must be covered up. Tape made of paper or fiberglass is used to cover the space between the wallboards. It is held in place with **joint compound**, which is like a kind of plaster.

New technologically advanced latex paints are easy to apply. They are easy to clean and last a long time. (Courtesy of PPG Industries Inc.)

Paint

Paint is the finish most often used to cover wallboard. Paints can be oil-based or water-based. Oil-based paints tend to last longer, but they have more of an odor when they are applied. They are also harder to clean up. They are thinned with turpentine.

Water-based (or latex) paints are thinned with water. While they are wet, water-based paints can be cleaned up with water.

Paint can be purchased as **flat**, **semigloss**, or **gloss**. Gloss refers to the ability of the paint to reflect light. Flat paints do not have a sheen. They are generally used for living rooms and bedrooms. The advantage of semigloss and gloss paints is that they can be wiped clean without harming the paint job. They are commonly used in kitchens and bathrooms where walls are subjected to wear and grime.

Paints consist of pigments and vehicles. The **pigment** gives the paint its color. It consists of chemical colors and natural minerals. Some common pigments are white lead (white), iron ores like ochre (yellow) and sienna (brown), iron oxides (red), lampblack (black), and chemical compounds of chromium and lead. The **vehicle** is the liquid portion of the paint. It carries the pigment particles.

Plywood paneling has surfaces made from hardwood veneers. Expensive hardwood grain patterns are sometimes photographically simulated. Paneling is used to beautify the inside walls of a home. (Courtesy of American Plywood Association)

INSTALLING UTILITIES

Before the inside walls are finished, plumbing, heating, and electrical wiring must be installed. These utilities are often taken for granted. They are there when you need them, but they are not visible because the pipes and wires are hidden behind the walls.

Utilities include plumbing, heating, and electrical systems.

PLUMBING

Domestic plumbing systems include the hot water supply, cold water supply, and drainage. These three systems are actually separate from each other. The main water supply feeds both the cold and hot water systems. The fresh water supply comes into the house. It is never allowed to mix with the waste water that leaves the house through the drainage pipes. The fresh water supply comes from individual wells or from reservoirs. Reservoir water passes through water treatment plants and is carried through underground pipes to the home.

The water supply in the home circulates through pipes under pressure. The cold water is carried directly to sinks, showers, tubs, toilets, washing machines, and hose outlets. The cold water also flows to a **hot water heater**. Hot water heaters burn oil or gas or use electricity to heat the water. Once the water is heated, the hot water circulates through a separate system of pipes. Some of the hot water gets piped directly to the faucets. This is referred to as **domestic hot water**. Often, **faucets** mix the cold and hot water.

Plumbing fixtures are fed by pipes installed behind the walls. (Courtesy of Kohler Company)

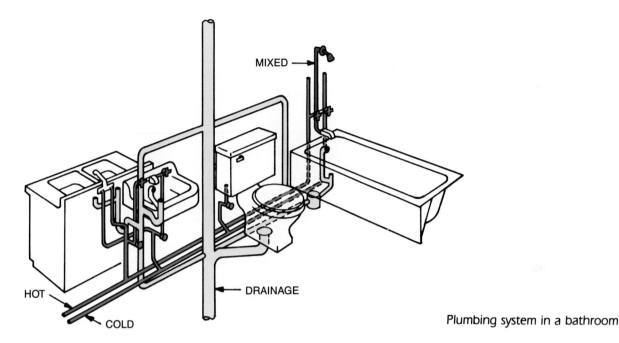

MIXED

HOT

COLD

DRAINAGE

Plumbing system in a bathroom

The pipes used to carry the domestic water supply are generally made from copper or from a plastic known as polyvinyl chloride (PVC). Copper pipe does not rust and resists corrosion. It has a smooth inner surface which permits water to flow easily. It can be easily bent and joined. To cut copper pipe, a **tubing cutter** is used. Copper pipes are joined together with copper **fittings**. Fittings come in a variety of forms. **Couplings** join two pipes. **Tees** provide a junction for three pipes. **Elbows** permit pipes to be joined at 45 or 90 degrees. **Unions** allow pipes to be disconnected.

Fittings are usually **soldered** onto the pipe. Solder is a metal alloy that is made from tin and lead. It melts at about 450 degrees Fahrenheit. **Soldering flux** is first applied to the joint. The flux keeps the copper clean (from oxydizing) while it is being heated. A propane torch is used to heat the pipe to a temperature that will cause the solder to flow. The solder fills the space between the pipes and creates a waterproof joint. **Compression fittings** are also used to join pipe. These fittings use a collar that tightens around the pipe and forms a watertight connection.

In recent years, PVC pipe has been used more and more. PVC is cheaper than copper. It will never rust or corrode. PVC pipe is joined with a chemical bonding agent or by compression fittings.

Drainage systems sometimes use cast iron pipe. The iron pipe is very stong and also resists corrosion.

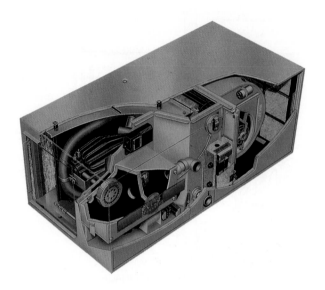

A cutaway view of a forced-air furnace (Courtesy of Lennox Industries Inc.)

HEATING SYSTEMS

Conventional heating systems use oil, natural gas, or electricity to provide heat. Alternative systems use wood- or coal-burning stoves, or solar heaters. Most systems use a furnace that warms air or water and circulates it to the areas to be heated. In a forced-air furnace, oil or gas is burned. The heat causes steel fins on a heat exchanger to get hot. Fans blow air over the heat exchanger through ducts (sheet metal or plastic pathways) that supply heat to the rooms.

Some furnaces heat water in a **boiler** to between 150 and 180 degrees Fahrenheit. A device called a circulator pumps the hot water through pipes to radiators. The water is then pumped back to the boiler for reheating.

ELECTRICAL SYSTEMS

Local communities insist that builders conform to the standards set by the *National Electrical Code (NEC)*®. The *NEC* has been developed by the National Fire Protection Association. It establishes safe methods of installing electrical wiring and equipment.

Electrical service is supplied to the home by power companies. Generally, power companies provide 120-volt and 240-volt service. **Voltage** is a measure of the force with

National Electrical Code® and *NEC*® are registered trademarks of the National Fire Protection Association Inc., Quincy, MA 02269.

Electricity is distributed from the power company to the home through overhead transmission lines. These must be repaired after damage from wind and rain storms. (Courtesy of Southern California Edison)

When several businesses share a building, separate watt-hour meters are installed for each. (Courtesy of Southern California Edison)

which electric current is pushed through wires. The 120-volt service powers lighting and most appliances. The 240-volt service is used to run stoves and heavy appliances, such as central air conditioners, that draw a lot of electric current.

The electric service enters the home through the service-entrance equipment. This includes a **watt-hour meter**, a **main disconnect** switch, and the **electrical panel**.

The watt-hour meter measures the amount of electrical energy consumed by all the electrical equipment in the house. Once every month or two, the power company sends a meter-reader out to read the meter. The company then sends you a bill for your electricity usage.

The main disconnect switch can shut off all the power supplied to the house. It is used in case of an emergency such as an electrical fire.

The electrical panel provides a way to distribute electricity to various branch circuits. The branch circuits supply power to each of the various rooms throughout the house. The panel contains **circuit breakers**. Each branch circuit has its own circuit breaker. Circuit breakers have replaced **fuses** in most homes. Circuit breakers control the amount of current that can safely pass through household wiring. If a short circuit occurs, the resulting high current causes the breaker to trip. The electric circuit is then temporarily disabled. Once the short circuit is corrected, the breaker can be reset.

Series and Parallel Circuits

Electrical devices are wired in **series** or in **parallel** circuits. A series circuit is one that provides only one path for electric current to flow. If anything interferes with the path, electric current stops flowing altogether. An example of a series circuit is the way lights are wired on some Christmas trees. On these trees, if one light is removed or burns out, the entire path is broken. All the lights, therefore, go out. Circuit breakers and fuses must be connected in series with the devices they protect. In series circuits, the source voltage is divided among all the electrical components.

A parallel circuit provides more than one path for current flow. Outlets in the home are wired in parallel. Each appliance receives its own source of voltage. Several appliances can be connected to the outlets at the same time. If one appliance is disconnected, all the others will continue to operate.

Electrical wire is covered with heavy insulation. The insulation prevents the wires from causing an electrical short circuit. Electrical wire used in residential construction is known by the trade name of **Romex** cable.

Almost all the cable used in homes today has an extra wire within it. This extra wire is used to connect all the metal outlet boxes together. This process is known as **grounding** the outlet boxes. The term **grounding** is used because the other end of the extra wire is connected directly to the earth. When electrical equipment is plugged into a grounded outlet box, it too is grounded.

Today many contractors use plastic boxes to house electrical outlets and switches. Plastic boxes reduce the possibility of an electrical short circuit.

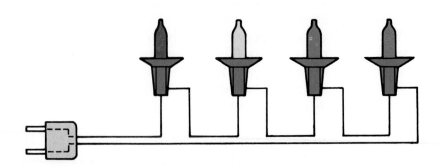

A series circuit has only one path for current flow.

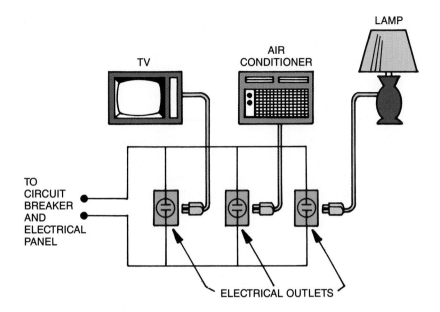

A parallel circuit has more than one path for current flow.

MANUFACTURED HOUSING

In the United States, much of the residential construction is done on site. Some parts such as trusses are manufactured in factories and delivered to the construction site. In some countries such as Japan and Sweden, there is a trend toward **manufactured housing**. In Japan, 15 percent of new homes are built in factories. In Sweden, more than 90 percent of new homes are built in factories.

These homes are not built **stick-by-stick** on site. They are produced in factories modeled after automobile plants. Walls are completely framed, sheathed, and insulated on assembly lines. Windows, plumbing, electrical wiring, and even kitchen and bathroom cabinets are installed. The interior finishing might even be completely done. An entire house could be manufactured and delivered to the site. All that is necessary is to place the house on its foundation, and connect the utility services.

Manufactured houses are built in modules (standard size sections) that are about 12 × 24 feet in size. A module can consist of one room or several rooms. All of the prefabricated modules are assembled at the site.

Manufactured housing does occur in the United States. You may have seen motel rooms, restaurants, and diners that have been prefabricated being transported by truck from the factory to their destination. The advantage of this type of construction is the great saving on labor cost.

Manufactured houses are built in a factory and delivered to the construction site.

Grounding Electrical Equipment

The reason all metal electrical equipment should be grounded is to protect people against electric shock. Some devices like electric drills have metal casings. One of the electrical wires carries a voltage of 120 volts to the motor inside the drill. This is called the **hot** wire. Consider what would happen if the insulation on the hot wire inside an electric drill became frayed. The hot wire then could touch the metal case.

Since the metal case conducts electricity, the case would now have 120 volts applied to it. A person who touched the metal case would also have 120 volts applied to him or her. If that person were standing on the ground, electricity would flow from the drill case, through the person, to the ground. The result would be a shocking experience! Electrical shocks can cause injury or death.

Electricity takes the path of least resistance. A person is not a very good conductor of electricity. A metal wire is a good conductor. If electricity has a choice between flowing through a grounding wire, and flowing through a person, it will flow through the wire. The round prong on electrical plugs carries the ground wire. It should never be cut off. When it is plugged into an electrical outlet, the ground wire provides an electrical path to the earth.

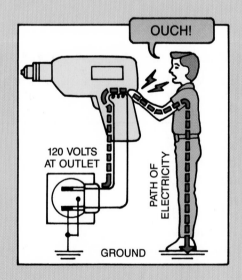

DRILL NOT GROUNDED

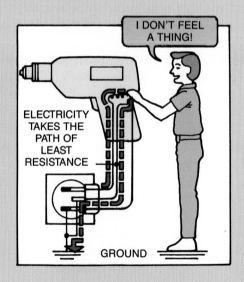

DRILL GROUNDED

COMMERCIAL STRUCTURES

Commercial structures are buildings that are used for banks, stores, and offices. These structures are usually occupied by more than one tenant, and are therefore larger than most residential structures. Commercial structures also have footings and foundations. The footings and foundations are generally made from reinforced concrete. The concrete must be very strong to support the weight of a large building.

Like most houses, commercial buildings are often framed superstructures. However, steel and concrete are used instead of wooden studs and joists. Steel **beams** are like the floor joists. They are horizontal supports that hold up the floors and the walls. Vertical **columns** made from concrete or steel beams are like the wall studs. They transfer the load on the beams to the ground. Since the columns on lower floors support more weight than those on upper floors, the lower columns must be very strong.

Assembling manufactured houses in a factory (Courtesy of Cardinal Industries, Inc.)

A finished module ready for transport (Courtesy of Cardinal Industries, Inc.)

Installation of a manufactured house (Courtesy of Cardinal Industries, Inc.)

Steel and reinforced concrete are used on commercial structures like the IBM building. (Courtesy of Turner Construction Company)

Installing steel beams and columns (Courtesy of Turner Construction Company)

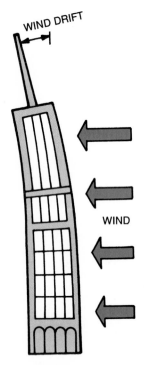

WIND DRIFT

WIND

The towers of The World Trade Center drift 3 feet in high wind, and up to 7 feet during a hurricane.

Wind effects influence the design and construction of skyscrapers.

The Effect of Wind on Tall Buildings

Wind is often greater at the top of skyscrapers than it is at ground level. Not only is the wind stronger at these heights, but its effect is greater on high buildings. Tall buildings are like long levers connected to a pivot in the ground. Consider how a small force applied to one end of a long lever can result in a large force at the other end. A pry-bar used to remove nails or lift heavy weight is a good example. When wind blows against the top of a tall building, the building sways. This is called **drift**. It refers to the number of feet the building moves from the vertical center.

The second tallest buildings in the world are the two World Trade Center towers in New York. These buildings are 1,350 feet high (110 stories). The World Trade Center must be able to withstand about 4,000 times the wind load as a 20-foot high house. In a strong wind, the building may drift three feet from center. Can you imagine how a person standing on one of the top floors must feel when that happens! In that case, a very large force is exerted on the foundation. The foundation must be very strong to counter the force of the wind. The foundation that supports the World Trade Center buildings rests in bedrock, 70 feet below the ground.

A great deal of consideration must be given to the effect of wind (the **wind load**) as it blows against the structure. The wind load of a building increases as the square of the building height. A 20-foot high building is twice as high as a 10-foot high building. Its wind load, therefore, will be four times greater than the 10-foot high building.

When engineers design buildings, they need to know how much wind will blow at the location. They usually study historical data from the past 50 years. The strongest wind that has occurred within the last 50 years is called a **50-year wind**. Buildings are usually designed to withstand the force of a 50-year wind. That means that the construction materials and design are chosen so that the building will not be damaged by a wind that is so strong that it only occurs once in every 50 years.

Why not design buildings to withstand 100-year winds (the strongest wind that is likely to occur in 100 years)? The reason has to do with costs. It would probably cost a lot more to construct a building that would withstand a 100-year wind than it would to repair damage due to a 50-year wind.

Generally, cost/benefit trade-offs are made in designing buildings. Costs need to be kept as low as possible while still providing safety for the occupants of the building.

Stiffening the Building

To reduce drift, buildings are designed to be stiff. To stiffen a building, builders use materials that do not bend. Such materials include concrete and specially constructed steel shapes.

If the outer walls of a skyscraper are stiff, the entire building acts like a huge hollow tube sticking out of the ground. The entire building, not just the inner core, resists bending.

One way of making the outside of the building stiff is to use concrete panels on the exterior. Another way is to build steel trusses into the building framework.

The Norwest Corporation headquarters in Minneapolis is designed as a concrete tube. The 950-foot high building uses high-strength concrete that costs half as much as steel. (Courtesy of Kenneth Champlin of Cesar Pelli and Associates)

Giant trusses stiffen the John Hancock Center in Chicago. (Courtesy of John Hancock Center)

SUMMARY

Modern construction technology requires a combination of practical knowledge, engineering principles, and management skill. Building construction involves building the footing, foundation, floors, walls and roof. It also involves installing utilities and insulation, and finishing the structure on the outside and the inside.

The footing is generally made from concrete. It acts as the base of the foundation and distributes the weight of the structure over a wider area of ground. The foundation supports the entire weight of the structure. It transmits that weight to the footing.

Walls are framed using studs made from 2×4 or 2×6 inch lumber. Walls transmit the load from the floors above them to the foundation. Walls also serve as partitions between the rooms.

Sheathing covers the exterior walls. It makes the structure more rigid. Sheathing closes in the structure and protects it against the weather. It also provides a surface upon which to attach exterior wall covering such as siding, shakes, or shingles.

The roof protects the entire house against wind, rain, and snow. It prevents heat from escaping in cold weather. It also beautifies the house. In the past, most roofs were framed. Today, roofs are made from prefabricated trusses. Trusses are carefully engineered. Because they are mathematically precise, they are built from thinner lumber than framed roofs.

Roofing materials such as asphalt shingles, wooden shingles, and shakes are used for weather protection. They also add to the appearance of the house.

Insulation helps to keep the inside of the house at a constant temperature. It isolates the inside of the house from the outside temperature. Insulation can be made from loose fiberglass. The fiberglass can also be made into blankets and batts. Rigid sheets of polyurethane foam are commonly used as an insulating material.

The outside of the house is finished with siding of wood or vinyl. Shingles and shakes are also used. The inside of the house is finished with gypsum wallboard. The joints are taped and spackled. Paint, wallpaper, or paneling then covers the wallboard.

Utilities include plumbing, heating, and electrical systems. Domestic plumbing systems include hot water and cold water supplies, and drainage. The drainage carries waste water. It is never allowed to mix with the fresh water systems. Plumbing pipes are made from copper, plastic, or iron. They are joined together with solder, special chemical bonding agents, or fittings.

Furnaces used in heating systems use oil, natural gas, or electricity to provide heat. A furnace warms air or water and circulates it to areas in the house to be heated.

The electrical system carries electricity from the electrical panel, to circuit breakers, and then to branch circuits throughout the house. Electrical outlets in the home are connected in parallel. If one appliance is unplugged, the others will continue to operate. All electrical equipment with metal cases should be connected to a good ground to safeguard against electrical shock.

Some kinds of structures are now being manufactured in a factory. These structures are made in standard size modules. They are transported by truck from the factory to their destination. Manufactured housing saves on labor costs.

Commercial structures such as banks, stores, and offices also require footings and foundations. Since these buildings are often quite tall, framing materials are made from steel and concrete, rather than wood.

Tall buildings must be built to stand up against strong winds. The outsides of these buildings are stiffened. Their foundations are generally quite deep.

(Courtesy of Holiday Corporation)

PREFAB PLAYHOUSE

Setting the Stage

He tore off the last wall, revealing rotted 2 × 4s. The house was so old that the foundation had settled into the ground in places, throwing the whole superstructure askew. The project in the next weeks will be to calculate the amount of wood needed to replace the frame, order it from the lumberyard, and replace the old framing boards.

Your Challenge

Design and construct a model prefabricated playhouse to be sold to the school or someone in your community. Use a reference book on residential construction to help you. (You may want to check with your customers for placement of windows, doors, and paint.)

Procedure

1. Divide your work crews into six teams (four wall teams and two roof teams).
2. Draw a framing layout for each wall and roof section. Studs and rafters will be 8" on center. Walls will be 4' × 4'. Roof pitch and length will be determined by the two roof teams. Draw a door or window in each wall section. Use the carpentry book for information on the correct layout of studs and rafters.

Walls

1. Determine the correct nail size to use for each nailing process. Attach the studs to the top and bottom plate of the walls, using two nails for each end of the stud.
2. Lay 4' × 4' covering on each wall section and trace the openings for doors and/or windows.
3. The teacher will help you cut each opening.
4. Nail each wall covering in place.
5. Determine how each wall section will be fastened to the other wall sections. Precut the corner pieces.

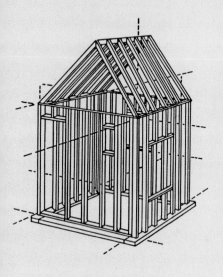

Suggested Resources

23 boards for framing—1" × 2" × 12'

2 boards for base—2" × 4" × 8'

1 board for ridge board—1" × 3" × 8'

2 lbs. cement coated nails

½ lb. common nails (assorted sizes)

3—4' × 8' sheets of homosote, ¼" plywood, or masonite

Rafter assembly jig

Reference books on modern carpentry

Roof

1. Using the reference book, lay out a rafter pattern with the correct angle for the ridge board and bird's mouth. Cut the rest of the rafters using this pattern.
2. Place each rafter in a rafter assembly jig provided by your teacher. This jig will hold all rafters for one side of your roof in place. Nail a 1″ × 3″ ridge board to each set of rafters. Nail a top plate in the bird's mouth.
3. Turn the roof section over and nail the covering material in place.

Delivery and Assembly

1. Draw a set of directions for assembly.
2. Assemble the walls at the site.
3. Nail the roof section in place from the inside of the building through the top plate at the bird's mouth. Then nail the two ridge boards together.

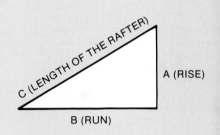

Technology Connections

1. You need to supply the purchaser with a snow-load specification for the roof you designed. You will find this information and charts in the reference book you used.
2. Describe some ways of making your roof stronger.
3. What are the advantages and disadvantages of on-site construction versus prefabricated construction?
4. Why are studs usually placed 16″ or 24″ on center for framing?

Science and Math Concepts

▶ The length of a roof rafter is determined by using the following formula:

$$C = \sqrt{A^2 + B^2}$$

▶ A roof that rises 4″ per foot of run will have a 4 in 12 slope.

INSTALLING ELECTRICAL SYSTEMS

Setting the Stage

Electrical systems are installed after the frame of the building has been constructed, but before the interior walls have been installed. Electrical systems are composed of several different circuits, each with different purposes, and located in different parts of the structure. All of the circuits must be connected to the service panel.

Suggested Resources

Per Group:

Romex with ground—size 14-2

One steel rectangular outlet box with internal wire clamps

One steel octagon outlet box with internal wire clamps

One SPST light switch

One porcelain ceiling lamp socket

One single switch plate or cover

One box hanger for 16" centers

Three red wire nuts

Three grounding clips

One light bulb

One heavy duty three-prong plug for solid cable

Wall frame (one per six groups)

Per Class:

Two lengths of 2" × 4" construction grade lumber—8' long

Seven lengths of 2" × 4" construction grade lumber—4' long

Twelve common nails—size 16d

Your Challenge

Install an electrical circuit in which a light fixture is controlled by a switch.

Procedure

1. Assemble the wall frame(s).
2. Mount the rectangular and octagonal outlet boxes as shown in the drawing.

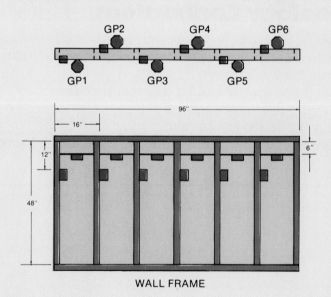

WALL FRAME

3. Cut 36" of Romex and attach the plug to one end.
4. Measure and cut the amount of Romex needed to connect the two boxes. Allow approximately 8" in each box to make connections.
5. Install the opposite end of the plug wire in the octagon box, allowing approximately 8" to make connections.

348

6. Install the cable to connect the boxes. Clamp all wires with the internal wire clamps.

7. Use a cable stripper to remove about ¾" of the sheathing (outside insulation) of the cable from the clamps to the ends of all wires. Be careful not to disturb the colored insulation.

8. Attach the ground (bare) wires to the outlet boxes with grounding clips.

9. Following the circuit diagram, connect the wires in the octagon outlet box with wire nuts.

10. Using needle nose pliers, form a partial circle with both stripped ends of the wires in the rectangular outlet box. Place the wires around the terminals of the switch so that the wire points in a clockwise direction. Tighten the terminal screws on the switch. No bare wire should be showing.

11. Test your circuit with a circuit tester, then ask your teacher to check your work. All grounds must be connected, and no bare wires should touch.

12. Attach the switch and light socket to the outlet boxes with the screws provided. Bend the wires in a "Z" shape as you push them into the boxes. Screw the switch plate to the front of the switch box. Put a light bulb in the socket.

13. Under the supervision of your teacher, plug your circuit into an electrical outlet and turn the light on and off.

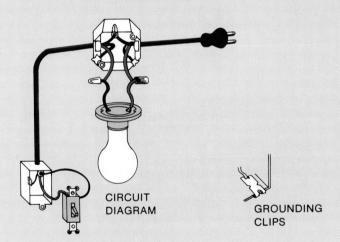

CIRCUIT
DIAGRAM

GROUNDING
CLIPS

Technology Connections

1. An electrical circuit in a structure is made up of several devices which have different purposes. What is the purpose of a duplex outlet receptacle? A switch? A light fixture outlet?

2. Many of the things in our homes work by electricity. How would your life be changed without electricity in your home?

3. What happens when someone tries to use more electricity in a structure than the electrical system was designed to provide?

Science and Math Concepts

▶ A *watt* is the unit of measurement that tells us how much electrical power is being used. To calculate watts, multiply volts × amps.

▶ Kilowatt hours (KWH) are how electricity is metered (sold). A *kilowatt hour* is 1000 watts used for one hour.

LASER CONSTRUCTION

Setting the Stage

There are many applications of lasers in the construction industry. A laser's most valuable asset is its ability to be directed anywhere, including over long distances. A helium-neon laser is used because its power is low enough to be safe, yet the beam is easily seen. A helium-neon (HeNe) laser is used to align underground pipes by projecting the beam in a straight line. The laser can also be used to level things such as floors, ceilings, and walls.

Your Challenge

Demonstrate the application of the laser in the construction industry.
1. Using six height adjustment jigs, align two sections of drain pipe with a laser. The pipe must be laid with a 1 degree pitch (⅛" per ft.).
2. Lay three courses of "concrete" blocks using clay for the mortar. (Use wood models to represent the concrete blocks.)

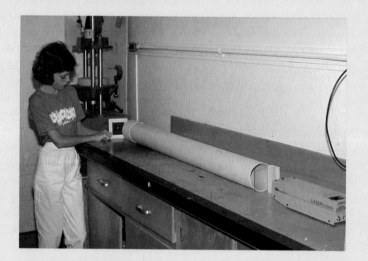

Suggested Resources

Helium-neon laser
Black 4" × 4" target
Height adjustment jigs
Acrylic target—⅛" × 4" × 4"
2—4" plastic drain pipes and couplings
Play-dough or plasticene clay
40 scaled wood blocks—8" × 8" × 16" (3/32 = 1")
Level—4-foot

Procedure

CAUTION: Never look directly into a laser light beam or into its direct reflection.

Drain Pipe Alignment

1. Place the laser at the point where you want to start the drain line. Lay one section of drain pipe in the jigs provided by your teacher. Using the acrylic target and a ruler, adjust the height of the laser to hit the center of the drain pipe.

2. Now align the other end of the drain pipe using a black target (back of the mirror). Remember, the pipe has to have a 1 degree pitch.
3. Add a second section of pipe, repeating the laser alignment process. (The laser will have to be adjusted lower on the first pipe to obtain the correct pitch with the second section of pipe.)

Concrete Blocks

1. Make a small jig by mounting a piece of acrylic on a small piece of wood. The height of the acrylic surface should be 1 1/2 times the height of the blocks you are laying. Scratch a reference line near the center of the acrylic jig. The measurement from the base of the jig to the scratched line will represent exactly one layer of blocks plus the mortar.
2. Determine how high to place the laser beam. To do this, place the laser one layer higher than the height of the layer you are working on.
3. Lay the first course of blocks (wood models) using clay for the mortar. Align each block as you lay it, using the acrylic jig for the laser target. Check with a builder's level. (You may use more than one acrylic jig as a target.)

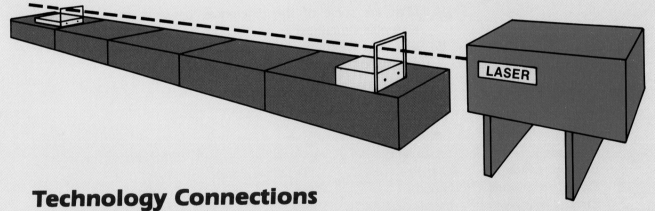

Technology Connections

1. Using a 4-foot level, tape measures, and chalk lines, try accomplishing the tasks you have just accomplished using the laser.
2. Can you design jigs and procedures to more easily accomplish the tasks? Could you have the laser turn a corner with the use of a mirror?
3. List the operations that have to be accomplished before a foundation wall is built and aligned with a laser.
4. List examples of structures and explain where laser alignment might be used.
5. Why are alignment and leveling important in the building of a structure?

Science and Math Concepts

▶ The angle of incidence is equal to the angle of reflection.
▶ The incident ray and the reflected ray lie in the same plane.

REVIEW QUESTIONS

1. If you were to build a house, how would you choose its location?
2. Describe the function of the footing and the foundation wall.
3. What are two purposes served by walls?
4. Draw a labeled sketch of a section of a framed wall. Show one window, the studs, the top and bottom plates, headers, trimmers, and jack studs.
5. Why is sheathing nailed over the entire wall section, including windows and door openings, when it is installed?
6. Why is a trussed roof stronger than a framed roof?
7. What material would you choose to use as roofing material? Explain why.
8. What is the function of insulation?
9. What is the purpose of the vapor barrier on the outside of the insulation?
10. Why is it necessary to provide separate systems for fresh water and drainage?
11. Explain how a home heating system works. Use a system diagram to help explain your answer.
12. Would circuit breakers be connected in series or in parallel with electrical outlets in a room? Explain your answer.
13. How does grounding electrical equipment guard against electric shock?
14. What are some of the differences between commercial and residential structures?
15. Why does wind affect tall buildings to a greater degree than it affects short buildings?

KEY WORDS

Asphalt	Frost line	*National Electrical Code*	Spackle
Bottom plate	Girder		Stud
Building permit	Header	Pitch	Subfloor
Circuit breaker	Insulation	Plumb	Top plate
Floor joist	Manufactured housing	Plumbing	Truss
Footing		Rafter	Vapor barrier
Foundation wall	Mason	Sheathing	Wind load
Framing	Mortar	Sill plate	

SEE YOUR TEACHER FOR THE CROSSTECH PUZZLE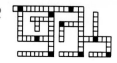

CHAPTER 12

MANAGING PRODUCTION SYSTEMS

MAJOR CONCEPTS

After reading this chapter, you will know that:

- A company's management team must coordinate the activities of different departments in the company to wisely use the company's resources.
- Architects design a building's shape and select basic materials for it.
- Engineers design a building's structure and its major systems.
- A general contractor organizes the work of many different people on a construction project.
- A company's marketing department decides what market the company will develop products for, what features those products should have, and the strategy to be used in selling them.
- A company that produces a product should provide service and technical information about the product after it has been sold.
- The survival and prosperity of a company depend on sound financial management.

MANAGING PRODUCTION SYSTEMS

The success or failure of a manufacturing or construction project depends to a large degree on how well it is managed. It is not enough to build a product well. It is important to build it within a schedule, within a budget, and safely. It is also important to be able to deliver it undamaged to the customer, and to service it if it breaks after it is put into use.

The coordination and proper use of the seven resources used in the manufacturing or construction system are part of the job of a company's **management.** Management also decides what products to make, and how to make potential customers aware of the products.

The general methods of managing a construction project are similar to the methods of managing a manufacturing project. The specific details of managing the two are quite different, though, due to the differences between construction and manufacturing. In this chapter, managing each of these systems is considered separately.

A company's management team must coordinate the activities of different departments in the company to wisely use the company's resources.

WHO MANAGES A CONSTRUCTION PROJECT?

Architects and Engineers

As was described in Chapter 10, a structure is designed by **architects.** Architects work directly for the **owner** of the building. Architects are responsible for the shape and form of the building, the choice of basic materials, and the many thousands of details involved in laying out the building.

Architects are also concerned with how the building will affect surrounding buildings, and how it will be affected by other new buildings. For example, will the building cast a large, permanent shadow on existing buildings? In most large cities, a view of the sky from existing buildings must not be blocked by a new building. Architects deal with this kind of problem by making buildings smaller at the top than at the bottom. This kind of design gives more light to the surrounding buildings.

Plans and drawings are being prepared for a new building in this architect and engineers' office. (Courtesy of The Turner Corporation, photograph by Michael Sporzarsky)

Architects design a building's shape and select basic materials for it.

Engineers are responsible for ensuring that the building is structurally sound. This includes selecting the size of each column and beam, and making sure that it will carry the load of the completed building and its occupants. Engineers often use a computer aided design (CAD) program to do this work.

Engineers design the building's structure and its major systems.

Engineers also design the major systems within the building. The major systems include the electrical system, the heating, ventilating, and air conditioning (HVAC) system, the fire detection system, and the plumbing system. As the building goes up, both engineers and architects must visit the site and inspect the work to make sure that the construction has been done according to the drawings.

Specialists are often hired to design some of the individual systems within a building. These specialists work with the owner, architect, and engineer. Specialists include security system specialists, fire system specialists, and communication system specialists. Although the specialists provide the design for each of these systems, the architect and engineer still have the overall responsibility for them.

Contractors

Once a building has been designed, it must be built by a team of construction specialists. Many different activities must be coordinated even on small construction projects. Very small projects are often managed by the owner. If a job is more complex, however, the owner usually has someone manage the project. A **general contractor** is a person or a company who accepts the total and complete responsibility for building a construction project. This includes managing the project, hiring people to work on the project, and purchasing the materials. Most small and medium size jobs are managed by a general contractor.

Specialists help architects and engineers in highly specialized areas such as this telephone and communication cabling installation. (Photo by Robert Barden)

The agreement between the owner and the general contractor is called a **contract.** It states what the contractor will do, and how much the owner will pay. The contract is often very lengthy, and includes the architect's and engineer's drawings. It tells in great detail exactly what will be done and when it will be done. It also usually tells who is responsible in the event of an accident, and who is responsible for obtaining permits to build the structure. It may also include penalties for failing to complete the job on time.

The general contractor can have its own employees do the actual work on the job, or it can hire another company,

A general contractor organizes the work of many different people on a construction project.

called a **subcontractor,** to do a portion of the work. On a large construction project, there may be dozens of subcontractors, each working in a specialty field, such as plumbing or carpentry. General contractors often use a combination of employees and subcontractors to complete a project. General contractors enter into legal agreements with their subcontractors. These agreements are called **subcontracts.**

It is very important that the work of each subcontractor and of the general contractor's employees is completely described in the subcontracts. All details must be described so that all of the work is done, without two people trying to do the same job, and without leaving anything undone. If anything is left out and is added later, there will be additional charges by the subcontractor. These additional charges are called **contract extras.** Contract extras on a poorly managed job can drive the cost of the project up to very high levels. These additional costs are sometimes called **cost overruns.**

Project Managers

On very large construction projects, a separate company, called a **project manager,** is hired. Project managers oversee the contracts, scheduling, material deliveries, and overall progress of the job. In this case, the general contractor is still responsible for much of the construction, but not for the

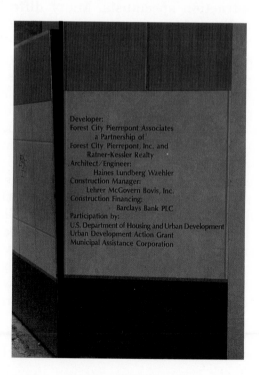

This sign posted at a construction site lists some of the companies and agencies that are involved in the planning, financing, and managing of the project. (Photo by Robert Barden)

management of all of the subcontractors. The project manager is independent from the general contractor. The project manager works directly for the owner. This is often necessary on large projects, such as skyscrapers, because they are so complex that special skills are needed to manage them.

One of the jobs of the project manager or general contractor is to make sure that the different subcontractors do not interfere with anyone else's work. For example, water pipes that are installed above a ceiling must not block access to electrical boxes or air conditioning vents. Before work starts, meetings are held with all of the subcontractors present. They jointly mark on a master drawing exactly where their systems are going. They provide enough detail in areas where problems may occur so that each system can be built independently, and not interfere with another. This process is called **coordination.**

In some areas of the country, all construction is done by workers who are represented by **unions.** These are labor organizations that bargain with employers to set wages and work practices for their members. Each union claims a carefully described portion of the work, called a **jurisdiction,** for its members.

On a unionized job, the general contractor or project manager must be especially careful to select the correct trade, or union, to do each job. Craftspeople in the carpenter's union, for example, will not do electrical work on a job. In some areas of work, it is not clear which union workers should do the work. This can cause a jurisdictional dispute. A project manager's skill at handling these negotiations often has a large impact on the success of the job.

SCHEDULING AND PROJECT TRACKING

The general contractor or project manager must carefully schedule all the work to be done on the job. It is important that the work be done in the right order to maintain safety on the job site and to protect systems from being damaged by later work. For example, a floor must be structurally sound and rough finished before interior work can start. Electrical or telephone cables cannot be installed until walls are framed.

With so many tasks and subcontractors on a single job site, special management tools are used to create and maintain the schedule. One of these tools is called a **Gantt chart.**

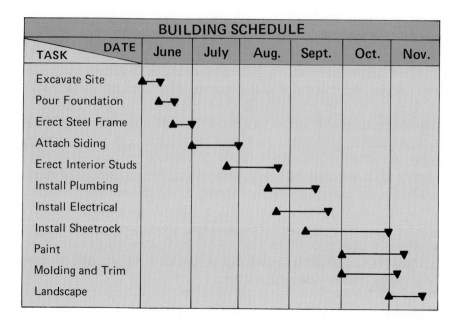

BUILDING SCHEDULE						
TASK \ DATE	June	July	Aug.	Sept.	Oct.	Nov.
Excavate Site	▲▼					
Pour Foundation	▲▼					
Erect Steel Frame	▲—▼					
Attach Siding		▲——▼				
Erect Interior Studs			▲——▼			
Install Plumbing			▲——▼			
Install Electrical			▲———▼			
Install Sheetrock				▲———▼		
Paint					▲———▼	
Molding and Trim					▲———▼	
Landscape						▲——▼

The Gantt chart is used to help managers schedule and monitor a project.

A Gantt chart is a kind of bar graph that shows when a particular activity is scheduled to start, and when it is to be completed. The chart shows where one project must wait to start until another is done. The total impact of schedule changes can easily be seen on the chart. The total number of work hours can also be shown on the Gantt chart. The chart is continually updated to show actual start dates and completion dates, and their impact on the total project. Gantt charts are used as a management tool in both construction and manufacturing systems.

Meetings, called **project meetings,** are held on a regular basis during the life of the job. At a project meeting, the owner, architect, engineer, project manager, general contractor, and some of the specialists discuss the progress of the project. Any question that is not resolved is called an **action item.** Written notes called the **minutes** of the meeting are made and sent to all participants. Action items are recorded in the minutes. Work done to deal with the action item is reported at the next meeting. In this way, problems are identified quickly, and they are dealt with by everyone who is concerned. Project meetings are usually held once per week or every other week.

PERMITS AND APPROVALS

Construction projects require the approval of local building and fire departments for many of the designs, work in progress, and completed systems. These **permits** and approvals are designed to protect the safety of workers and the public nearby. They also ensure the safety of the people who will occupy the building.

Before the construction is started, a building permit must be issued. In order to obtain a building permit, plans and drawings of the proposed building must be submitted to the local government agency responsible for issuing permits. The builder must also have the approval of the **zoning board.** Zoning includes keeping residential areas free of industry. It also involves examining the impact of a large new building on existing roads, parking facilities, nearby schools, and so on. Sometimes, a builder has to agree to provide new road entrances or do other additional construction in order to obtain zoning approval for a new building.

In commercial buildings, a **life safety plan** must also be prepared. A life safety plan is an analysis of the stairways, exit doors from all areas, and other building design features that affect the safety of the people inside the building in the event of a fire or other disaster.

As construction proceeds, permits for many of the activities at the site must be obtained. For example, if a building in a city is next to a sidewalk, the sidewalk may have to be closed to pedestrians for their safety. A special permit is required to do this. In addition, a new walkway with a protective roof may have to be built to protect pedestrians from falling tools or small debris. If a crane is needed to lift large parts or equipment into place, a permit must be obtained for it. If the crane sits in the street, blocking traffic, a special permit must be obtained to close the street. Other special permits must be obtained to dig up the ground around the building to get access to electrical, water, sewer, and gas services.

These are some of the permits required to be posted at a construction site. (Photo by Robert Barden)

Special structures sometimes have to be made to protect the public from falling tools and debris. (Photo by Robert Barden)

Some of the major systems that have to
be tested in a new building.
(Photo by Robert Barden)

When the building is almost complete, the major systems of the building are tested to ensure that they were installed properly. The fire department may require that they witness the testing of the fire detection or fire suppression system. A fire suppression system may be a sprinkler system such as those used in general office and living areas, or a Halon system which is used in computer and communication areas. Halon is a chemical used to fight electrical fires. Elevators usually must be inspected and approved before they can be used. Not all building systems have to be tested by city officials, but the building owner has them tested to be sure that all systems in the new building are working.

After all systems have been tested, and the building is complete, the local building department issues a **Certificate of Occupancy,** or CO. The building may not be occupied by any tenants until the CO has been issued.

BUILDING MAINTENANCE

After the building has been completed, the owner has different management tasks to perform. If the owner will rent the building, advertisements must be placed or a rental agent must be assigned. If the building is for rent, it is important to have a renter move in as soon as possible, so that the owner starts to get paid. The owner has to make payments on construction loans or a mortgage. If the building sits empty, the owner has no income to offset the expenses.

Ongoing maintenance includes keeping all building systems operating properly and obtaining the necessary regular inspections (fire, elevator, health, and so on). It also involves keeping the building in a safe, neat condition so that it continues to attract new renters and keeps existing renters. If the owner's company occupies the building, the building must be kept safe and neat for the employees' health and well-being.

MANAGING MANUFACTURING SYSTEMS

Managing manufacturing systems includes identifying appropriate products to manufacture and sell. It also includes developing and producing these products at the lowest possible cost, selling them, and servicing them. While each of these areas is handled by specialists, the overall effort must be coordinated by a general manager, a management team, or a company president. Each of these activities will be examined.

Marketing

Market research, research and development, and making the prototype have each been briefly described in Chapter 9. It is the job of the marketing department to identify what features the product must have to make it attractive. To do this, they must identify who the potential customers are. This is called the **market.** They must then decide exactly which products to make. Some companies adopt a strategy of making the same product that other companies make, but making it better or less expensively, or doing a better job of making the public aware of it. Other companies try to analyze the existing market and identify instances of products that people need that no other company is making. This analysis is called a **market gap analysis.**

The marketing department also tries to identify how much the product can cost and still sell well. Based on these data, they make an estimate of how many of the product will be bought by the public in one year (the **total market**). They also estimate what percentage of the public will buy the company's product, rather than a product from another company. This is called **market share.**

The total market times the market share is the company's potential sales for the product. If this is large enough, management will develop the product. This is often done in an area of the company called **Research and Development** (R&D).

Research and Development (R&D)

Research is the search for new materials, processes, and techniques that may be useful to a company's business. While the research is done in fields that are likely to provide new discoveries that will be helpful to the company, it is not

A company's marketing department decides what market the company will develop products for, what features those products should have, and the strategy to be used in selling them.

Research is the search for new materials, processes, and techniques. (Courtesy of Cetus Corporation)

directed at a specific product. **Development** is the use of these new discoveries and previously known methods and techniques to solve a specific problem. Development is most often directed at a specific product or group of products.

The end result of development is a **prototype.** A prototype is a test version of the product. It is used to test consumer reaction and to learn the problems of manufacturing the product. After the prototype has been built, it is test marketed. The prototype is shown to potential customers to get their reaction to it. This feedback is then used to improve the product before it is produced in large quantities.

Production

The product is then changed to make it easier to produce and to add the features that the test marketing showed were needed. This step is called **production engineering** or **productizing.** After a product has been productized, it can be put into production.

During production, quality must be monitored constantly to ensure that the products are being made correctly. **Quality control,** or QC, is the monitoring of quality

Models and prototypes are built so that a product can be tested before it is manufactured. (Courtesy of General Motors Design)

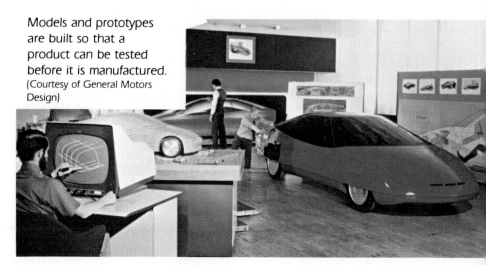

Quality control inspections at several points in the manufacturing process ensure a high-quality product. (Courtesy of Grumman Corporation)

throughout the production process. Incoming materials that will be used in the manufacturing process are examined. Partial assemblies are examined at various stages in the manufacturing process. The final product is then inspected before it leaves the plant. Companies often have reputations for good quality products that can be traced directly back to their quality control procedures.

Other management tasks are to make sure that the manufacturing plant is operated safely and that laws and regulations are followed. These include safety rules established and policed by **OSHA,** the Occupational Safety and Health Administration. Rules published by OSHA are meant to keep workers safe while on the job.

The Environmental Protection Agency, **EPA,** establishes rules and regulations to keep the air and water surrounding manufacturing and construction jobs clean and free from harmful wastes. The EPA has regulations covering the amount of emissions permitted from smoke stacks, what may and may not be dumped down a sink drain, and the proper disposal of toxic wastes.

As in construction, some factories are unionized. Union contracts set wages and work rules for some of the workers in these plants. Management is involved in negotiating the contract. Then management must plan jobs to use the proper class of employee for each job.

Servicing

The manufacturer must have a plan to deal with products that need routine maintenance or that have broken after they have been sold. Most products now come with a guarantee or warranty stating the manufacturer's intent to repair or replace a defective product within a certain period of time after the sale. The time period varies greatly, from none (no guarantee) to five years or more. After the warranty period, the manufacturer may regard the product either as a **throwaway** or a repairable product. If it is a throwaway product, the customer must buy a new one to replace a broken one. Throwaways are usually small and inexpensive.

A company that produces a product should provide service and technical information about the product after it has been sold.

The manufacturer must establish a plan for dealing with repairable products. Sometimes, the customer has to mail the defective product back to the manufacturer. The manufacturer must then have a repair department that can provide cost-effective, rapid repairs. This creates jobs that range in skill level from entry level to highly skilled, depending on the products to be maintained and repaired.

Other options include having another company perform the maintenance under a contract or having the dealer or distributor who sold the product provide the maintenance. Not all dealers and distributors have a repair department for doing this kind of work.

SELLING THE PRODUCT

The marketing department sets the strategy or the plan for selling the product. The sales department actually goes out and sells it. Depending on the type of product that is to be sold, a company may choose one or more ways of selling the product.

Direct sales refers to having salespeople who are employees of the company selling products directly to customers. If the company's customers are widespread, the salespeople may be located in regional offices that serve different parts of the country. A company's sales volume in a given area has to be large enough to justify opening an office in that region.

A company that does not have enough sales volume in a certain area to warrant hiring salespeople and opening an office may hire an independent **sales representative** or "rep" organization. Sales representatives may sell the products of several manufacturers. They develop a customer

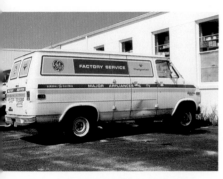

Authorized repair service like this form part of a company's servicing plan.
(Photo by Robert Barden)

base in their area and usually offer products that are similar but not competing with each other. When a sales representative takes an order, it is usually an order made directly with the factory. The manufacturer delivers the product to the customer, and the customer pays the manufacturer directly. The manufacturer pays the sales representative a **commission,** or percentage of the sale, after is has been paid by the customer. The sales representative sometimes waits a long time before being paid for a sale.

Dealers sell products to customers by taking orders for the products and placing orders with the manufacturer. The manufacturer then ships the products to the dealer who delivers them to the customer. The dealer charges the customer more for the products than the dealer is charged by the factory. Dealers also keep a small amount of the products on hand to sell.

Companies that sell large quantities of products often use **distributors.** Distributors buy large quantities of products from the manufacturer at a low price, stock them in a warehouse, and sell them directly to customers for a higher price. Distributors sometimes use catalogs or mail order advertising to sell the products. Others have showrooms where customers can inspect the products before buying them. Some manufacturers operate their own showrooms, making the money (and dealing with the problems) that a distributor would make by selling the products.

Distributors like this buy products in large quantities from manufacturers and sell them to the public at showrooms and warehouse buildings. (Photo by Robert Barden)

FINANCING THE MANUFACTURING SYSTEM

While the features, quality, and sales of the products are very important, the survival of a manufacturing company (and other companies) lies in successful **financial management.** Sound planning together with ways to monitor and control the company's finances are necessary for the company to survive, grow, and prosper.

The survival and prosperity of a company depend on sound financial management.

Before a business is started, a **business plan** is prepared. The business plan states the goals and objectives of the business, the strategy and methods that will be used to achieve the goals, and the financial requirements for the business. One of the most important parts of a business plan is the **cash flow analysis.**

The cash flow analysis predicts how much money will have to be spent each week or each month of operation and how much income is expected in the same periods. If there is more income than there is money spent in a certain period of time, the business is said to be operating at a **profit.** If the expenses are more than the income, the business is operating at a **loss.**

Most **start-ups,** or new businesses, are expected to run at a loss for a period of time while their products are being developed and their customer base established. During this period, there is little income. The company must start with enough money to pay its bills while it is in this growth phase. This money can come from the personal savings of the people starting the company, a **venture capitalist,** or a loan. A venture capitalist supplies money to a new or growing company in exchange for part ownership in the company.

Another way for a company to raise money is to sell shares of stock to the public. A share of stock is a piece of ownership in the company. When people buy stock in a company, they become **shareholders.** A company that makes a profit may distribute some of the profit to the shareholders. If the company does well, the price of its stock may rise, providing a profit to those shareholders who choose to sell their stock. If the company does poorly, however, the price of the stock may go down. Shareholders could then loose some or all of the money they have invested. The sale of stock is regulated by the Securities and Exchange Commission (SEC).

SPREAD SHEET: SIX MONTH CASH FLOW						
	JAN	FEB	MAR	APR	MAY	JUNE
INCOME, DOLLARS	2400	3000	4500	4000	5000	5500
EXPENSES						
Rent	850	850	850	850	850	850
Telephone	116	173	164	175	204	227
Salaries	575	1270	2350	2350	2525	2610
Materials	255	478	763	652	940	985
Shipping	18	22	35	25	38	43
Insurance	125	125	125	125	125	125
TOTAL	1939	2918	4287	4177	4682	4840
INCOME–EXPENSES	461	82	213	−177	318	660
CASH TO START 4000						
CASH IN BANK	4461	4543	4756	4579	4897	5557

A spread sheet shows the cash flow of a business. If an expense or income item changes, the effect of the change on the whole business can be seen quickly.

In order to keep cash needs as small as possible, the company tries to complete its development work as quickly as possible. This will allow it to sell products and get income from them more quickly. During manufacturing, managers try to manufacture the products as quickly as possible for the same reason. They also try to have the parts necessary for product assembly arrive just before they are needed. (See "Just In Time Manufacturing" in Chapter 9.)

A powerful scheduling tool that managers in manufacturing systems use to help keep production times to a minimum is the **PERT chart.** PERT stands for Program Evaluation Review Technique. It was a scheduling technique developed by the U.S. Navy for use on the Polaris missile and submarine project, the most complex manufacturing project undertaken to that time.

In a PERT chart, each step necessary for the completion of a project is identified. The fact that some steps must be done before others is shown. The time needed to complete each task is also shown. The total effect on the job schedule of any late completion dates can be seen right away. Thus, a manager is able to shift resources to the most time-critical tasks.

A company usually has to pay its suppliers before it gets paid for its final products. Keeping the smallest possible inventory (stock) for the shortest possible time will keep cash needs as low as possible. Inventory is measured in how many times it completely changes in one year, called turn. A company that can completely manufacture a product and sell its inventory in sixty days has a turn of about 6 times per year. (365 days per year divided by 60 days equals about 6 turns per year.)

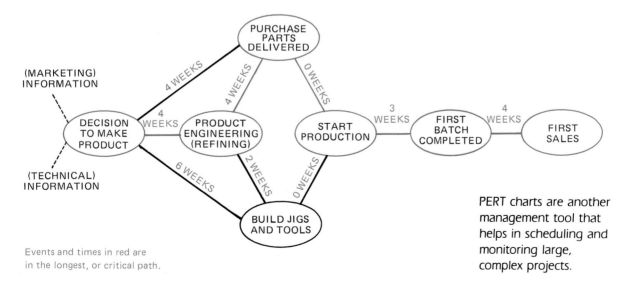

Events and times in red are in the longest, or critical path.

PERT charts are another management tool that helps in scheduling and monitoring large, complex projects.

SUMMARY

The success or failure of a manufacturing or construction project depends to a large degree on how well it is managed. The coordination and proper use of the seven resources used in the production system are part of the job of a company's management. It must coordinate the activities of different departments in the company to wisely use the company's resources.

Architects design a building's shape and select basic materials for it. Engineers design the building's structure and its major systems. Specialists in different fields supplement the skills of architects and engineers.

A general contractor organizes the work of many different people on a construction project. General contractors hire subcontractors to do different parts of the work. On very large construction jobs, separate companies called project managers are hired to oversee contracts and coordinate construction activities.

Scheduling complex jobs often requires special scheduling tools. One such tool is the Gantt chart. The Gantt chart is a kind of bar graph that shows the start and completion dates for different activities needed on the job. Progress is reviewed regularly at project meetings.

Permits and approvals are necessary before the construction project starts, while it progresses, and at completion. Some of these permits are zoning approval, a building permit, sidewalk closure permits, street closing permits, and street opening permits. When the construction is complete and the building and its safety systems have been inspected, a Certificate of Occupancy is issued.

A company's marketing department decides what market the company will develop products for, what features those products should have, and the strategy to be used in selling them. It also tries to identify how much the product can cost and still sell well.

Research is the search for new materials, processes, and techniques that may be useful to a company's business. Development is the application of these new developments and previously known methods and techniques to a specific product. Quality control is the monitoring of quality throughout the production process.

A company that produces a product must develop a method of supporting the product after it has been sold. This support is called servicing.

A company may use one or more methods of selling its products. These methods include direct sales, using sales representatives, selling through dealers, and selling through distributors.

Sound planning coupled with a way to monitor performance and to control the company's finances are necessary for the company to survive, grow, and prosper. This planning and management strategy is listed in the business plan. Cash is the lifeblood of the company. A company can raise money from the savings of the people who formed the company, venture capitalists, and selling stock in the company.

In order to keep a company's cash needs to a minimum, it is important to speed development to bring products to market as quickly as possible. It is also important to build products as quickly as possible (within the bounds of good quality) and to keep inventories as low as possible. PERT charting helps to achieve this objective.

(Courtesy of Turner Construction Company)

(Courtesy of Corning Glass Works)

ONE, IF BY LAND . . .

Setting the Stage

The manufacturing industry usually doesn't put a new product on the market "overnight." It can take many years of planning, research and development, production, and marketing to establish a new company or introduce a new product. There are seven resources of technology: people, information, materials, tools and machines, energy, capital, and time. These resources play an important role in the decisions that industry must make. If a wrong decision is made, it can be a *very* expensive mistake.

Your Challenge

Organize and establish a company that will manufacture lanterns using an assembly line to make the manufacturing process more efficient.

Procedure

1. Elect the executive board (president, vice president, secretary, and treasurer) of the new company known as *Lanterns, Inc.* The foreperson, safety supervisor, advertising and publicity staff, and other workers will later be "hired" by the board.
2. Determine the responsibilities of each executive board member.
3. Develop as a group a list of what resources are probably needed to start *Lanterns, Inc.* This list should include people, information, materials, tools and machines, energy, capital, and time.
4. Using the chalkboard or large sheet of paper and 3" × 5" cards, develop a flowchart of the *steps* in making a lantern.
5. Using the preceding step as a guide, develop another flowchart of the *assembly line* needed to efficiently mass-produce the lanterns. This chart should include the machines, tables, tools, work stations, quality control checkpoints, etc. needed.
6. Everyone in the class should contribute to the design and fabrication of any jigs and fixtures needed for the assembly line. Accuracy is especially important at this stage.
7. The executive board of *Lanterns, Inc.* should interview "prospective employees" for the company's job openings. A standardized job application should be completed by each applicant.

Suggested Resources

Safety glasses and lab aprons
Heavy gage tin plate
Steel or brass rod—³⁄₁₆" in diameter
Copper tubing—¾" in diameter
Spot welder
Propane soldering torch
Solder and flux
Squaring shear
Notcher
Bar-folder
Box and pan brake
Slip-form rolls
Variable speed drill
Drill bits

8. The assembly line positions should be rotated periodically. When people switch jobs in the assembly line, they may have to be "retrained" for their new positions.

9. The safety supervisor's responsibility includes making sure students wear safety glasses and any other protective equipment during the manufacturing process. The safety supervisor also helps to ensure that all workers are following the safety instructions described by the teacher.

Note: Refer to the drawings for the following steps.

10. Using the squaring shear, cut 3 ¾" × 13 ½" pieces of heavy gage tin plate for the body of each lantern.

11. Use the notcher to cut out the corners and the 90° "V" cuts on the edges of the body.

12. The bar-folder is now used to bend the ¼" hems to the *inside* of the finished lantern.

13. The ⅜" laps are bent on the bar-folder to a 90° angle in the same direction as the ¼" hems were bent.

14. Complete the bending of the lantern body using the box and pan brake. Be careful not to crush any of the laps during this step.

15. The 1" × 6" corner posts (two for each lantern) are cut from heavy gage tin plate with the squaring shear. One-quarter inch of *one* long edge of each corner post is folded over into a hem on the bar-folder. The bar-folder is also used to make the final bend, a ⅜", 90° bend, in the long direction.

16. The corner posts can now be spot welded to the lantern body. *Be absolutely sure all of the lantern angles are exactly 90° before welding.*

17. Using a template, trace the shape of the lantern top onto tin plate. Cut out the top using tin snips, aviation snips, or the notcher. Make sure there are no sharp burrs on the edges of the lantern top. File or sand as necessary. Scribe the bending lines onto the layout of the lantern tops.

18. Bend the lantern tops to a pyramid shape using a box and pan brake. Clamp and spot weld the tops together. Slightly bend the lower "rounded" part of the lantern top so it fits firmly on the lantern body.

19. Make sure the completed top fits properly on the lantern body. Weld it in place.

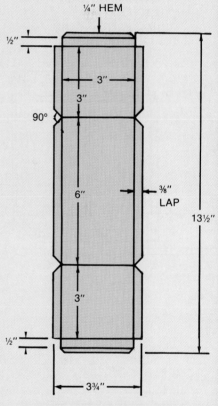

LANTERN BODY LAYOUT

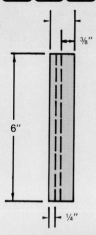

CORNER POST LAYOUT
(MAKE 2)

20. Drill two ¼" holes 1" down from the point of the top. These holes should be made on opposing sides of the "pyramid" shape. The handle ring will be inserted through these holes.

21. To make the handle ring for the lantern, cut a 9" long piece of steel or brass rod with a hacksaw. File both ends round.

22. Use the slip-form rolls to bend the rod into a circle. Insert the completed handle ring through the ¼" holes drilled in the lantern top.

23. Cut ¼" × 4" strips of tin plate for the "X" designs used in the lantern windows. A total of eight ¼" × 4" strips are needed per lantern. Spot weld these strips into an "X" maintaining a 90° angle at the intersection. Four X's are needed for each lantern. Weld these to the inside of the lantern.

24. To make the candle holder, cut a 1" long piece of ¾" diameter copper tubing with a tubing cutter. Remove any burrs, shine on the buffing wheel, and open-flame solder this piece to the center of the bottom of the lantern. Wash off the excess soldering flux.

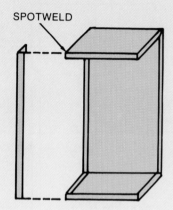

SPOTWELD

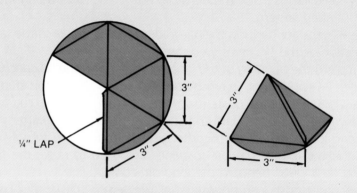

¼" LAP

LANTERN TOP LAYOUT

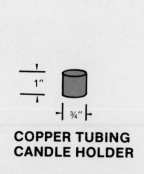

1"

¾"

COPPER TUBING CANDLE HOLDER

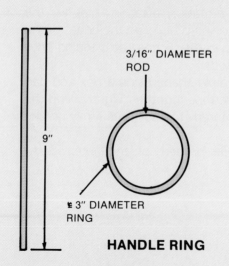

9"

3/16" DIAMETER ROD

3" DIAMETER RING

HANDLE RING

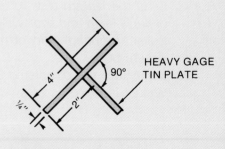

90°

HEAVY GAGE TIN PLATE

4"

2"

¼"

WINDOW CROSSES
(MAKE 8)

Technology Connections

1. Why is it important to make a *prototype,* or "original" working model, of the project being mass-produced before going into full production on the assembly line?
2. Why is accuracy important when manufacturing most products?
3. What is the purpose of quality control checkpoints?
4. How do mass production techniques differ from the craftsman techniques?
5. Why is advertising so important to the manufacturing industry? Make a list of the different ways manufacturers can advertise a product.
6. Match the advertising slogans:

1. Soup is good food!	Allstate Insurance	_____
2. 99 and 44/100% pure!	New York City	_____
3. The wings of man	Campbell's Soup	_____
4. When it rains, it pours!	General Electric	_____
5. We bring good things to life!	U.S. Army	_____
6. You're in good hands . . .	Greyhound Bus	_____
7. Over 30 billion sold!	Eastern Airlines	_____
8. Be all that you can be!	Ivory Soap	_____
9. The Big Apple	McDonald's	_____
10. Leave the driving to us!	Morton Salt	_____

Science and Math Concepts

▶ A spot welder fastens metal by passing electricity through it. The heat needed to melt the metal is caused by *resistance* to the flow of electricity.

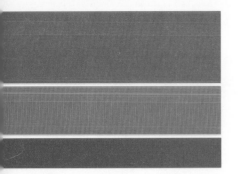

REVIEW QUESTIONS

1. What are the tasks and duties of management in a production system?
2. What do architects do on a construction project? What do engineers do?
3. Describe the role of a general contractor at a construction site. What are subcontractors?
4. Why are construction contracts so detailed and long?
5. What role does a project manager play on a construction project?
6. What is coordination? Describe what might happen if there were no coordination among subcontractors.
7. Draw a Gantt chart of your class activities for the next five weeks. Ask your teacher for information to do this.
8. Is the management job done when the building construction has been completed?
9. What is the difference between marketing and sales?
10. What is the difference between research and development?
11. Is quality control important to sales and marketing? Why or why not?
12. Name four ways that a manufacturer could sell his products.
13. What does a business plan show? When should it be prepared? When should it be updated?
14. Draw a PERT chart for a project that you have undertaken or will undertake in class.
15. It takes 90 days for a manufacturer to produce and sell a product. What is the turn on the product? (Be approximate.)

KEY WORDS

Action item	Direct sales	Overruns	Research and
Architect	Distributor	Owner	development
Business plan	Engineer	Permits	Sales
Cash flow analysis	EPA	PERT chart	representative
Certificate of	Gantt chart	Productizing	Shareholder
occupancy	General	Profit	Subcontractor
Commission	contractor	Project	Throwaway
Contract	Loss	manager	Union
Coordination	Marketing	Prototype	Venture capitalist
Dealer	OSHA	Quality control	Zoning

SEE YOUR TEACHER FOR THE CROSSTECH PUZZLE

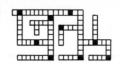

CAREERS IN PRODUCTION

Production technology includes construction and manufacturing. Workers in the construction trades build, repair, and modernize homes and other kinds of buildings. They also work on a variety of other projects, including airports, mass transportation systems, roads, recreation facilities, and power plants. Most manufacturing workers work in manufacturing plants, although some jobs involve sales and considerable travel.

OCCUPATION	FORMAL EDUCATION OR TRAINING	SKILLS NEEDED	EMPLOYMENT OPPORTUNITIES
ARCHITECT—Plans and designs attractive, functional, safe, and economical buildings.	College degree (bachelor's degree in architecture) as well as three years of experience in an architect's office.	Engineering design and managerial skills. Knowledge of building materials and modeling techniques.	Above average. Rapid growth in construction of non-residential structures will increase the demand.
JOURNEYMAN—A member of the building trades, such as:	On-the-job training; two- to four-year apprenticeship programs. A high school education including courses in basic mathematics, applied science, electricity and electronics, mechanical drawing, and construction technology.	Sketching; reading drawings; must layout, measure, cut, shape, and fasten materials; use hand and power tools safely. Must know about building codes and regulations.	Average. As the population grows, more journeymen will be needed to help build and maintain structures.
Carpenter—Builds framework, frames the roof and interior partitions.			
Concrete mason—Places and finishes concrete.			
Electrician—Installs, assembles, and maintains electrical systems.			
Plumber—Installs and maintains water and heating systems.			
Roofer—Installs and repairs roofs.			
MANAGER AND ADMINISTRATOR—Plans, organizes, directs, and controls an organization's major functions.	College degree and management training. Top managers often have a master's degree in business administration.	Determination; self confidence, high motivation; strong decision-making, organizational, and interpersonal skills.	Opportunities will increase faster than the average occupations as business operations become more complex.
MANUFACTURING SALESPERSON—Most manufacturers employ sales people who market products to consumers in other businesses.	For technical products, college degrees in scientific or technical fields. For nontechnical products, college degrees in liberal arts or business administration.	A pleasant personality and appearance, and the ability to get along well with people are important.	Lower than average. Many large firms and chain stores buy direct from manufacturers, but this is a large occupational field with many yearly openings.

Data from *Occupational Outlook Handbook, 1986-87*, U.S. Department of Labor.

Equipment and Supplies

(This activity can be carried out by using any one or a combination of the following materials — wood, metal or plastic.)

Standard lab equipment, including either hand or power tools to cut stock to length and drill holes in the selected materials

(Optional) equipment for testing of geometric shapes)

Standard drill press

Bathroom scales

Lengths of stock

Wood — may be as small as popsicle sticks or as large as wood lath

or

Metal — approximately 26-gauge, in ½" strips for small structures or thin wall electrical conduit for large structures

or

Plastic — ⅛" sheet acrylic in ½" strips for small structures, or ½" plastic water pipe for large structures

Fasteners — Small bolts or other fastening devices to attach the ends of the components together

Miscellaneous materials to construct jigs and fixtures used in the manufacturing of the components

PRODUCTION COMPANY

Objectives

When you have finished this activity, you should be able to:

■ List three geometric shapes and list two advantages and two disadvantages of each.

■ Set up a manufacturing company to design, test, and manufacture components to be used in a construction company.

■ Establish a construction company which will use the components manufactured in the class to build structures.

■ Distinguish between manufacturing and construction industries and list ten items produced by each.

■ Better understand the importance of each individual's work in order for an organization to be successful.

Concepts and Information

As we look around, we see a large variety of objects that have been constructed from a variety of materials. We also find an infinite number of sizes and shapes in our surroundings. Upon more careful study, notice that all of these shapes are made up of curved or straight lines, which are parts of circles, rectangles, or triangles. During this activity, you will learn more about those geometric figures that are made up of straight components. These shapes are used to construct a

EXAMPLES OF STRAIGHT-SIDED GEOMETRIC FIGURES

TRIANGLES (THREE-SIDED FIGURES)

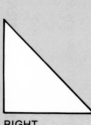 

RIGHT   ISOSCELES   EQUILATERAL

QUADRILATERALS (FOUR-SIDED FIGURES)

RECTANGLE   SQUARE   TRAPEZOID

variety of structures, such as roof and floor trusses, bridges, geodesic domes, and towers. Also, you will have the opportunity to be part of a manufacturing or construction company that performs the design, fabrication, and construction of a structure.

(Courtesy of Marriott Corporation)

Whatever a production company decides to manufacture or construct, it must meet certain standards of quality and still make the product within the price range of the customer. In the production of large structures, one of the major concerns is to be able to build the structure with the least amount of materials. As larger-sized items are produced, the factors of material costs, weight, and ease of production all play a part in how efficient the end product will be. Therefore, engineers design items that will have adequate strength but use a minimum amount of materials. This can be seen in the framing of homes, construction of bridges, electrical transmission towers, and automobiles.

Often the product made by one company is used as an input for another company. For example: a logging company harvests trees, which are the output of the landowner but are the input for the logging company. The logs are then sold to a sawmill (output of the logging company, input for the sawmill), which in turn manufactures lumber (output of the sawmill, input for the construction company). The lumber is sold to a construction company, which constructs homes that are sold to the homeowner.

Sometimes companies are organized into various divisions. Each division produces a service or part that will be used by another division of the same company.

Activity

In the following activity, your class will set up a production company with a design and engineering division, a manufacturing division, and a construction division. The design phase of the entire project will be the responsibility of the design and engineering division. This division will prepare the specifications and drawings for the components to be made by the manufacturing division. The construction division will then use these components to construct various structures. By working together, the company will make a profit so it can remain in business.

Procedures

1. With two or three other students, design a variety of straight-sided geometric shapes using provided materials, such as popsicle sticks and fasteners.
2. Test the geometric shapes with a drill press and a bathroom scale, or other material-testing apparatus to determine the strength of each shape. Choose the shape you'd like to use for your structures.

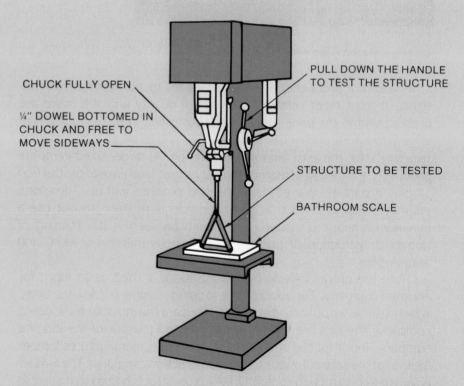

CHUCK FULLY OPEN

PULL DOWN THE HANDLE TO TEST THE STRUCTURE

¼" DOWEL BOTTOMED IN CHUCK AND FREE TO MOVE SIDEWAYS

STRUCTURE TO BE TESTED

BATHROOM SCALE

CAUTION: MAKE SURE DRILL PRESS IS UNPLUGGED! DO NOT APPLY MORE THAN 50 LB. BE CAREFUL TO AVOID SPLINTERS SHOULD THE STRUCTURE BREAK DURING THE TEST. GOGGLES SHOULD BE WORN DURING THE TEST.

3. Divide the class into design teams of three to five members each. Design and sketch a minimum of four structures that can be constructed from geometric shapes fastened at the ends to each other.

4. As a class activity, select one, two, or three structures to be built. Each structure should be built from multiples of the same geometric shape.

5. Set up a company with three divisions: a design and engineering division, a manufacturing division, and a construction division. Select a vice president of each division, a company president, and other company and division officials as is necessary to carry out needed business and make company and division decisions.

One suggested format for the company is as follows:

Component Production Company:
  Company President
  Vice President of Engineering and Design
  Vice President of Manufacturing
  Vice President of Construction
  Manager of Personnel Dept.
  Financial Manager
  Manager of Materials Dept.
  Tool and Machine Supervisor

(Courtesy of Montana Power Company)

6. Build a prototype or small model of the structure to make sure the components will work properly.
7. Design and make detailed drawings and specifications of the components to be produced.
8. Under the direction of the financial manager, determine how to obtain capital to purchase the materials needed to build the components.
9. Design and build jigs and fixtures to be used in the manufacture of the components.
10. Set up a production line so the components will move from one station to the next with the least amount of wasted time and motion so the line is efficient.

(Courtesy of Montana Power Company)

11. Establish quality control devices that will ensure a quality product. The product must meet the specifications originally agreed on at the beginning of the production run.
12. Operate a trial run to make sure all jigs, fixtures, and materials are in the proper working order to assure a smooth run.
13. Complete the production run.
14. Stockpile the components with proper identification. Deliver to the construction division for assembly.
15. Write detailed instructions on how to assemble the components into desired structures.
16. Construct the structures and inspect during construction to make sure each structure meets original design specifications.

Review Questions

1. List the seven resources that are required by any production company in order for it to be successful.
2. List several inputs of the engineering and design division of your company. Do the same for your manufacturing division and your construction division.
3. What is the processing phase of the manufacturing and construction divisions?
4. List the outputs of the manufacturing and construction divisions of your company.
5. Determine those activities that provide feedback for your company.
6. Discuss what could happen to a company if any one of the employees did not do the best possible job.
7. Make suggestions on how the manufacturing division of your company could have produced the components more efficiently.
8. Make suggestions on how the workers in the construction division may have been able to do their jobs more efficiently.
9. Study your surroundings and list at least ten different items that are built in such a way that a triangle is used in the design to add strength and stability to the device or structure.
10. List the major differences between a constructed item and one that is manufactured.

SECTION 4

ENERGY,
POWER, AND
TRANSPORTATION

CHAPTER 13

ENERGY

MAJOR CONCEPTS

After reading this chapter, you will know that:

- Work is equal to the distance an object moves, multiplied by the force in the direction of the motion.
- Energy is the ability or capacity for doing work. It is the source of the force that is needed to do work.
- Kinetic energy is energy in motion. Potential energy is energy stored in an object due to its position, shape, or other feature.
- Potential energy can be converted into kinetic energy, and vice versa.
- The principle of conservation of energy states that energy cannot be created or destroyed, but it can be changed from one form to another.
- Energy sources can be classified as limited, unlimited, or renewable.
- Most of the energy used by the United States currently comes from limited energy sources.

INTRODUCTION

Energy is one of the seven resources needed in a technological system. The history of technology has been largely tied to the availability and development of energy. For thousands of years, until the beginning of the Industrial Revolution, energy was supplied mainly by muscle, wind, and water. Mills that ran on water needed to be located near a waterfall or water race. Pumps that ran on wind needed to be in locations where the wind blew reliably.

As new energy sources, such as coal and oil were developed, industries that were heavy users of these resources grew up around them. Glass factories were built near natural gas wells. Steel mills sprang up near coal and iron deposits.

Newer industries, such as electronics manufacturing and electronic communications, use less energy than steel mills and glass factories. Because of this, they can locate farther away from energy sources. The use of electricity as an energy source has also given industry new freedom to locate away from an original energy source. Energy can now be used far from its source to perform useful work.

Industries often grew near the source of energy needed to make them run. Steel mills are often located near coal and iron deposits. (Courtesy of American Iron and Steel Institute)

WORK AND ENERGY

Work

If your teacher were to tell you to read the next two chapters in this book for homework, you might say, "That's a lot of work!" It would certainly be a lot of mental work. But in science, the word *work* has a very precise meaning.

If a force pushes or pulls against an object, and the object moves, then **work** is done. The amount of work done is the product of the force, multiplied by the distance that the object moved in the direction of the force. Work is measured in foot-pounds.

Work is equal to the distance an object moves, multiplied by the force in the direction of the motion.

Two horses hitched to the opposite ends of a carriage pull in opposite directions for an hour. Neither one wins the tug-of-war, and the carriage does not move. Even though the horses have exerted a great deal of force (and have worked up quite a sweat), no work was done because the carriage did not move.

384

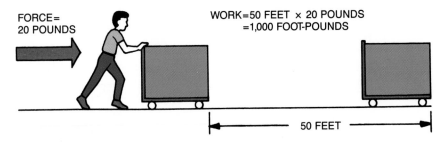

FORCE=
20 POUNDS

WORK=50 FEET × 20 POUNDS
=1,000 FOOT-POUNDS

50 FEET

Work is done only when a force moves an object over a distance.

Energy

Energy is the ability or capacity to perform work. It is the source of the force that is necessary to do work. A greater amount of energy is needed to do a large amount of work than to do a small amount of work. Energy can take many forms. Much of technology is concerned with harnessing energy from different sources to do useful work for people. In order to harness energy, people have made machines that change energy from one form to another, or that apply energy to a specific task.

Kinetic energy is energy in motion. A rag falling into a pail of water has kinetic energy. When it hits the water, it will move some of the water out of the pail (causing a splash). A bullet fired out of a gun has kinetic energy, as does a moving bicycle. The amount of kinetic energy that an object has depends on its weight and how fast it is going.

Potential energy is energy that is stored in an object due to its position, shape, or other feature. The elastic band in a slingshot that is pulled back, ready to hurl a rock, has

Energy is the ability or capacity for doing work. It is the source of the force that is needed to do work.

Kinetic energy is energy in motion. Potential energy is energy stored in an object due to its position, shape, or other feature.

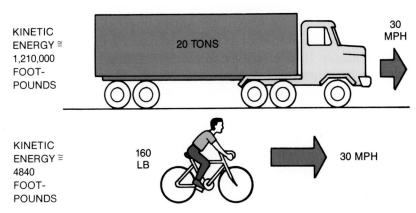

KINETIC
ENERGY ≅
1,210,000
FOOT-
POUNDS

20 TONS

30
MPH

KINETIC
ENERGY ≅
4840
FOOT-
POUNDS

160
LB

30 MPH

If two objects are moving at the same speed, the larger object has more kinetic energy.

POTENTIAL ENERGY
STORED IN SPRING

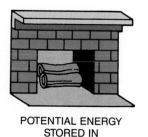

POTENTIAL ENERGY
STORED IN
MOLECULES
OF WOOD

POTENTIAL ENERGY
STORED IN POSITION
ABOVE FLOOR

Potential energy is energy that is stored in an object due to its position, shape, or other feature.

Potential energy can be converted into kinetic energy, and vice versa.

The principle of conservation of energy states that energy cannot be created or destroyed, but it can be changed from one form to another.

potential energy ready to be used. It was put there by the person who pulled back the band. A rock held over muddy ground has potential energy. It was put there by the person who lifted the rock up. When the rock is dropped, it will do work by pushing the mud aside as it sinks into the mud. Potential energy can be stored in other ways. Gasoline has stored chemical energy. A magnet has stored magnetic energy.

Potential energy can be changed into kinetic energy. For example, when you let go of the elastic band in the slingshot, the potential energy in the elastic is given to the rock which is sent flying through the air. Conversely, in order for an object to have potential energy, kinetic energy must have been used to lift it, stretch it, or change it in some other way.

Energy from different sources can also be changed from one form to another. For example, chemical energy stored in gasoline is changed to thermal energy (heat) and light energy when the gasoline is burned. If it is burned in an engine, the engine converts the thermal energy into mechanical energy. Converting energy from one form to another is one of the major tasks of technology. Energy conversion stems from one of the most important principles of science, the principle of **conservation of energy**. This principle states that energy cannot be created or destroyed, but it can be changed from one form to another.

The gravitational kinetic energy of the falling water is being converted to electrical energy.
Conversion of energy from one form to another is one of technology's most important tasks.
(Courtesy of United States Department of Energy)

Energy in Our Modern Society

Our society uses a large amount of energy each day. We use energy to light our homes at night, to heat them when it is cold, and to cool them when it is hot. We use energy to get to and from school and work. We use energy when we pick up the telephone to call our friends. Large amounts of energy are needed to construct new buildings and in industrial processes. Energy is needed to operate the equipment on farms to make them more productive.

The energy used in this country comes from many sources. It is used in many ways. The technological tools that people have made to solve problems require energy. People then must develop new technological ways of providing that increased energy.

Energy in the United States

The total amount of energy used each year in this country is about 80 **quads.** A quad is one quadrillion (1,000,000,000,000,000) BTUs. BTU stands for British Thermal Unit. It is the amount of energy needed to raise one pound of water one degree Fahrenheit. A quad is equal to the energy given off by burning about 20 million gallons of gasoline every day for a year.

The pie chart on the left shows how different energy sources contribute to the United States' annual energy needs. The pie chart on the right shows the four major uses of energy in the United States.

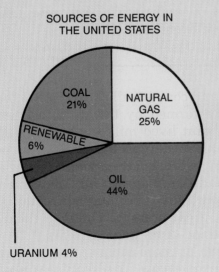

SOURCES OF ENERGY IN THE UNITED STATES

COAL 21%
NATURAL GAS 25%
RENEWABLE 6%
OIL 44%
URANIUM 4%

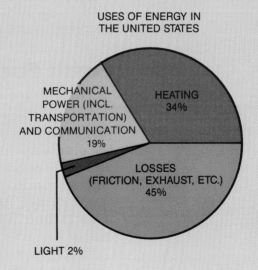

USES OF ENERGY IN THE UNITED STATES

MECHANICAL POWER (INCL. TRANSPORTATION) AND COMMUNICATION 19%
HEATING 34%
LOSSES (FRICTION, EXHAUST, ETC.) 45%
LIGHT 2%

TYPES OF ENERGY SOURCES		
Limited	Unlimited	Renewable
Coal	Solar	Wood
Oil	Wind	Biomass Gasification
Natural Gas	Gravitational	Biomass Fermentation
Uranium	Tidal	Animal Power
	Geothermal	Human Muscle Power
	Fusion	

Energy sources can be classified as **limited, unlimited,** or **renewable.** Limited energy sources, such as oil, are those of which we have a fixed supply. They are sources that we can run out of if we continue to use them. Unlimited resources, such as the sun, are energy sources that we have more of than we can ever use, at least in a practical way. Renewable energy sources, such as wood, are those that can be replaced as they are used. These sources need to be managed carefully so that we do not use more than we can replace.

Energy sources can be classified as limited, unlimited, or renewable.

LIMITED ENERGY SOURCES

As you can see from the pie chart of energy sources, most of the energy used by the United States currently comes from limited energy sources. It takes so long for nature to make these energy sources, that as we use them up, there is no hope for them to replenish themselves. It takes millions of years for natural action to make coal, oil, and natural gas. Various estimates show that we have only a few hundred years' supply left of oil and natural gas, and 1,000 years' supply of coal at our current rate of use.

Most of the energy used by the United States currently comes from limited energy sources.

Fossil Fuels

Oil, natural gas, and coal are all called **fossil fuels.** This is because they were formed by the fossilized remains of plants and animals that lived millions of years ago. After the plants and animals died, their bodies did not decay completely because they were covered over by fallen trees, leaves, and mud. Additional layers and movement of the earth carried the partially decomposed remains further underground. There, great pressure changed them into oil, gas, or coal over a period of millions of years.

Molecules are two or more atoms joined together by a chemical bond. The molecules in fossil fuels are held together with high-energy bonds. When fossil fuels are burned, the molecules break down into simpler molecules. As they break down, they release much of the energy in the high-energy bonds. This is the energy that we get from fossil fuels.

The more energy that is released, the better the fuel is. The high-energy bonds in gasoline, which is a refined form of oil, release a great deal of energy when the gasoline burns. That is why gasoline is so widely used as a fuel.

When fossil fuels are burned, complex molecules with high-energy bonds are changed into several simpler molecules, and some of the energy stored in the bonds is released. These changes are called **chemical changes** since the chemical structure of the fuels has been altered. This is sometimes called **chemical energy**.

Coal was one of the first fossil fuels to be used widely during the industrial revolution. Coal can be mined from the surface of the land in a process called **strip mining.** Strip mining ruins the surface of the land because very large holes are dug. In many locations today, operators of strip mines are required to fill in the holes. They must also restore the area by replanting trees and grass when they have finished using the mine.

Below the surface, coal may exist in different forms at different depths. To reach this coal, miners have to dig tunnels into the earth, dig the coal out of the ground, and transport it back to the surface. Dangers in this kind of work include cave-ins, gas in the mines, and long-term

Coal is transported in large quantities, by barges, trucks, or trains. (Photo by Jeremy Plant)

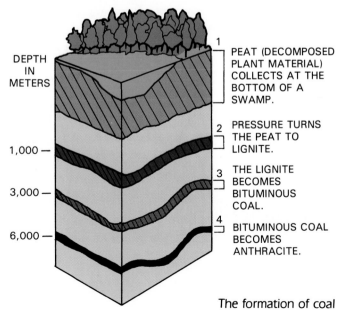

DEPTH IN METERS

1,000 —

3,000 —

6,000 —

1 PEAT (DECOMPOSED PLANT MATERIAL) COLLECTS AT THE BOTTOM OF A SWAMP.

2 PRESSURE TURNS THE PEAT TO LIGNITE.

3 THE LIGNITE BECOMES BITUMINOUS COAL.

4 BITUMINOUS COAL BECOMES ANTHRACITE.

The formation of coal

health problems from breathing coal dust every day for many years. Modern equipment and safety standards are helping to make this work safer.

In the early part of this century, coal was widely used to heat homes. Coal was stored in a coal bin in a basement or outside the house, and burned in a stove or central heater to provide warmth. Coal has since been replaced by oil, gas, and electricity for home heating. It is still burned by heavy industries, such as steel, to produce heat for industrial processes.

Coal is not as easy to transport as oil. It must be moved in large quantities by train, barge, or truck to its destination. Oil can be moved through pipelines. In addition, burning

Technology in the Coal Mines

Technology has made the job of extracting coal from the ground faster and safer. Coal that is near the surface is extracted by strip mining. In strip mining, very large earth movers, shovels, and trucks are used to dig the coal from just beneath the surface and place it in waiting railroad trains. In this country, strip mines must be reclaimed (the surface replaced, and trees and shrubs replanted) when the mining operation has been completed.

In mines that recover coal from deep underground, two shafts are dug. One is used to transport workers to and from the coal seam. The second is used to lift coal to the surface. Large machines called continuous miners and longwall miners are used to loosen coal from the wall of the mine and convey it to waiting trucks or carts. These machines are used together with other machines that hold up the mine roof while the mining operation continues.

This heavy shovel and truck is being used for strip mining. (Courtesy of United States Department of Energy)

Longwall miners are used in underground coal mines. (Courtesy of United States Department of Energy)

coal produces sulfur dioxide, which pollutes the air. Yet coal is the limited energy source of which we have the greatest reserve remaining. Engineers are working to solve these problems so that we can take better advantage of coal.

To transport coal more easily, it can be crushed and mixed with water to form a slurry. The slurry can then be pumped through pipelines in its liquid form. Attempts are also being made to burn coal more cleanly. Impurities can be removed by cleaning the coal carefully before it is burned. Pollutants can also be chemically removed while the coal is burning. After the coal has burned, smoke can be cleaned by devices called scrubbers. Scrubbers are installed in smokestacks. Because of advances in these areas, coal is expected to become an even more important energy source in this country.

Oil, or "black gold," is the energy source we depend upon the most. About half of the energy we use comes from oil, but remaining supplies are much more limited than coal. Oil is an attractive source of energy because it can be extracted, stored, and transported easily. It can also be refined into many other useful fuels. Some of these fuels are home heating oil, diesel fuel, gasoline, and jet fuel.

Oil is in such demand that oil drilling wells are erected wherever supplies exist. Wells are dug even in harsh climates such as the northern coast of Alaska. They are also set up at difficult construction sites such as ocean floor fields in order to satisfy the need. Try to imagine your daily life without gasoline, heating oil, jet fuel, or diesel fuel.

Oil is extracted from the ground by wells such as this . . . (Courtesy of United States Department of Energy)

. . . and carried by pipeline to waiting tankers that will carry it to refineries at distant locations. (Courtesy of National Society of Professional Engineers)

The High Technology Search for Oil

Oil and gas form only where the right combination of rock types and structures exist. People who look for oil search for a combination of these underground structures and rocks. In their search, they use highly sensitive magnetometers, seismographs, and gravimeters. Magnetometers record magnetic fields; rocks that are near oil fields are generally not magnetic.

Seismographs measure vibrations in the earth created by setting off a small explosive just beneath the surface. By measuring the time delay and strength of the echoes from underground rock formations, explorers can get a picture of them.

A gravimeter measures the earth's gravity. Large heavy rock formations underground will change the gravimeter's readings slightly at the surface. Explorers use this as another tool to map underground formations.

Special photographs taken from satellites are also used to identify land formations that are likely to yield oil. Computer enhancement of these photographs, or images, provides great detail. Most often, multiple color displays are used to aid in identifying patterns.

A satellite image.
(Courtesy of
Litton Industries, Inc.)

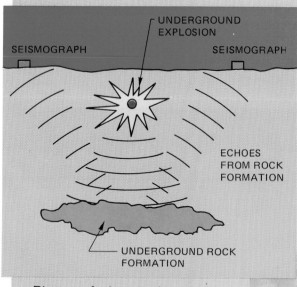

SEISMOGRAPH

UNDERGROUND
EXPLOSION

SEISMOGRAPH

ECHOES
FROM ROCK
FORMATION

UNDERGROUND ROCK
FORMATION

Diagram of seismograph use.

People go to great trouble to extract oil from the earth. For example, this offshore rig is drilling for oil under the seabed. (Courtesy of DuPont Company)

Natural gas is used for home heating and cooking. Natural gas is also used as a fuel in the glass industry. Huge gas-fired furnaces melt glass in large quantities. Gas is also used by other manufacturing industries for heating and processing materials.

Nuclear Fuels

Uranium is another important limited energy source. Unlike oil, gas, and coal, uranium is not a fossil fuel, and it is not burned to obtain energy. Uranium is a **nuclear,** or **atomic** fuel.

The great scientist, Albert Einstein, developed a theory that energy can be changed into matter, and matter can be changed into energy. His famous mathematical formula, $E = mc^2$, shows that a tiny amount of matter can be changed into a tremendous amount of energy. (In the formula, E stands for energy, m stands for matter, and c stands for the speed of light, 186,000 miles per second.)

In **nuclear fission,** an atom with a very heavy, unstable nucleus, such as Uranium-235, is bombarded by tiny particles called **neutrons**. When the nucleus is hit by a neutron, the nucleus splits into two smaller nuclei, more neutrons, and energy in the form of heat and light. The new neutrons bombard other nuclei in the vicinity. They cause the nuclei to split and release even more neutrons. If there are enough unstable nuclei, and there is no control of the neutrons, the chain reaction can produce enormous energy

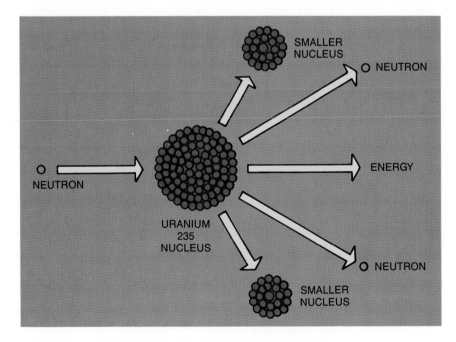

When a large atomic nucleus is split, a very small amount of its matter is converted into an enormous amount of energy.

The atomic bomb produces incredible amounts of energy from nuclear fission. (Courtesy of Los Alamos National Laboratory)

The energy released by nuclear fission is harnessed to generate electricity in this nuclear power plant. (Courtesy of United States Department of Energy)

in an explosion. This is the principle behind the atomic bomb. If the fuel used has a smaller number of unstable nuclei, and there is a way of controlling the amount of neutrons, the atomic energy can be harnessed. This is what happens in an atomic power plant.

A relatively small amount of atomic fuel can generate an enormous amount of energy. The use of atomic energy, however, involves some risks not encountered with the use of fossil fuels. The fuel and the waste products from a nuclear fission plant give off high-energy particles (called **radiation**). This radiation can cause burns and radiation sickness in people. It can be fatal to people who are exposed to it at high enough levels. These radioactive materials must be handled with great care. Safety precautions must be taken to ensure that people are not accidentally exposed to the radioactive materials. Care must also be taken that no radioactive gases escape into the atmosphere from the nuclear site.

Another major problem with using nuclear energy is that the waste products remain radioactive for thousands of years. No one has yet found a safe way to dispose of this radioactive waste. Despite these problems, hundreds of power plants, ships, and research facilities using nuclear energy are in operation worldwide.

UNLIMITED ENERGY SOURCES

Some energy sources are so vast in size or replenish themselves so quickly that they are practically unlimited. For example, energy from the sun is called unlimited, even though the sun will burn itself out in a few billion years. It is very attractive to use energy from a virtually limitless source. But it is sometimes difficult, inconvenient, impractical, or very expensive to harness these sources of energy. They are, however, playing an increasingly important role in our energy supply as we become more mindful of the limits of our more popular energy sources, such as oil and gas.

Solar Energy

Every fifteen minutes, the sun provides enough energy to meet the entire world's energy needs for one whole year. We depend on solar energy for the heat and light necessary to support life on earth. Many of the other forms of unlimited energy that will be described actually come from solar energy. For example, the heating effects of the sun cause the air movements that result in **wind energy.** And the sun's gravitational pull, along with the moon's and other planets', cause ocean movements that can be harnessed as **tidal energy.**

The sun's energy can be used by people in many ways. In **active solar** systems, the heat of the sun is used directly to perform a specific task. An example of an active solar system is solar collectors mounted on the roof of a house to heat water. In some solar collectors, specially shaped mirrors, called **parabolic reflectors,** focus the light (and heat) from the sun at one spot. This increases the temperature at that one spot. In very large installations, many mirrors can form a very large parabolic reflector that can generate very high temperatures at one spot.

Passive solar designs of houses take into account the savings that can be obtained by positioning walls, windows, and doors to take advantage of the sun's heat, or to avoid it in warm climates. Most modern buildings take passive solar design into account in some way, as described in Chapter 11.

An aerial view of a solar electricity-generating plant. (Courtesy of Southern California Edison)

Many buildings built today incorporate both active and passive use of solar energy. (Courtesy of General Electric Company)

NASA is developing this Power Extension Package (PEP). It will use a large array of solar cells to convert sunlight into electrical energy. Most satellites and other space vehicles use solar cells in one way or another. (Courtesy of NASA)

The Solar Challenger's only power source was sunshine. About 16,000 solar cells changed the sun's rays into electricity to power the plane's motor. (Courtesy of DuPont Company)

Solar cells, or **photocells** (short for photovoltaic cells), turn light into electricity. By putting many of them together, enough electricity can be generated to do useful work. In remote areas that are not served by power lines, solar cells are used to charge batteries that operate radio relay stations, radio telephones, and other devices that need electricity. Solar cells are used on most satellites and space vehicles. They provide prime power, or charge on-board batteries. One possible space project that has received some attention is a solar-powered satellite. It would have a huge array of solar cells to collect the sun's energy. The energy would then be sent to the earth using a microwave beam.

Wind Energy

Wind is air that moves because of the sun's unequal heating of the earth's surface. Wind has been used as an energy source for agriculture and transportation since ancient times. For centuries, people have built machines that harness the wind to pump water and grind grain. More recently, the wind has been used to turn generators to generate electricity.

Until the invention of the steam engine, ships used the wind as their major source of energy. Sailing ships opened up new parts of the world to explorers and stimulated world trade. In some parts of the world, such as the Far East, wind-powered vessels are still widely used in commerce. Today, sailboats are used mostly for recreational travel in this country.

Wind-powered electric generators do not generate electricity when the wind isn't blowing, or when it is blowing below a certain speed. Because of this, wind-powered generators are only practical in locations where the wind blows briskly most of the time. Large wind-powered generators create a loud noise. They can also cause interference with television signals, so they are located away from residential areas.

Experiments are being done to make more efficient use of the wind. A novel wind-powered ship is described in Chapter 15. Other developments have produced windmills that generate electricity in very low winds, and diesel-powered freighters with sails added.

This Darrieus windmill looks like an eggbeater. Wind blowing from any direction makes the blades turn. (Courtesy of the Department of Energy)

Gravitational Energy

In earlier times, water wheels provided power for grinding grain and milling flour. Today, the power from falling water is used to turn modern versions of the water wheel, called **turbines.** The turbines are connected to generators that produce electricity. In some locations, huge dams have been constructed to provide power from falling water. This type of energy production is called **hydroelectricity.** Gravity also causes tides in the oceans and rivers. **Tidal energy** is being used on an experimental basis to generate electricity.

Energy from falling water can generate enough electricity to supply the needs of entire cities.
(Courtesy of Kajima Corporation)

Water falling through a penstock turns the blades of a turbine. The turbine is connected to a generator, which produces electricity.

This geothermal electric power plant generates 900 million watts. (Pacific Gas and Electric)

Geothermal Energy

At the center of the earth, there are hot molten gases and minerals. About three miles below the surface of the earth, the temperature is around 600° Fahrenheit. Geysers like "Old Faithful" at Yosemite National Park result from steam and hot water escaping through cracks in the earth's surface. It is possible to use the steam produced within the earth's core to turn turbines, which can generate electricity.

Only a very small percentage of our energy now comes from geothermal reserves. New technological methods of reaching these reserves could turn geothermal energy into an important energy source in the decades ahead.

Fusion

Fusion is another energy source that comes from the conversion of matter into energy. In fusion, two atomic nuclei are forced together, resulting in one new nucleus plus a tremendous amount of energy. A huge amount of energy is needed to force the two nuclei together, but under the right conditions, more energy is given off than is used in the process. An atomic bomb is needed to start the process in a hydrogen bomb, or fusion bomb. It is calculated that the fusion of two atoms requires temperatures of over 100 million degrees Fahrenheit.

The researchers at this Tokamak fusion reactor at Princeton Plasma Physics Lab are trying to make controlled fusion power a reality in our lifetimes. (Courtesy of United States Department of Energy)

Proposed commercial fusion power-generating plants will use deuterium and tritium as fuel. These are commonly found in sea water. The waste products of the fusion process are not radioactive and do not pollute the atmosphere or the surrounding water. Many people expect fusion reactors to become the main way of generating electricity. They will replace the need for oil or coal in generating electricity. A commercial fusion reactor has not yet been built, but research continues to be very promising.

RENEWABLE ENERGY SOURCES

Renewable energy sources are those that can be replenished rapidly enough so that, with careful management, they appear to be limitless. If they are not replenished, renewable energy sources are used up, and appear to be limited.

Human and Animal Muscle Power

One of the earliest sources of energy was muscle power, both human and animal. Initially, human muscle power was used for many small tasks. As people learned to work together cooperatively, larger tasks could be accomplished with human power.

Early in the development of civilization, people learned how to harness oxen, donkeys, and other animals to machines. The machines could then plow the land, pump water,

In China, human power is an important source of energy. Bicycles outnumber private automobiles by almost a million to one. (UN Photo 152,715/John Isaac)

and perform other tasks. In Egypt, Greece, Rome, and Mexico, there are very large stone buildings that were built thousands of years ago. These were built using human and animal power as the primary energy source for hauling, lifting, and positioning huge building blocks. In some parts of the world, human and animal power are still important sources of energy.

Biomass

Biomass refers to vegetation and animal wastes. Biomass can be processed and serve as a major source of renewable energy. Three basic biomass processes are used to produce energy. The first is **direct combustion** (burning) of the waste products. Burning wood is a biomass process. In some parts of the world, where wood is not plentiful, dried animal wastes are burned to provide heat in homes.

The second, used more in this country, is **gasification.** In this process, methane gas is derived from the biomass. The methane is then collected and stored to be used as a fuel.

The third biomass process is **fermentation.** Fermentation uses microorganisms to turn grain into alcohol and carbon dioxide gas. The alcohol can then be stored and used as a fuel or it can be used as an additive to other fuels to make their supplies last longer. An example of this is **gasohol,** which is a mixture of gasoline and alcohol.

Wood

Wood is a form of biomass. Wood was burned as a primary energy source until the Middle Ages. A cord of wood (an amount 4 ft. × 4 ft. × 8 ft.) can provide the same amount of heat as about 200 gallons of oil. Some woods are better than others for heating. Hardwoods contain more heat energy per cord than softwoods. Oak is a very good wood to use for firewood. Around 1600, a shortage of firewood developed. Great forests were destroyed as people burned wood without thinking of the consequences. In some parts of this country, wood is becoming popular again for home heating as prices for other heating fuels climb.

Wood is renewable through forest management. Trees from a particular area are cut down to be processed into firewood, paper, and other products. That area is then replanted with small trees, and left alone for enough time for the trees to grow again.

SUMMARY

Work is only done when a force moves an object. The amount of work is equal to the distance the object moves multiplied by the force in the direction of the motion. Energy is the ability or capacity to do work. It is the source of the force that is necessary to do work.

Kinetic energy is energy in motion. The amount of kinetic energy that an object has depends on the object's weight and its speed. Potential energy is energy stored in an object due to its position, shape, or other feature. Potential energy can be converted into kinetic energy, and vice versa.

The principle of conservation of energy states that energy cannot be created or destroyed, but it can be changed from one form to another. Thus, chemical energy stored in fuel oil can be changed into thermal (heat) energy. Thermal energy can boil water to make mechanical energy in the form of steam pressure. Steam pressure can turn a turbine to make rotational energy that can drive a generator to make electrical energy.

Energy sources can be classified as limited, unlimited, or renewable. Most of the energy used in the United States comes from limited energy sources. The major limited energy sources are fossil fuels and nuclear fuels used in nuclear fission reactors.

Fossil fuels such as oil, natural gas, and coal were formed over millions of years. They are the result of pressure and heat on the partially decomposed remains of plants and animals. Because this natural process takes so long, and we are using fossil fuels at such a rapid rate, we could run out of these energy sources in a relatively short amount of time.

In fossil fuels, energy is stored in the high-energy bonds that hold the molecules together. When the fuel is burned, these molecules break down into simpler molecules, releasing some of the energy stored in the high-energy bonds. The more energy that is released, the better the fuel.

We have the largest remaining reserves of the fossil fuel coal. Problems in using coal include mining it, transporting it, and the pollution it creates when it burns. Technologists are working on ways to reduce these problems.

About half of the energy we use comes from oil. Oil is refined to form other fuels. These fuels include home heating oil, diesel fuel, gasoline, and jet fuel. Natural gas is used for home heating and cooking.

Nuclear fuels give up energy when small amounts of matter are converted into enormous amounts of energy through nuclear fission. Fission occurs when an atomic nucleus is split into smaller nuclei, neutrons, and energy. Strict safety precautions must be observed during the handling of fuel and waste products, and during the operation of nuclear plants, so that people are not exposed to radiation.

Solar energy can be used in building designs to heat water or living spaces through active or passive solar systems. Solar cells convert the sun's light directly into electricity. They are used in remote locations and in space vehicles.

Historically, wind energy has been widely used in transportation and agricultural systems. It is now being used to generate electricity. Gravitational energy is harnessed when water flowing over a dam turns a generator or a grinding wheel. The earth's geothermal power has been tapped to generate electricity. Atomic fusion is a promising unlimited supply of energy. When it is fully developed, it will produce large amounts of energy from materials found in sea water.

Renewable energy sources must be managed properly so that our use of them does not exceed our ability to replenish them. Renewable sources include human and animal muscle power and biomass processes. Biomass processes include direct combustion, gasification, and fermentation.

(Courtesy of United States Department of Energy)

HOT DOG!

Setting the Stage

The NASA training plan was simple. They took us by helicopter to the middle of the desert and waited to see if we would make it back. We were given mirrored thermal blankets, some water, a knife, a first-aid kit, and a radio transmitter. We were only to use the radio in an emergency. No matches, no food, and no wood!

We had no trouble catching small animals, but not one of us wanted to eat the meat raw. Now we had to find a way to make our catch edible.

Suggested Resources

Safety glasses and lab apron
Tin plate—26 gauge, 22¼" × 12"
Pine—¾" × 8" × 40"
Band iron—⅛" × ¾" × 32"
Steel rod—⅛" × 14"
Galvanized steel—22 gauge, 1" × 5"
12 round head wood screws—⅝ #4
2 round head machine screws—1¼", ¼ #20
2 wing nuts—¼ #20
Mirrored mylar
Spray adhesive
Finishing supplies
Jig or band saw
Drill press and drills
Diacro bender
Bar folder and box and pan brake
Sheetmetal hole punch—⅛"
Squaring shear
Hack saw
File
Belt sander

Your Challenge

Construct a device that can cook a hot dog using solar energy.

PARABOLIC REFLECTOR

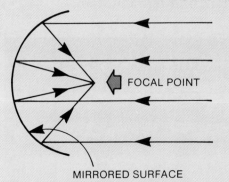

FOCAL POINT

MIRRORED SURFACE

Procedure

1. Be sure to wear safety glasses and a lab coat.
2. If you have not been told how to use any tool or machine that you need to use, check with your teacher BEFORE going any further.
3. Lay out a parabola with a 6" focal length on a large sheet of graph paper. Mark the *focal point* and complete the layout. Cut out the parabola with scissors and use it as a template.

4. Trace the parabola onto ¾" pine and carefully cut it out with a band saw. You will need two wood pieces for the cooker.

5. Smooth and trim the edges of the pine to final size using a belt sander. *The edge with the parabolic curve must be very accurate.*

6. Drill a ³⁄₁₆" diameter hole in the focal point and a ¼" hole for the stand in each piece.

7. Sand, stain, and finish the wood.

8. Cut a piece of tin plate about 22¼" × 12". Bend a ½" hem on the two shorter ends.

9. Locate five holes ⅜" in from the edge and equally spaced along the two long edges of the tin plate. Punch or drill ⅛" holes at these ten points.

10. Cut a piece of mirrored mylar 24" × 14". Using spray adhesive, cement the mylar to the tin plate on the side opposite the hems. Trim to size with scissors.

11. Fasten the tin plate to the wood pieces using screws. Start in the center and work your way toward the ends. Two students helping each other will make this step much easier.

12. Cut the 32" long cooker stand from the ⅛" × ¾" band iron. Drill ¼" diameter holes ¾" in from both ends. Bend to shape with a diacro bender.

13. Clean, prime, and paint the cooker stand.

14. Attach the stand to the cooker with 1¼" machine screws and wing nuts.

15. Cut a piece of galvanized sheetmetal 1" × 5" for the focusing sight. Mark the bending lines 1" in on both ends. Lay out and punch the three ⅛" holes.

16. Using wood screws, fasten the focusing sight parallel to the center line of the cooker.

17. Cut a 14" length of ⅛" steel rod. File one end to a rounded point. Bend 1½" of the opposite end to 90 degrees.

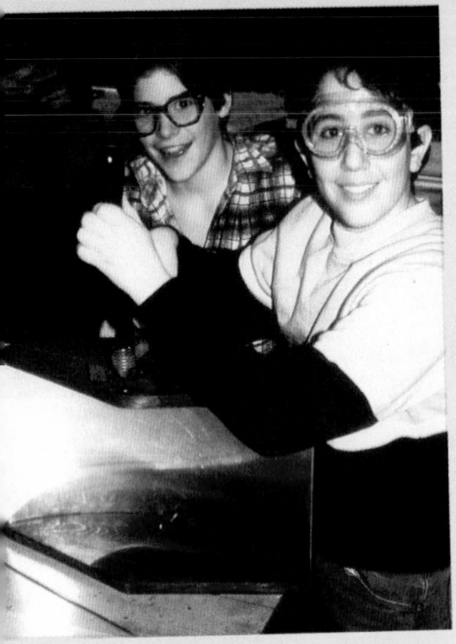

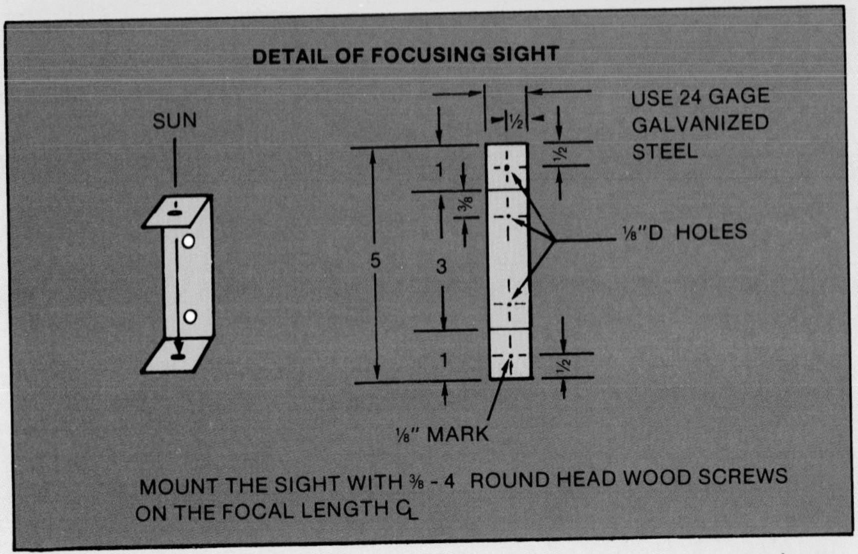
DETAIL OF FOCUSING SIGHT
SUN
USE 24 GAGE GALVANIZED STEEL
½
1
½
⅜
5
3
⅛"D HOLES
1
½
⅛" MARK
MOUNT THE SIGHT WITH ⅜ - 4 ROUND HEAD WOOD SCREWS ON THE FOCAL LENGTH CL

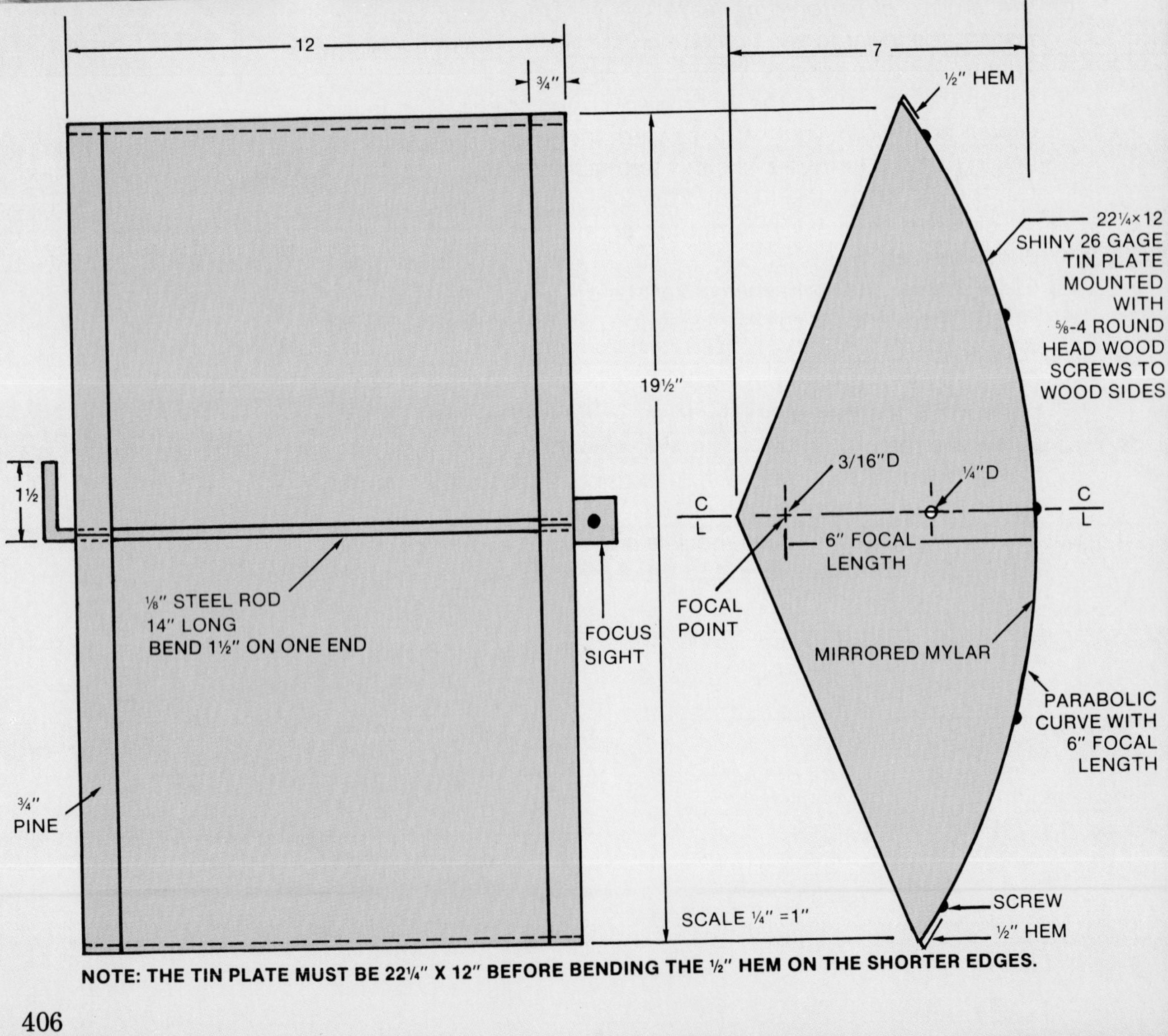
12
¾"
7
½" HEM
22¼×12 SHINY 26 GAGE TIN PLATE MOUNTED WITH ⅝-4 ROUND HEAD WOOD SCREWS TO WOOD SIDES
19½"
1½
3/16"D
¼"D
C L
C L
6" FOCAL LENGTH
⅛" STEEL ROD 14" LONG BEND 1½" ON ONE END
FOCUS SIGHT
FOCAL POINT
MIRRORED MYLAR
PARABOLIC CURVE WITH 6" FOCAL LENGTH
¾" PINE
SCREW
SCALE ¼" =1"
½" HEM

NOTE: THE TIN PLATE MUST BE 22¼" X 12" BEFORE BENDING THE ½" HEM ON THE SHORTER EDGES.

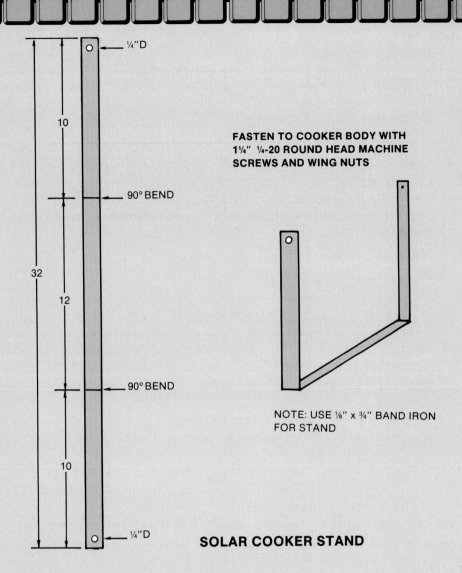

FASTEN TO COOKER BODY WITH
1¼" ¼-20 ROUND HEAD MACHINE
SCREWS AND WING NUTS

NOTE: USE ⅛" x ¾" BAND IRON
FOR STAND

SOLAR COOKER STAND

Technology Connections

1. What are some advantages of using solar energy? What are some disadvantages?
2. How is the heat from the sun concentrated by the solar cooker? Why must the parabolic reflector be as smooth and reflective as possible?
3. White objects (like marshmallows) do not cook very well on the cooker. Why not?
4. What effect would increasing the size of the reflector have on the time it takes to cook a hot dog? Why?

Science and Math Concepts

▶ The *focal point* is the point to which a lens or mirror converges parallel rays of light.

▶ A *parabolic reflector* is a type of concave mirror that focuses incoming parallel rays of sunlight to a focal point.

A PENNY FOR YOUR THOUGHTS!

Setting the Stage

Energy that is stored is known as *potential energy*. A stretched rubber band has enough potential energy to fly through the air for some distance. A boulder at the top of a hill also has great potential energy. As the boulder rolls down the hill or the rubber band flies through the air, the potential energy becomes *kinetic energy*. Kinetic energy is energy in motion.

Your Challenge

Design a device that can toss a penny into a 2' diameter target placed 15' away. Use only the resources listed below. The device must sit on the ground and have some sort of trigger mechanism.

Procedure

NOTE—Read the entire procedure before you begin.

1. Be sure to wear safety glasses and a lab coat.
2. Make a sketch and plan of your device before beginning to build it.
3. Computer aided drawing (CAD) is recommended after the first sketch is made. If CAD is not available, a full-scale drawing on graph paper is usually helpful.
4. If you have not been told how to use any tool or machine that you need to use, check with your teacher BEFORE going any further.
5. Carefully cut, bend, and assemble your penny-toss device.
6. Be careful! Resources are limited, so use the ones you have wisely.
7. Remember—hot glue is HOT!
8. You may be judged on originality, neatness, and construction techniques as well as accuracy.
9. A competition for the greatest throwing distance may also be held—67'3" is the record so far!

Suggested Resources

Safety glasses and lab apron
3 pieces of any hardwood—¼" × ¼" × 3'
1 piece of sheetmetal—6" × 9" (22–24 gauge)
1 piece of string—3' long
3 paper clips
Rubber bands
Unlimited use of fasteners (nails, screws, rivets, etc.) and shop tools and machines (saws, hammers, files, drills, etc.)
Hot glue gun and glue sticks

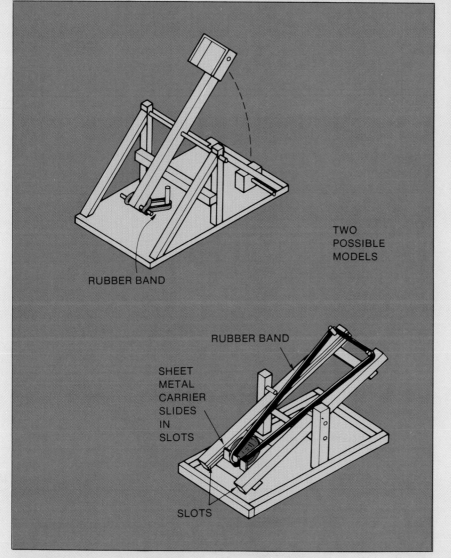

TWO POSSIBLE MODELS

RUBBER BAND

RUBBER BAND

SHEET METAL CARRIER SLIDES IN SLOTS

SLOTS

Technology Connections

1. Solving technological problems requires skill in using all seven resources. These resources are people, information, materials, tools and machines, energy, capital, and time.
2. Where did your penny-toss device get its energy?
3. What types of devices were used in the middle ages to throw or shoot stones or arrows?
4. What do you think would happen to the throw distance, or range, of the penny-toss device if you used something heavier than a penny?
5. What can you do to increase the range of the penny-toss device?
6. Why does the penny continue to fly after it has left the penny-toss device?

Science and Math Concepts

▶ *Potential energy* can be defined as stored energy or energy that an object has because of its position or condition.
▶ *Kinetic energy* can be defined as energy that matter has because of its motion.
▶ Newton's first law of motion: If an object is at rest, it tends to stay at rest. If it is moving, it tends to keep on moving at the same speed and in the same direction. This is known as *inertia*.

MUSCLE POWER

Setting the Stage

Through our muscles, our bodies have potential and kinetic energy. In this activity you are going to convert some of the stored or potential energy of your muscles into kinetic energy by pedalling a bicycle. Now this seems easy, but let's make it a bit more challenging. Let's test how much muscle power you have by then converting this kinetic energy into electrical energy via a generator! Who can produce the most electricity?

Your Challenge

Using the bicycle/generator set-up provided by your instructor, find out just how much power in the form of electrical output you can produce.

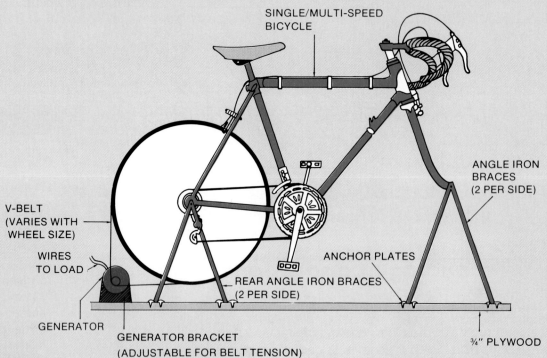

SINGLE/MULTI-SPEED BICYCLE

ANGLE IRON BRACES (2 PER SIDE)

V-BELT (VARIES WITH WHEEL SIZE)

WIRES TO LOAD

ANCHOR PLATES

REAR ANGLE IRON BRACES (2 PER SIDE)

GENERATOR

GENERATOR BRACKET (ADJUSTABLE FOR BELT TENSION)

¾" PLYWOOD

Suggested Resources

Bicycle/generator set-up
Voltmeter (if not built in)
Ammeter (if not built in)
Lights and switches
Lots of muscle power!

Procedure

1. Check the bicycle and make sure it is attached securely to the frame.
2. Check the tension and placement of the belt on the bicycle rim and generator.

3. Make sure all switches to the lights are turned off.
4. Taking turns, each student will mount the bicycle.
5. Begin pedaling and maintain a steady speed.
6. Turn on light number 1, and note the resistance to your pedalling. The light should light up. (Be sure that no one touches any exposed wires or terminals)
7. Have your instructor or another student quickly measure and record your voltage and current output.
8. Keep on pedaling and turn on light number 2. The second light should light up.
9. Have your instructor or another student measure and record your voltage and current. See how bright you can get the lights! Pedal faster!
10. If you still have some energy left, turn on light number 3 and see how bright you can get them all to light. Also take readings of voltage and current at this time.

LAYOUT OF LOAD/TEST BOARD

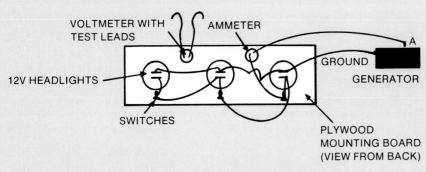

VOLTMETER WITH TEST LEADS

AMMETER

A

GROUND

GENERATOR

12V HEADLIGHTS

SWITCHES

PLYWOOD MOUNTING BOARD (VIEW FROM BACK)

NOTE: Use 12v generator available from local salvage yard. There are different types of generators. Consult your local service center for particular wiring information pertaining to your unit. Lights are wired with switch controls for each to increase load. Plywood mounting board may be attached to ¾" base.

Technology Connections

1. Why did the pedaling get harder after each light was turned on?
2. If a multi-speed bicycle is used what is the effect of gearing on the output of the generator?
3. What would be the mechanical advantage of this system? *Hint:* Measure the circumference of the bicycle wheel and generator pulley. What would happen if we were to change the circumference of either?
4. Who produced the highest voltage and current readings, or the brightest lights? Was he or she the biggest person in class? Do you think that size makes a difference?

Science and Math Concepts

▶ Your muscles have potential (stored) energy which can be converted into kinetic (motion) energy.
▶ Power transmission took place via a belt driven system.
▶ The generator (mechanical energy) produced electrical output that was created by your energy input (human muscle power).
▶ You produced work because the force of your leg muscles moved the pedals of the bicycle.

REVIEW QUESTIONS

1. A girl pushes on a bicycle with 10 pounds of force, moving it 100 feet. How much work has she done?
2. A boy pushes on a car with 25 pounds of force, but is unable to move it. How much work has he done?
3. While playing pinball, a boy draws back the spring-loaded shooter to fire a ball bearing into play. Describe how and at what points energy is converted between kinetic and potential energy.
4. If one million foot-pounds of gravitational energy are stored in the water behind a dam, how much electrical energy will be generated if all of the water flows over the dam and turns the generator? Why? (Assume that there are no losses in the generator.)
5. Is solar energy really unlimited? Why do we call it an unlimited energy source?
6. How is energy stored in a fossil fuel?
7. Name three major fossil fuels.
8. Why is new emphasis being put on the development and use of coal as an energy source?
9. a) In what ways is nuclear fission different from nuclear fusion?
   b) Why is fusion classified as an unlimited energy source, while fission is classified as a limited energy source?
10. Draw a sketch of a solar-powered car. Why might this be an impractical vehicle?
11. Is wood a practical fuel in your area? Why or why not?

KEY WORDS

Active solar	Fission	Kinetic energy	Quad
Biomass	Fossil fuel	Nuclear fuel	Radiation
Conservation of energy	Fusion	Parabolic reflector	Solar cell
Energy	Gasohol	Passive solar	Work
Fermentation	Gasification	Potential energy	
	Hydroelectricity		

SEE YOUR TEACHER FOR THE CROSSTECH PUZZLE

CHAPTER 14

POWER SYSTEMS

MAJOR CONCEPTS

After reading this chapter, you will know that:

- Power is the amount of work done during a given period of time.
- An engine is a machine that uses energy to create mechanical force and motion.
- A transmission is a device that transfers force from one place to another or changes its direction.
- Modern engines convert the potential energy stored in a fuel to mechanical force and motion.
- Both external and internal combustion engines convert the potential energy stored in the molecular bonds of a fuel into heat. The heat expands a gas, moving a piston.
- Newton's third law of motion states that for every action, there is an equal and opposite reaction.
- A generator converts rotary motion into electrical energy.
- An electric motor converts electrical energy into rotary motion.

WHAT IS POWER?

In Chapter 13, different sources of energy were described. In industry and in our homes, these sources of energy are harnessed to perform useful work. The machines that use energy sources to provide power that performs useful work are called **power systems.** Power systems are found in automobiles, jet airplanes, and the tape drive of a cassette recorder.

In the previous chapter, the words *work* and *energy* were shown to have very precise meanings in science and technology. The word **power** also has a definite meaning. In a technical sense, power is the amount of work done during a given period of time. If a certain amount of work is done in ten hours, then ten times the power is used to do the same work in one hour. Power measures how quickly we do a given amount of work.

Power is measured in **horsepower.** One horsepower is equal to 550 foot-pounds per second. It is also equal to 33,000 foot-pounds per minute. Another measure of power is the **watt.** Watts are used in the metric system of measurement, which uses meters rather than feet to measure length. One horsepower is equal to 746 watts.

Power is the amount of work done during a given period of time.

(Courtesy of Lockheed-Calfornia Company)

Power systems come in many shapes and sizes, and perform many different kinds of jobs.

(Courtesy of Fruehauf)

(Courtesy of New York Power Authority)

Power is needed to light our cities, to provide us with transportation, to cook our food, and to build buildings. (Courtesy of Perini Corporation)

POWER SYSTEMS

Power systems generally have two major parts: an **engine** and a **transmission.** An engine is a machine that uses energy to create mechanical force and motion. Examples of modern engines are the engines found in automobiles and trucks, aircraft jet engines, electric motors in appliances, and atomic reactors on submarines.

An engine is a machine that uses energy to create mechanical force and motion.

A transmission, or drive, is a device that transfers force from one place to another or changes its direction. A transmission may also change the force by increasing it, decreasing it, or dividing it into smaller parts. Examples of transmissions include automobile transmissions, belts and pulleys, and chains and gears. A particular transmission is selected to match the needs of an engine used for a specific job.

A transmission is a device that transfers force from one place to another or changes its direction.

Simple engines were used by people as early as 600 BC. These engines harnessed the energy of animals and the wind to perform tasks such as pumping water, grinding grain, and lifting loads. Later, the energy of running water was harnessed for grinding grain and sawing logs.

One disadvantage of these simple engines is that they have to be located near the energy source (e.g., near a running stream or where the wind blows regularly). Another disadvantage is having to rely on an energy source that is not always available. While they are still in use in some places, these simple engines have generally been replaced by modern engines that change the potential energy found in different fuels into mechanical force and motion. These fuels are often selected for their portability as well as their high energy content. When a portable fuel is used, the engines can be placed where they are needed.

Wind and running water were the earliest sources of energy for engines.

Modern engines convert the potential energy stored in a fuel to mechanical force and motion.

ENGINES

Modern engines convert the potential energy found in different fuels to mechanical force and motion. Different tasks for engines have lead to the development of a variety of engine types and fuels. These different engine types include **external combustion engines, internal combustion engines, reaction engines, electric motors,** and **nuclear reactors.**

External Combustion Engines

In external combustion engines, the fuel is burned in one chamber, heating a liquid or a gas in another chamber. As the gas is heated, it expands. Expanding gas builds up pressure, which applies a force to a movable piston or turbine. Force applied to a piston produces a back-and-forth (reciprocating) motion, while force applied to a turbine produces a rotational (circular) motion.

The steam engine that fueled the industrial revolution is an early example of an external combustion engine. A diagram of a simple steam engine is shown in Chapter 15. In the steam engine, a fuel such as wood or coal is burned under a large container of water. As the water is heated, it boils, turning into steam. The expanding steam pushes against a piston, driving it forward and then back. The motion of the piston depends on the position of a slide valve. A flywheel and connecting rod are used to convert the reciprocal motion of the piston into rotary motion. The flywheel also builds up **momentum** (that is, once it has been set in motion, it tends to keep turning) giving smoothness to the rotary motion.

During the industrial revolution, steam engines were used to provide power to factories, as well as to move trains and boats. In a factory, the flywheel of the steam engine

In a steam turbine engine, the force of the expanding steam is converted into rotary motion by the turbine. (Photo courtesy of New York Power Authority)

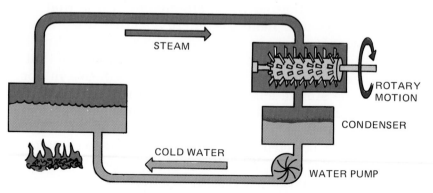

STEAM

ROTARY MOTION

CONDENSER

COLD WATER

WATER PUMP

was connected by canvas belts to a long shaft that ran the length of the factory, high overhead. Where power was needed, a wheel was mounted on the shaft, and a canvas belt put on the wheel. At the factory floor, the machine that had to be run also had a wheel. The other end of the canvas belt was on the machine's wheel. The canvas belt was normally loose so that the shaft near the roof would turn, but the machine on the factory floor would not. When power was needed, the machine operator would pull on a rope connected to an **idler wheel,** putting pressure on the canvas belt. This pressure would make the belt tight, turning the shaft of the machine at the floor.

A single steam engine could thus power a large number of different machines in a factory. In this kind of power system, each operator had control over whether his machine was getting power. The canvas belts, overhead shaft, and idler wheels formed the transmission, or drive, for the power system.

In railroad locomotives powered by steam engines, the drive wheels of the locomotive are connected to steam pistons. In steam-powered ships, the flywheel of the engine is connected to the ship's propeller.

Internal Combustion Engines

In 1876, the first successful internal combustion engine was invented in Germany by Nicolas Otto. It was successful because it was very small. It was also more than three times as efficient as steam engines of the day. In the modern internal combustion engine, fuel is burned in a closed chamber (the combustion chamber), becoming an expanding gas. As in the steam engine, the expanding gas pushes a piston, causing reciprocating motion. The piston is connected by a rod to a crankshaft, which converts the reciprocating motion into rotary motion.

The internal combustion engine has come into widespread use. It is used in nearly all cars and trucks. It is also used in piston engine airplanes, many railroad locomotives, and ships. Small engines are used in lawn mowers, go-carts, portable generators, leaf blowers, snow throwers, and outboard engines. In these small engines, the **two-stroke cycle,** which is a variation of the four-stroke cycle, is used.

In both the external combustion engine and the internal combustion engine, the energy that is used is the potential energy stored in the molecular bonds of the fuel (see Chapter 13). When the fuel is burned, the energy is released in the form of heat. The heat expands a gas, causing motion of a

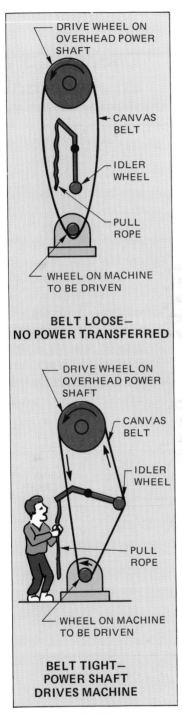

BELT LOOSE— NO POWER TRANSFERRED

BELT TIGHT— POWER SHAFT DRIVES MACHINE

The machine operator in this old factory would turn his machine on by pulling on a rope connected to an idler wheel, tightening the canvas belt.

Energy Conversion in an Automobile Engine

An automobile engine is called an internal combustion engine. The gasoline (chemical energy) burns, producing heat. The heat is then converted to mechanical energy. Since there are four parts to the engine cycle, it is referred to as a four-stroke cycle.

During step one (the intake stroke), the piston moves downward, creating suction. A mixture of gasoline and air enters the cylinder through an open intake valve. During step two (the compression stroke), the piston moves up and compresses the fuel. At this instant, the spark plug fires, igniting the gas-air fuel mixture. The heat from the explosion causes the gases to expand, pushing the piston down. This third step is called the power stroke. During step four, the piston moves back up and the exhaust valve opens. The burned gases are forced out. The cycle then repeats.

The gasoline is a source of potential energy. As it burns, it is converted to heat energy. The heat energy causes the engine parts to move. As the parts move, they develop kinetic energy.

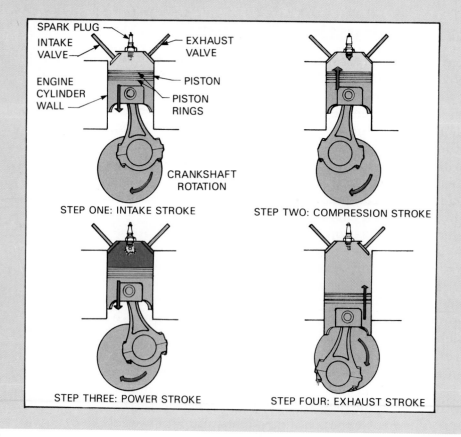

SPARK PLUG
INTAKE VALVE
EXHAUST VALVE
ENGINE CYLINDER WALL
PISTON
PISTON RINGS
CRANKSHAFT ROTATION

STEP ONE: INTAKE STROKE

STEP TWO: COMPRESSION STROKE

STEP THREE: POWER STROKE

STEP FOUR: EXHAUST STROKE

Everywhere we look, there are internal combustion engines.

piston. When the piston or its flywheel causes an object (the **load**) to move, work is done. In this way, the engine delivers power.

The external combustion engine is different from the internal combustion engine in several ways. In an external combustion engine, the fuel is separate from the gas that is expanding. Supplies of the fuel and the gas are needed to keep the engine running. A steam engine thus needs to have ready supplies of both fuel (wood, coal, and so on) and water. In the internal combustion engine, the burning (exploding) fuel supplies the gas that expands, so only the fuel is needed. Because of this, the fuel used in an internal combustion engine is picked to burn quickly, expanding the gas rapidly, providing a large force on the piston. The fuel used in an external combustion engine is picked to burn more slowly while releasing enough heat to make the water boil.

Other types of internal combustion engines are in use today. One of these is the **diesel** engine.

The diesel engine operates similarly to the four-stroke engine, except that no spark plug is used to ignite the fuel and air mixture. In a diesel, air is compressed by a piston. As it is compressed, it gets hotter. As the piston reaches the top of its stroke, a spray of diesel fuel is injected into the cylinder. The heated air is hot enough to make the diesel fuel explode without needing a spark. Diesel engines are

Both external and internal combustion engines convert the potential energy stored in the molecular bonds of a fuel into heat. The heat expands a gas, moving a piston.

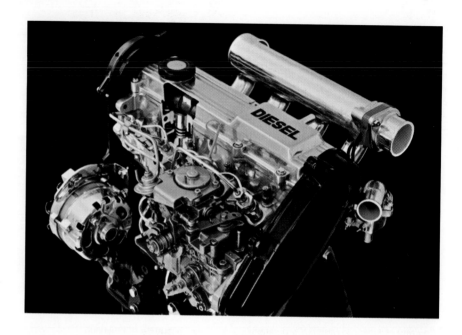

Diesel engines do not require spark plugs to operate. They provide long life with low maintenance. (Courtesy of Ford Motor Company)

thus simpler than gasoline engines. They are used in large trucks, locomotives, and ships. They are known for their low maintenance and long life, but they generally do not have the quick response that gasoline engines have.

Reaction Engines

Reaction engines rely on Newton's third law. Newton's third law states that for every action, there is an equal and opposite reaction. This law can be demonstrated if you lean against a wall and push against it with both hands. The action of pushing against the wall (action) will push you away from the wall (reaction). Two examples of reaction engines are **jet engines** and **rocket engines.** Diagrams of both jet engines and rocket engines appear in Chapter 15.

Newton's third law of motion states that for every action, there is an equal and opposite reaction.

In a jet engine, air is pushed into a combustion chamber by a compressor. There, jet fuel (similar to kerosene) is sprayed into the chamber and mixed with the air. The

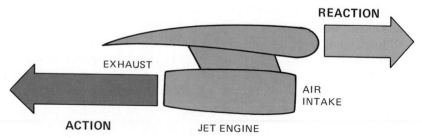

A jet engine is a good example of Newton's third law.

Jet engines have come into widespread use because they provide large amounts of power and are very reliable. (Courtesy of Boeing Corporation)

The nuclear reactor aboard this submarine supplies all of its power needs, including propulsion and electricity. (Courtesy of Lockheed Horizons)

Rockets do not need outside air to operate. This rocket is launched underwater from a submarine. (Courtesy of Lockheed Horizons)

mixture ignites and burns rapidly. The burning fuel expands, rushing out the exhaust nozzle. The gases rushing out the nozzle (the action) cause the engine to move forward (the reaction). Jet engines are used on most modern airliners and fighter planes due to the large amounts of power they generate and the low maintenance required for them.

Rocket engines operate on the same principle as jet engines, but rocket engines do not use the air around them to support combustion. Rocket engines carry their own supply of oxygen. The oxygen supply is usually in the form of liquid oxygen or other liquid oxidizer. Rocket engines are used for rockets that are going very high in the atmosphere where the air is thin, or into space where there is no air.

Nuclear Reactors

Nuclear fission reactions were described in Chapter 13. In a nuclear fission reaction, tremendous amounts of heat energy are released when large atomic nuclei are split into smaller atomic nuclei. In a reactor, this heat is used to boil water, making steam. The expanding steam is used to turn a turbine. On a ship, the turbine can drive the propeller to move the ship. The steam can also turn a turbine connected to a **generator** that converts rotary motion into electrical energy. The reactor thus supplies the ship with motion and all of its electrical needs.

A generator converts rotary motion into electrical energy.

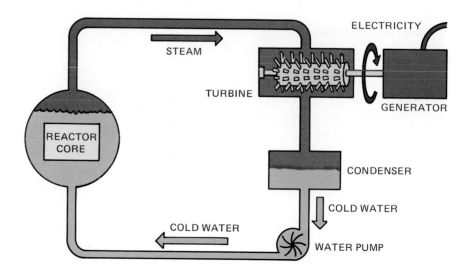

In a nuclear power plant, heat from the nuclear fission reaction boils water, making steam. The steam turns a turbine that is connected to a generator. The generator converts the rotary mechanical motion to electrical energy.

Atomic power plants use the heat of the atomic fission reaction to boil water. The resulting steam then drives a turbine that is connected to a large electrical generator. The electrical energy generated is sent to the surrounding community over an electrical power transmission system consisting of poles, wires, transformers, and other electrical components. These components are the same as those used with non-nuclear (coal-fired or oil-fired) power plants.

How Power Plants Make Electricity

There are many different kinds of electric power plants in use today. All of them use some form of energy to turn the shaft of a generator. A generator changes mechanical rotary motion into electrical energy.

In a generator, a cylinder called a **rotor** spins within a surrounding stationary housing called a **stator**. The rotor has large permanent magnets attached to it. The stator has coils of wire, called **poles,** positioned at regular intervals near the rotor. As the rotor turns, the moving magnets repeatedly change the magnetic field

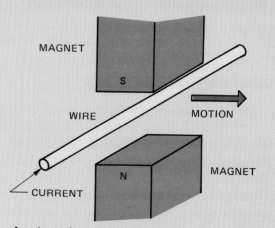

A voltage is generated across a wire whenever a magnetic field moves past it.

surrounding each of the poles on the stator. As the magnetic field changes around any piece of wire, a voltage is generated. In the generator, a voltage is thus generated by the motion of the magnetic field near the coils of wire.

The generator shaft can be turned by any one of a number of energy sources. In a hydroelectric power plant, the shaft is turned by the rush of falling water from a natural drop or from a man-made dam. In a coal-fired or oil-fired power plant, heat from the burning coal or oil boils water into steam. The expanding steam turns a turbine that is connected to the generator shaft. In a nuclear plant, heat from the atomic reaction is used to boil water, producing steam that turns a turbine. In the solar plant described in Chapter 13, reflected sunlight heats a container of water, boiling it to produce steam. The steam is used to turn a turbine that is connected to the shaft of a generator. Wind-powered generators use propellers to capture the wind and turn it directly into rotary motion that turns the generator.

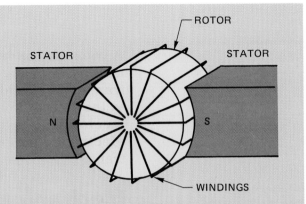

In this generator, coils of wire on the rotor repeatedly move past the magnet poles, producing an electric voltage across the coils.

These turbine generators are driven by steam from a nuclear reactor. (Courtesy of New York Power Authority)

Electric Motors

Electric motors are electromagnetic devices that change electrical energy into rotary motion. They operate in the opposite way of a generator. Small motors are found in appliances, toys, and small machines. These are often called **fractional horsepower** motors because they deliver less than one horsepower. Much larger motors are used to move subways and trains, and elevators in buildings.

Electric motors are easy to control, provide instant response to a request for mechanical power, and do not use energy when they are not doing useful work. However, they must either carry their power source with them (e.g., a battery), or they must be connected by wires to a power source.

An electric motor converts electrical energy into rotary motion.

TRANSMISSIONS

A transmission transfers, or carries, force from one place to another or changes its direction. Transmissions are used to transfer force and motion from the engine of a power system to the object to be moved (the load). Transmissions may be **mechanical, hydraulic** (fluid), **pneumatic** (air pressure), **electrical, magnetic,** or may use some other method to transfer the force.

Mechanical Power Transmission

Mechanical transmissions, sometimes called **mechanisms,** may use very simple parts or they may combine many simple parts into a complex device. In mechanical transmissions, parts with different physical shapes are joined together to transfer the force. Some of these parts are gears, pulleys, cams, levers, and linkages.

Hydraulic transmissions use water or other liquids to transfer force from one location to another. Liquids do not compress (get smaller) under **pressure.** Pressure is the force on a liquid or object divided by the area over which it is applied. Because a liquid does not compress, the pressure is the same at all points throughout the liquid. Since the pressure is the same everywhere within the liquid, a piston pushing into a tube of liquid at one end will push a second piston out at the other end of the tube. Using different size pistons, the amount of force exerted by one piston will be different from the amount of force exerted by another.

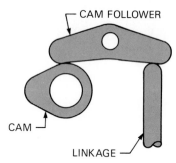

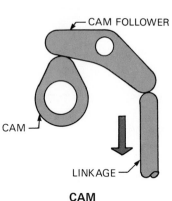

Simple parts with different physical shapes are used alone or in combination to transfer force in a transmission. (Photo courtesy of Prime Computers, Inc., Natick, MA, 1984 Annual Report)

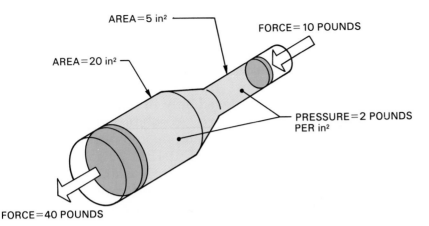

AREA=5 in²

FORCE= 10 POUNDS

AREA=20 in²

PRESSURE=2 POUNDS PER in²

FORCE=40 POUNDS

Pressure throughout the liquid is the same because the liquid will not compress. Pistons of different sizes can be used to obtain larger or smaller forces from this kind of transmission.

Electrical Power Transmission

Electrical power is transmitted by networks of wires that carry electric current from the place where it is generated to the places where it will be used. At the user locations, motors and other electrical equipment change the electrical energy into force, motion, and other useful activities.

Alternating current, or ac, is produced by generators at power plants. As the current is carried over wires from the power plant to the user locations, some electrical energy is converted to heat by the resistance in the wires. Electrical energy converted to heat in this way is lost because it is no longer available at the user location.

Wires carrying high-voltage electricity lose less to heat than low-voltage lines carrying the same power load. Therefore, high voltage is sent on wires that carry electricity over long distances. Devices called **transformers** are used to convert the high voltage to a lower voltage ("step-down") near the user locations.

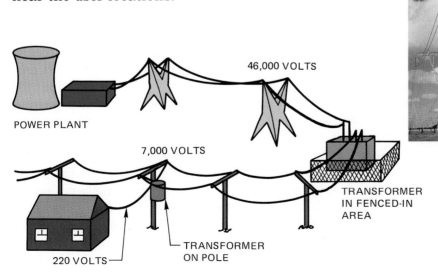

46,000 VOLTS

POWER PLANT

7,000 VOLTS

TRANSFORMER IN FENCED-IN AREA

TRANSFORMER ON POLE

220 VOLTS

Transformers are used to change ac voltages in an electric power transmission system. Power transmitted at high voltage suffers less losses. (Photo courtesy of New York Power Authority)

A pole-mounted transformer. Even the "low" voltage used in power transmission, and the power coming into homes is enough to cause serious injury or death if someone touches it. You should never climb a power pole or go near power lines. (Courtesy of Allegheny Ludlum Corporation)

Recent developments in **superconductivity** hold great promise for power transmission systems. A superconductor is a wire or other conductor that carries electric current with no resistance. Because there is no resistance, there is no heat loss as current flows through the wire. Until 1986, superconductivity was achieved only in materials that were cooled to nearly **absolute zero.** Absolute zero is the temperature at which all molecular motion stops—about 459°.

In 1986 and 1987, researchers at IBM Zurich Labs, Bell Labs, and other research facilities in Houston, Tokyo, and elsewhere, developed a group of new materials that become superconducting at much higher temperatures. As materials become superconductors at temperatures nearer air temperature, it becomes more and more feasible to build superconducting electric transmission lines that will not lose power over long distances. Transmission lines that do not lose power will result in less expensive electricity.

Power transmission systems from several power generating stations or from several power companies are often joined together in a **grid.** Because they are connected, when one power station doesn't have enough capacity to serve its customer load, another power station can supply some of its unused capacity.

CONTINUOUS VERSUS INTERMITTENT POWER SYSTEMS

Superconducting wire, tap, and cable can carry electric current with no resistance. Researchers are now developing ways to allow superconductors to function at increasingly higher temperatures. (Courtesy of Intermagnetics General Corporation)

Power systems get their energy from many different sources. An internal combustion engine can run on gasoline, diesel fuel, gasohol, or propane. The shaft of an electric generator can be turned by water falling over a dam, a propeller driven by the wind, or a turbine turned by steam from a nuclear reactor, a coal-fired boiler, or a solar collector. It can also be turned by energy obtained as a by-product of another system. For example, a generator can be turned by a bicycle wheel or a generator can be turned by a turbine in the exhaust of a jet engine.

Some of these energy sources are available on a continuous basis, while others are available only part of the time. An internal combustion engine runs as long as fuel is supplied to it. A wind driven generator, however, produces electricity only when the wind blows. A hydroelectric power plant only produces electricity when there is sufficient water behind the dam to turn the turbines.

Power systems that operate from intermittently available (available from time to time) energy sources are sometimes designed to store some of the energy so that power can be made available when the primary energy source is not available. Sometimes, energy in these systems is stored during a period of light demand, so that it can be used to help meet the need during a period of heavy demand. For example, some hydroelectric power stations use some of the power they generate during the night when demand is low to pump water back up behind the dam. During the next day, when the demand is high, there is then more water to fall over the dam, producing electricity to meet the demand.

Another way that energy is stored by a power system is in batteries. Remotely located radio relay stations, for example, are sometimes powered by solar cells. The solar cells charge a battery while the sun is shining. At night, or during very cloudy weather, the solar cells do not produce electricity. During this time, the equipment runs on power supplied by the battery. When the sun comes out again, the solar cells recharge the battery.

In a car, the battery is used to start the car. The battery is then recharged by a generator called an **alternator.** While the engine is running, most of the electric needs of the car are supplied by the alternator. When the engine is not running or when the alternator cannot supply all the electric power that the car needs, the battery supplies the power.

This desert telephone is being powered by solar cells. (Courtesy of Woodfin Camp and Associates/R. Azzi)

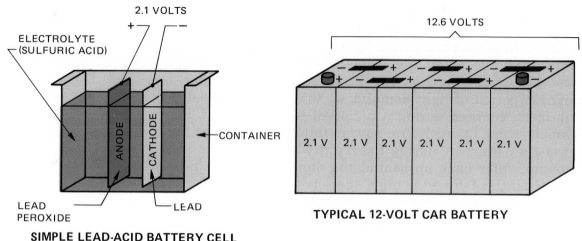

SIMPLE LEAD-ACID BATTERY CELL

TYPICAL 12-VOLT CAR BATTERY

A battery cell provides electric power released in the chemical reaction between the electrodes and the electrolyte. A battery is made by combining cells together.

SUMMARY

Machines that use energy sources to provide power that performs useful work are called power systems. Power is the amount of work done during a given period of time. Power is measured in horsepower or watts.

Power systems generally have two major parts: an engine and a transmission. An engine is a machine that uses energy to create mechanical force and motion. A transmission, or drive, is a device that transfers force from one place to another or changes its direction. A transmission may also change the force by increasing it, decreasing it, or dividing it.

Modern engines change the potential energy found in different fuels to mechanical force and motion. Different engine types include external combustion engines, internal combustion engines, reaction engines, electric motors, and nuclear reactors.

In an external combustion engine such as a steam engine fuel burns in one chamber and water boils in another chamber. The steam from the boiling water expands, pushing a piston forward and back. The reciprocal motion of the piston is changed to continuous rotary motion by a flywheel.

In an internal combustion engine, the fuel is made to explode in a chamber, pushing a piston. The piston is connected to a crankshaft and flywheel, which produce rotary motion. In a gasoline engine, a spark plug is used to explode the fuel. In a diesel engine, the air is compressed until it is hot enough to explode the fuel. No spark plug is needed in a diesel engine.

Reaction engines produce motion based on Newton's third law of motion. Newton's third law states that for every action, there is an equal and opposite reaction. Examples of reaction engines are the jet engine and the rocket engine. A jet engine burns jet fuel mixed with air, making a rapidly expanding hot exhaust gas. The exhaust gas rushing out the back of the engine pushes the engine forward. A rocket carries its own supply of oxygen to burn the fuel. Rockets can be used at very high altitudes where the air is thin, or in space where there is no air.

In a nuclear reactor, the heat energy released in an atomic fission reaction is used to boil water and make steam. The steam can then be used to turn a turbine that can directly propel a ship or turn an electric generator.

A generator is a machine that converts rotary motion into electrical energy. An electric motor is a device that converts electrical energy into rotary motion.

Mechanical transmissions use simple parts such as gears, cams, levers, pulleys, and linkages connected together to transfer force from one place to another. Hydraulic transmissions use water or other liquids to transfer force.

Electric power is transmitted over wires attached to poles or buried underground. To keep losses low, power is sent long distances at high voltage. Superconductivity holds great promise for reducing losses in electric transmission. Using superconductive wires will lower power costs. Near the customer, the voltage is reduced by devices called transformers. Even the "low voltage" entering a home is dangerous. You should avoid touching it in any way.

In some power systems, part of the power that is generated is stored for use when the energy source is temporarily unavailable. In other power systems, part of the power is used to replenish the energy source that supplies it. Batteries are often used to store electric energy. Batteries convert chemical energy into electric energy.

(Courtesy of United States Department of Energy)

TROUBLESHOOTING

Setting the Stage

You've tried all of the deep pools in the river and have two good trout to bring home after an all-day excursion. Your fishing partner pulls twice on the starter, and off you go. Within a couple of minutes, however, the engine dies, then spurts back to life. It dies again, spurts, and finally stops altogether. What can be wrong with the engine?

Your Challenge

Engines do not run because of problems in one of three subsystems. Three subsystems have to be present for an internal combustion engine to operate: ignition, compression, and fuel.

1. You will be given three identical 0.049 glow-plug engines. Each engine will have a mechanical problem and will not run. Isolate the problem by identifying in which of the three subsystems the problem could exist.
2. Demonstrate how to check ignition, compression, and fuel on a two-cycle or four-cycle engine used for transportation or recreation.

Suggested Resources

3—0.049 glow-plug engines mounted on wooden test blocks

Small two- or four-cycle engine

Prime bottle for 0.049 engine (acrylic glue bottle)

Pump can for priming two- or four-cycle engine

Compression gauge

Spark checker or used spark plug

Gloves for starting and adjusting 0.049 engine

Safety glasses

Procedure

Glow-plug Engine

Here is how to check each subsystem. You then have to decide which subsystem or subsystems are faulty.

1. Ignition
   a. Look for the glow of the plug or listen for burning fuel with the cylinder open.
   b. Unscrew the glow plug and hook it to the battery. The plug element should have a dull red glow.
2. Compression
   a. Loosen the glow plug two turns.
   b. Spin the propeller.
   c. Now tighten the glow plug and spin the propeller. It should be harder to spin the propeller with the plug tight.
3. Fuel
   a. Place a small amount of fuel directly on the top of the piston. Spin the propeller and the engine should fire. If it does, richen the fuel/air mixture by turning the needle valve counterclockwise.

b. You may have too much fuel. To check for this, remove the glow plug and hook it to a battery. If fuel has to be burned off before the plug glows, too much fuel is present. Screw the needle valve clockwise for a leaner mixture. Another method is to blow into the cylinder when the cylinder is open. If the engine fires every time you do this, turn the needle valve clockwise (leaner).

Two- or Four-cycle Engine

Now put this knowledge to use on a two- or four-cycle engine used for transportation or recreational vehicles.

1. Ignition

Ignition is checked by grounding a plug to the engine block. (Make sure no fuel is present.) This plug has to have the electrode bent for a 1/4" gap for magneto ignition. Up to 7/16" gap is used on some electronic ignition systems. You may have to design and construct your own spark checker for these systems.

2. Compression

Place a compression gauge in the spark plug hole. A reading of 70–80 lbs. minimum should be in each cylinder. Readings in each cylinder should be similar.

3. Fuel

a. You can easily bypass the complete fuel system by using a metal pump can and dispensing a small amount of fuel directly into the air intake for the carburetor. If the engine runs for only a few seconds, you have a fuel problem.

b. Check for too much fuel by unscrewing the spark plug and examining the plug. If the plug is wet with fuel, there is too much fuel present or you may have problems with one of the other systems.

Technology Connections

1. By practicing on an engine, using the preceding suggestions, you will be able to attribute the engine problem to one of the three subsystems (fuel, compression, and ignition). These are the only systems that have to function to operate the engine for two minutes. Many other subsystems are present to maintain engine operation (for example, lubrication, cooling, transmission, electrical, charging).

2. Name and describe three non-fossil-fuel-burning power systems that could be used for transportation vehicles.

Science and Math Concepts

▶ For combustion to take place, three elements have to be present: fuel, air, and some form of ignition.

▶ Eighteen thousand volts are required to produce a spark 1/4" long in dry air at atmospheric pressure.

▶ Heat engines obtain their name from the basic principle on which they operate. They convert heat energy into usable power in the form of motion. Heat engines include all types of steam, gasoline, diesel, jet, and rocket engines.

MY HERO!

Setting the Stage

Location: aboard the crippled starcraft Delphi. Captain's log shows stardate 2011—emergency entry number 3.

As a result of gyroscopic failure, we have lost all electrical power aboard our ship. The power failure has caused a release of potentially explosive hydrogen gas into the ship's atmosphere.

To regain electrical power we must find a way to start the gyroscope spinning again without igniting the hydrogen gas. All available Delphi crew members have been given this most urgent problem to solve.

Your Challenge

Find a way to restart the gyroscope before a spark blows up the ship.

Procedure

1. Be sure to wear safety glasses and a lab coat.
2. Remove the cover from the film container.
3. Locate the center of the cover. Drill a small hole just large enough to push the end of the swivel through.
4. Hot glue the swivel in place on the inside. Take care to seal the hole completely. Make sure the swivel and top rotate freely.
5. Tie the piece of string to the free end of the swivel.
6. Cut the straw in half.
7. Drill two holes in the film container directly opposite each other and slightly smaller than the straw.
8. Curve the straw as close to a *C* shape as possible without kinking or closing off the tube.
9. Insert the curved straw about ¼" into the holes and hot glue in place. Again, take care to seal the hole completely.
10. Place about 1 teaspoon of baking powder in the container.
11. Hold the container in a deep sink, bucket, or tub.
12. Fill the rest of the container with warm water and immediately snap the cover tightly in place.
13. Hold the string a few inches above the swivel in one hand. Rapidly shake the container with the other hand and then let it hang free.
14. Be sure to keep the container in the sink or tub.
15. Remember . . . you may have saved the starship from disaster, but you still have to clean up the mess you made!

Suggested Resources

Safety glasses and lab apron
Plastic 35 mm film container
Small diameter plastic cocktail straw (the skinnier the better)
Small swivel
1 piece of string—15" long
Drill and small drill bits
Hot glue gun and glue stick
Baking *powder* (NOT baking SODA)
Cup of water
Teaspoon
A DEEP sink, bucket, fishtank, or washtub

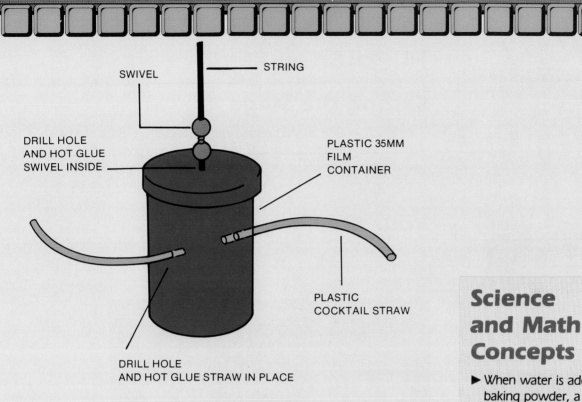

SWIVEL

STRING

DRILL HOLE
AND HOT GLUE
SWIVEL INSIDE

PLASTIC 35MM
FILM
CONTAINER

PLASTIC
COCKTAIL STRAW

DRILL HOLE
AND HOT GLUE STRAW IN PLACE

Technology Connections

1. The spinning cylinder you made is based on Hero's Engine, invented in the second century B.C. by Hero of Alexandria. The original engine used steam as the power source.
2. Where did your Hero's Engine get its power? What type of materials processing would this be considered?
3. Why did the cylinder continue to spin even after the fuel was used up?
4. If your engine was discharged in a weightless environment with no friction or air resistance, what would happen?
5. How is a jet plane similar to Hero's Engine?
6. People process materials by forming, separating, combining and conditioning. We used the separating and combining processes to make Hero's Engine. Can you explain where?
7. Glues and adhesives are used in combining processes. What process would nails and screws be used for? Scissors? Saws?

Science and Math Concepts

▶ When water is added to baking powder, a chemical reaction releases carbon dioxide gas (CO_2).
▶ Newton's third law of motion: When one object exerts a force on a second object, the second object exerts an equal and opposite force upon the first. (The cylinder and straws are one mass, and the exhaust gases are the other mass.)
▶ Remember Newton's first law of motion: If an object is at rest, it tends to stay at rest. If it is moving, it tends to keep on moving at the same speed and in the same direction. This is called *inertia*.
▶ Resistance to motion caused by one surface rubbing against another is called *friction*.

433

REPULSION COIL

Setting the Stage

All motors use a magnetic field to convert electrical energy into mechanical energy. Electrical inductance and electromagnetism are used by transformers to increase or decrease voltages. Someday, super-conducting magnets will be used to lift and propel high-speed trains.

Your Challenge

Build an electromagnetic repulsion coil that can lift an aluminum ring into the air using low-voltage alternating current (AC).

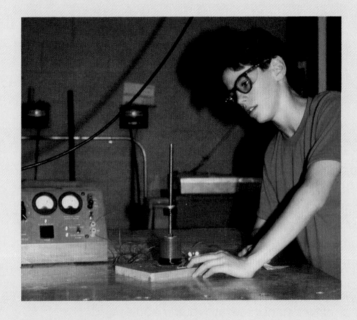

Suggested Resources

Safety glasses and lab aprons
Magnet wire (copper, enameled)—#22
Soft iron rod—⅜" diameter, 12" long
Aluminum washer—½" inside diameter
Fiber or plastic tube
Soft iron wire—#22
Bell wire—#20
3 soft iron washers—⅜"
Spring brass or momentary contact switch
Soldering iron and solder
Threading die (⅜"-16 NC)
2 soft iron nuts—⅜"-16 NC
Wood base—¾" thick
Rubber, plastic, or wood legs for the base—½" high
Black electrical tape
Round head wood screws—½" #4
Assorted machines, supplies, and tools as needed

Procedure

1. Be sure to wear safety glasses and a lab coat.
2. Cut a 6" × 9" wood base from ¾" stock. Sand and finish as desired. Install ½" high legs (rubber stick-ons work fine)
3. Drill a ⅜" hole in the center of the base.
4. Cut a 12" long ⅜" soft iron rod with a hacksaw. File both ends.
5. Thread 4½" of one end of the ⅜" rod with a ⅜"-16 NC die. (Use cutting oil and take your time.)
6. Cut 2¼" lengths of the soft iron wire.
7. Thread a ⅜"-16 nut all the way up on the ⅜" rod.
8. Cut a 2½" long piece from the fiber or plastic tube.
9. Slide a ⅜" washer over the threaded end of the rod.
10. Slide the fiber tube on next.

434

11. Fill the gap between the fiber tube and the rod with the 2¼"
    pieces of soft iron wire.
12. Slide the second ⅜" washer on next. (Careful . . . don't lose the
    wires!)
13. Insert the rod through the hole in the base and secure with a ⅜"
    washer and nut.
14. Wind at least 3 layers of #22 magnet wire over the fiber tube.
    (Be neat.)
15. Make and install the switch from the spring brass or use a
    commercial momentary contact switch (springs open automatically
    when you let go).
16. Hook up all the components with bell wire.
17. Place the ½" aluminum washer over the soft iron rod.
18. Apply 6 volts AC to the circuit.

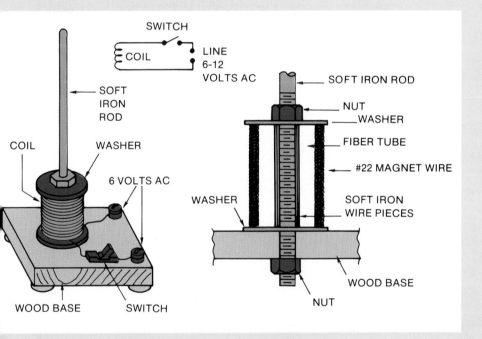

Technology Connections

1. Electronic circuits are made up of *components*. Each component
   has a specific function in the circuit. What is the function of the
   switch in the repulsion coil circuit? The coil?
2. Why is it important to use low voltage (6–12 volts AC) in this
   circuit? What will happen if too much voltage is applied?
3. Why can this device be used on *alternating current (AC)* only?
4. What is *induction*?

Science and Math Concepts

▶ When electricity flows
  through a wire, a
  magnetic field builds up
  around the wire. A *coil*
  of wire concentrates the
  magnetic field. A metallic
  object brought near the
  coil may have an electric
  current *induced* into it
  even without direct
  physical contact.
▶ When *alternating
  current* (AC) passes
  through a coil, the
  magnetic lines of force
  around the coil expand
  and collapse very
  quickly.
▶ A *transformer* uses two
  coils close to each other
  to increase or decrease
  AC voltages. The primary
  coil of a transformer
  induces an electric
  current into the
  secondary coil. A
  transformer will not
  work with *direct current
  (DC)*.

REVIEW QUESTIONS

1. Define power. How is it different from work?
2. A girl pushes a twenty-pound lawn mower 500 feet in ten minutes. How much work has she done? How much power was needed to do the work?
3. Name the two major parts of a power system. Define them.
4. Name four different kinds of engines.
5. Draw a simple diagram of a steam engine. Why are steam engines not used as often now as they used to be?
6. Draw a simple diagram of an internal combustion engine. Indicate how a gasoline engine is different from a diesel engine.
7. Describe the operation of a four-stroke engine.
8. State Newton's third law of motion. How does it apply to the operation of a jet engine or a rocket engine?
9. Define the terms engine, transmission, electric motor, and electric generator.
10. Describe the operation of a nuclear power plant.
11. Why is electric power sent at high voltage when it must be sent for a long distance? What device is used to change its voltage near the customer?
12. How will superconductivity be used in electric power transmission?
13. Why do hydroelectric plants sometimes pump water back up over the dam?
14. What is a battery? How does it operate?

KEY WORDS

Absolute zero	Four-stroke cycle	Jet engine	Rocket engine
Alternating current	Fractional horsepower	Load	Rotor
Battery	Generator	Momentum	Stator
Diesel engine	Horsepower	Nuclear reactor	Superconductor
Electric motor	Hydraulic	Pneumatic	Transformer
Engine	Idler wheel	Poles	Transmission
External combustion engine	Internal combustion engine	Power	Two-stroke cycle
		Power system	Watt
		Pressure	
		Reaction engine	

SEE YOUR TEACHER FOR THE CROSSTECH PUZZLE

CHAPTER 15

TRANSPORTATION

MAJOR CONCEPTS

After reading this chapter, you will know that:

- A transportation system is used to move people or goods from one location to another.
- Modern transportation and communication systems have helped to make countries interdependent.
- The availability of rapid, efficient transportation systems has changed the way we live.
- Transportation systems convert energy into motion.
- Steam was the first important source of mechanical power for transportation systems.
- Steam engines have been replaced by internal combustion engines, turbine engines, or electric motors in most transportation systems.
- Objects float if their buoyancy is more than their weight.
- Intermodal transportation systems make optimum use of each type of transportation used in the system.
- Most transportation systems use vehicles to carry people and/or cargo.
- Systems without vehicles are sometimes used to move people and materials from one place to another.

EXPLORING OUR WORLD

The first fast forms of transportation were used to communicate with people in faraway places. (Courtesy of The Museum of Modern Art/Film Stills Archive, W. 53rd Street, New York City)

The need to move about has existed since humans first walked the earth. Prehistoric people made camps and hunted nearby. They brought the animals they killed back to camp to share with others who had stayed behind to work in the camp.

As farming cultures developed, people needed a means for moving the harvested crops from the fields to storage areas. They often used a sled pulled by an ox or other animal or even people. Around 3500 B.C., the wheel was invented in Sumeria (in the Middle East). With the wheel, people were able to move larger loads from place to place.

As civilization developed, people recognized that rapid transportation could supply a means of communication with faraway places. Messengers on horses and in horse-drawn chariots delivered messages to the farthest reaches of the known world.

The early Phoenicians and Scandinavians used boats to explore the outer reaches of their known worlds and to trade with other cultures. A great age of exploration started in the late 1400s. During that time, European sailors explored and mapped much of the world.

The steam engine was introduced during the Industrial Revolution of the 1700s and 1800s. Powered by steam engines, trains moved people and goods rapidly across vast distances on land. In his novel, *Around the World in Eighty Days*, Jules Verne described a journey around the world in eighty days, a fantastic feat for that time. Flying in the Concorde Supersonic Transport (SST), today's traveler could make the same trip in much less than eighty hours.

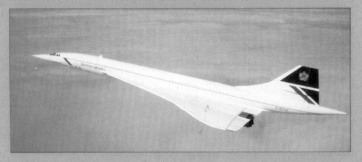

Today's traveler can travel around the world in eighty hours with plenty of time left over to sightsee along the way. (Courtesy of British Airways)

Using new modes of transportation, people are now exploring nearby space. We expect someday to be exploring the rest of our solar system and beyond. While the "Age of Exploration" is generally applied to the time when European explorers mapped the world, history is filled with stories of exploration. Perhaps the most exciting period of exploration is still ahead, as we invent new transportation technology to carry us further into today's frontiers—space and under the sea.

(Courtesy of NASA)

(Courtesy of Lockheed Corp.)

New types of vehicles enable us to explore the harsh environments of space and the deep sea.

THE WORLD IS A GLOBAL VILLAGE

Efficient, inexpensive transportation and communication have brought together civilizations that were previously cut off from each other. Transportation technology enables countries to trade goods with each other. We are able to taste foods from other countries, purchase goods manufactured in other countries, and sell goods made here to people in other parts of the world. Because of improved transportation, the economies of most countries are dependent on the economies of other countries.

Modern transportation has made it possible for more and more people to travel. This increase in **tourism** is not only interesting to the traveler, but also promotes international goodwill and understanding of different cultures and customs.

Modern transportation and communication systems have helped to make countries interdependent.

439

CREATING A NEW STYLE OF LIVING

The availability of rapid, efficient transportation systems has changed the way we live.

At the turn of this century in the U.S., the average citizen traveled 400 miles per year. In 1986, the average driver in this country traveled 12,000 miles per year. Where we once worked very close to where we lived, now we frequently must **commute**, or travel on a regular basis, to get to work.

Suburbs, or outlying areas surrounding cities, have developed. From the suburbs, many residents travel to work in the city each day and return to their homes each night. The travel time may be as long as two hours in each direction. Commuters who use a train often spend the time reading the paper, doing work in preparation for the day's activities, or socializing with other commuters. Commuters who drive must concentrate on driving. Car pools of several people who regularly ride together, and who rotate drivers, allow the passengers to use the travel time productively.

RESOURCES FOR TRANSPORTATION SYSTEMS

The way in which resources are used makes one type of technological system different from another. This is true in the case of transportation systems, as well.

Energy

The earliest transportation systems used natural sources of energy. The wind, animal power, and human power moved people and cargo from one place to another. In these systems, energy was converted into motion using simple tools such as sails, wheels, ropes, and pulleys.

Transportation systems convert energy into motion.

Transportation systems use energy that has been stored in different forms. Automobiles, for example, use energy stored in chemical form as gasoline. Subways and many surface trains use electricity. Some experimental buses use energy stored in a spinning wheel. The part of the transportation system that converts the energy into motion is generally called the **engine** or motor.

Electric motors convert electricity into rotary motion. The shaft of the motor is connected through gears or belts to the wheels of the vehicle. As the motor turns, the wheels turn, and the vehicle moves forward or backward. The gears

Many trains use electric power that comes from overhead wires. (Photo by Jeremy Plant)

Many kinds of vehicles must carry enough fuel with them to last the entire trip. (Courtesy of U.S. Navy)

or belts that connect the motor to the wheels are called the **transmission** or **drive** (sometimes called the drive train). The advantages of electric motors are that they are small, easy to control, and don't pollute the area around the vehicle. The disadvantage is that the source of electricity must either be provided along the track or roadway (through overhead wires or a "third rail") or must be carried in the vehicle in the form of batteries. Batteries are heavy and must be recharged fairly frequently.

Gasoline is one of the most efficient ways to store and transport energy. Gasoline burns easily, giving off a large amount of energy for its volume. Because it is a liquid, it can be stored and moved in tanks that can be carried on or in a vehicle. Because it burns so easily, however, it must be used with a great deal of caution. Gasoline engines are called **internal combustion** engines because the burning of the gasoline is totally enclosed within the engine.

Gasoline is a petroleum-based fuel (energy source). That means that it is made from oil that has been pumped from the ground. Diesel fuel, propane, and jet fuel are other petroleum-based fuels. Airplanes, trucks, and ships use petroleum-based fuels. Each of these vehicles must carry its energy source with it. Energy stored in petroleum fuels is a convenient way to do this.

People

Transportation systems are often designed by people to move people. In addition to designing and building transportation systems, people often form an integral part of them. Many systems rely on people to perform important functions within the system. For example, motormen drive trains. Drivers control cars and trucks. Pilots and crews fly airplanes, and astronauts fly space vehicles. In transportation systems that move people and goods from one side of the country to another, people sell the tickets, make the schedules, clean the vehicles, maintain the equipment, and purchase supplies.

Information

Transportation systems require knowledge and information from many specialists in order to run smoothly and efficiently. The specialists include the operators of the vehicles (pilots, drivers, engineers, etc.), the ticket agents, maintenance staff, and system managers. A great deal of technical, managerial, and political information is needed when designing the systems.

People who drive cars, pilot planes, or sail ships require a constant flow of information to guide them on their journey. Information about their current position, their expected course or route, their current speed, and the operation of their vehicles is crucial to the safety of their passengers and their ability to arrive at the proper destination on time. Many technological tools are used to provide this information. These tools include electronic road signs, radar, two-way radios, and on-board computers.

Materials

Transportation systems often include **vehicles** (containers that hold the people or goods being moved) and roadways. Sometimes, the roadways are very specialized, as with railroad tracks or canals. No roadways are used with airplanes and ships, but ports for loading and unloading people and cargo are required (airports and seaports). The materials used to make vehicles and roadways are determined by the type of transportation system.

Airplanes must be made of very strong, yet lightweight materials. Aluminum and titanium are often used, but new **composite** (see Chapter 8) materials are being used in some airplanes. These composite materials are made of fibers mixed with epoxies. They are stronger than metals. Cars and trucks are also made of metal, but weight is not as critical as in airplanes. Steel is the most often used metal. Other materials, including aluminum, fiberglass, plastics, and composites, are being used where weight is important (for better fuel economy, for example).

Materials such as concrete and asphalt are used to construct roadbeds. Roadbeds must be strong enough to hold the large weights of trailer trucks without yielding. They must be made in such a way that the freezing winter cold

This advanced plane must use composites that are stronger than metal to withstand stresses during flight. (Courtesy of Grumman Corp.)

and the summer heat don't crack the roadway. This can be quite a challenge, particularly when heavy loads are riding on the roadway. Many roadways need constant attention and rebuilding.

Tools and Machines

The tools and machines used in transportation systems include the vehicles, conveyor belts, pipelines, and support equipment (automatic controllers and maintenance equipment) required to move people and goods from place to place. Many transportation systems also require special tools and machines to maintain the primary tools and machines. For example, automotive engine analyzers are used to maintain cars. Automatic test equipment is used to check out airplanes, and a service tower is used for the space shuttle.

Tools and machines include the support equipment for vehicles. (Courtesy of International Business Machines Corp.)

Capital

Because of the cost of vehicles and the need for ports or roadways, a great deal of capital investment is required for many transportation systems. In addition to the original cost of building the facilities and vehicles, there is an ongoing cost to maintain them. Maintenance is important for the smooth operation of the system and for the safety of the passengers. Safety is generally regarded as the top priority in systems that carry people.

The cost of vehicles is often borne by companies in the private sector, while the cost of the port facilities and roadways is often borne by governments or public agencies. This is because the cost of building a road system or a port is so great that few, if any, companies could afford it. Also, such facilities need to be used by many people to get the best possible use out of them. Roadways are built and maintained using money collected through taxes or tolls.

Time

In a transportation system, time usually depends on the distance to be traveled and the technology chosen. Times range from seconds (moving parts from workstation to workstation along an assembly line) to weeks (a long ocean voyage) to years (interplanetary space travel). If travel time is one of the objectives, an appropriate technology can be selected in the design of a transportation system.

Mass transit systems move many thousands of people on a regular basis. Time is so critical to the smooth operation of these systems that the vehicles move on a rigid schedule. Schedules must be maintained so that people traveling on one route can meet trains, planes, or buses traveling on different routes in order to get to their final destinations. If arriving passengers are late, they will miss their connections and will not arrive at their final destinations on time.

TYPES OF TRANSPORTATION SYSTEMS

Transportation systems differ in many ways, but they also have basic similarities. Most systems use vehicles to carry people or goods, but some, such as pipelines and conveyor belts, may not. Most systems are made up of subsystems. For example, an energy conversion subsystem, a suspension subsystem, a steering or guidance subsystem, and others form the transportation system. These subsystems can be put together in different combinations to form different kinds of transport vehicles and systems.

For example, a diesel engine placed in a floating hull makes a marine transportation system (a boat). The same diesel engine placed in a vehicle with wheels makes a land transportation system (a truck or a train). One way to classify transportation systems is by the environment

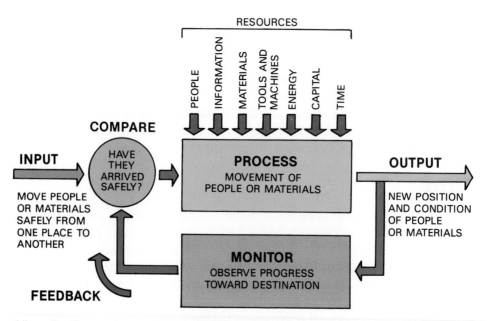

Like all other systems, transportation systems can be represented with system diagrams.

through which they are carrying people and goods (land, sea, air, space).

Many transportation systems must be integrated with other types of transportation systems to be truly useful to passengers or shippers. For example, a person traveling by airplane first needs to get to an airport. The trip to the airport may be by bus, taxi, private car, or train. When the airplane lands, there is once again a selection of ground transportation to be made. In some cases, people going on cruises must fly to the embarkation point of the ship. It is important to remember that different types of transportation systems work together to solve a transportation problem.

LAND TRANSPORTATION

The earliest form of land transportation, after walking, was riding on animals. It was a logical next step, then, to have animals drag heavy loads placed on sleds. The invention and development of the wheel improved both the load that could be carried and the speed at which it could be moved. In some societies today, animals and people are still used to provide the energy for transportation.

In the Malagasy Republic, people-powered push-pushes provide transportation. (Photo by Michael Hacker)

Steam-Powered Vehicles

Shortly after the invention of the steam engine, attempts were made to use it to propel various vehicles. The first vehicle to be moved by mechanical power was built in 1769 by Nicolas-Joseph Cugnot. The vehicle was a tractor powered by a steam engine. The tractor moved at a little more than 2 miles per hour. Unfortunately, its weight and simple steering system made it difficult to control. The first tractor crashed into a wall. A second version failed to become a commercial success, and interest in steam-powered vehicles shifted largely to trains and boats.

The first **railroad** to use all steam engines was opened in England in 1830. The engine could pull the small train at speeds up to 30 miles per hour. The railroad provided both passenger service and cargo (freight) service. Railroads began appearing in the U.S. at about the same time. They quickly came into widespread use.

As more and more people used and worked on railroads, safety became an important issue. Soon after the Civil War, several important inventions made railroads much safer.

Steam was the first important source of mechanical power for transportation systems.

Steam Engines

Steam engines are **external combustion engines**. The fuel is burned in an open chamber. Steam engines burn wood, coal, oil, or other fuel to heat a boiler containing water, creating steam. The steam pressure is used to push a piston. The piston's reciprocating (back-and-forth) motion is converted into rotary motion. The rotary motion can then be transferred to wheels. Steam locomotives need to stop frequently to take on new supplies of water and fuel, which they must carry with them.

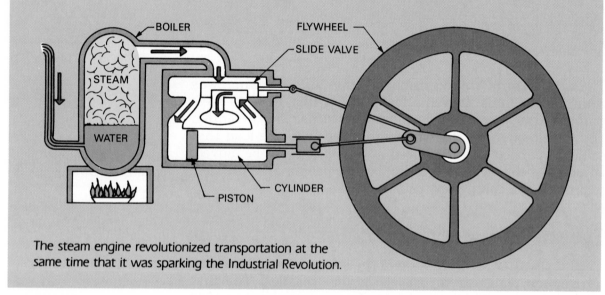

The steam engine revolutionized transportation at the same time that it was sparking the Industrial Revolution.

Many obstacles had to be overcome to build the railroad through the rugged west. (Courtesy of Oregon Historical Society)

One of the most important was the air brake, invented by George Westinghouse. With the air brake, a trainman could use one control to apply the brakes to all of the cars on the train at the same time. The air brake is still used on trains today, as well on trucks and buses. The air brake was so important that, in 1893, a law was passed requiring its use on all trains.

In 1862, President Lincoln signed the Pacific Railroad Act, authorizing the construction of a railroad between the Missouri River and Sacramento, California. The Union Pacific was to build west from the Missouri River, and the Central Pacific was to build east from Sacramento. Each railroad was awarded right-of-way to place the tracks on, ten square miles of public land, and up to $48,000 in loans from the government for every mile of track laid. In 1869, the two met in Promontory, Utah. The eastern and western

A steam-powered
logging engine
(Photo by Jeremy Plant)

Modern diesel engines
(Photo by Jeremy Plant)

A turbotrain (Photo by Jeremy Plant)

parts of the country were joined. The growth of the railroads made it much easier to travel across the country, aiding greatly in the settlement of the American west.

Steam engines were constantly being improved. At the same time, smoother, more reliable track was developed, maintenance service was improved, and passenger comfort became important. Trains that could travel at speeds of 100 miles per hour were in service by the late 1800s. Special trains such as the Orient Express (London-Istanbul) and the Twentieth Century Limited (New York-Chicago) gave a few privileged passengers very rapid, luxurious trips.

Gasoline-Powered Vehicles

Late in the 1800s, several people started using a new type of engine, the internal combustion engine, to power carriages. The engine had been invented by Nickolas Otto of Germany in 1876. The gasoline-burning internal combustion engines were lightweight and could be used on small, maneuverable carriages. A relatively small amount of gasoline powered the vehicles for many miles. The new "automobiles" quickly gained popularity.

The small size and fuel economy of the internal combustion engine led to its replacing the steam engine in many vehicles. During the late 1800s and the early 1900s, several manufacturers successfully built steam-powered cars. While

A modern luxury car (Courtesy of Ford Motor Company)

Cutaway view of an automobile (Courtesy of Saab)

A modern truck used to move large amounts of freight economically (Courtesy of Fruehauf)

Henry Ford at the wheel of his first car (Courtesy of Ford Motor Company)

they could go very fast (a Stanley Steamer set a world's record of 122 miles per hour in 1906), they needed a fresh supply of water every fifty miles. By 1920, no steamers were left in production.

The performance, safety, and comfort of the automobile improved rapidly. Many small companies sprang up. They offered cars with unique features for different uses, from sporty "raceabouts" to family sedans. The automobile was soon considered to be an essential by American families. Millions had been sold by the 1920s. Since that time, many of the smaller car manufacturers have gone out of business or were bought by the larger ones. Cars that were once the very best automobiles, such as Pierce-Arrow, Duesenberg, Stutz, and Franklin can now only be seen in museums. The automobile industry today is dominated by a few very large companies. It is America's largest industry.

Diesel-Powered Vehicles

Another kind of internal combustion engine, the **diesel** engine, also came into widespread use during this time. A diesel engine is very similar to a gasoline engine, except that it has no spark plugs. To get the fuel and air mixture to explode (providing the power to the engine), the piston squeezes the mixture even more tightly than in a gasoline engine. When any gas is squeezed (put under pressure), it gets hotter. When the fuel and air mixture is under enough pressure, it will become hot enough to ignite by itself, without a spark plug. Diesel engines are very good for carrying heavy loads at a constant speed. They also provide good mileage and long life, with little maintenance. They are used in cars, trucks, buses, locomotives, boats, and construction machinery.

On the railroads, the change from steam engines to diesel locomotives went very quickly after World War II. In 1949, the last steam locomotive built in the U.S. for main-line service was delivered. Only several hundred steam engines remain in this country today. Most are in museums or on outdoor display, but some are used for excursion trains and tourist attractions. Diesel engines replaced steam engines very quickly because they obtain more mileage from fuel, they don't bellow large clouds of smoke, and they require much less maintenance.

Electric Vehicles

Another kind of energy source, electricity, had been used for trains and cars almost since the earliest steam-powered vehicles. An electric car was first run in 1839. It used batteries to store energy and electric motors to turn the wheels. Although electric cars ran very well, they could not go very far before the batteries had to be recharged. Electric cars of today suffer from this same problem, even though battery technology has improved greatly.

Electric vehicles that get their power from overhead electric lines or from a third rail have been much more successful. Buses and trolleys that got electricity from overhead power lines were a common sight in many cities. Much of the track on the nation's railroads is electrified, particularly in densely populated areas, where the clean, nonpolluting electric engines are preferred to diesel engines. The subways that run underground in some of our largest cities are all electric-powered.

Steam engines have been replaced by internal combustion engines, turbine engines, or electric motors in most transportation systems.

This electric-powered subway runs both underground and above ground in San Francisco, California. (Courtesy of the Bay Area Rapid Transit District)

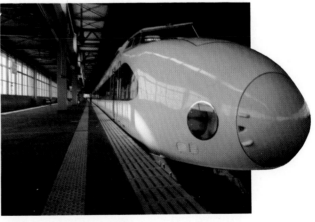

The electric-powered Japanese "Bullet Train" provides passenger service at 130 miles per hour. (Courtesy of Dave Bartruff)

WATER TRANSPORTATION

Streams, rivers, and oceans have made natural transportation routes from ancient times to this day. Large cities are often found at or near natural harbors on the seacoast or along rivers because of the travel and trade along these waterways. People have used water to carry cargo and move about since ancient times. They have used different forms of energy to move their boats.

Use of Natural Resources

The earliest energy source used was human muscle. Using poles, paddles, and oars, ancient people moved small boats on short trips close to shore. People also learned how to harness the wind to move their boats. They built large boats that used both wind and human power. Early civilizations, such as the Egyptians, the Phoenicians, and the Romans, traveled in ships that used dozens of rowers and sails at the same time.

As people's sailing skills improved, they set out on longer and longer journeys of exploration. The ancient Phoenicians, based in the Mediterranean, traded with the occupants of what is now Great Britain. The ancient Romans traded with the Far East. It is believed that early Scandinavian explorers visited North America. From the 1400s through the 1800s, there was a great wave of exploration by Portuguese, Spanish, English, and French explorers. They used two-masted and three-masted sailing ships that could carry enough provisions to sail for long distances. During this time, North and South America, Africa, and many of the Pacific islands were explored, mapped, and settled by immigrants. Throughout this time, the energy source for water transportation came from natural resources (human, animal, and wind power).

Sailing ships explored and traded with all parts of the world. (Courtesy of the U.S. Navy)

The use of wind power in a modern ship is demonstrated by the Alcyone,

built by famed ocean researcher Jacques Cousteau. (Photo Courtesy of The Cousteau Society, a member-supported environmental organization)

What Makes a Boat Float?

Early boats were made of wood. Today, boats are made of many different materials, including steel, fiberglass, and cement. A solid piece of most of these materials would sink, but boats made of them float. The reason that a boat floats is stated in a scientific observation called Archimedes' Principle.

Archimedes was an ancient Greek mathematician and physicist. He observed that any object put into water had a certain amount of **buoyancy**. Buoyancy is caused by the upward force that water (or any fluid) exerts on any object in the fluid. The amount of upward force (buoyancy) is equal to the weight of the water that was displaced (pushed aside) by the object. If the buoyancy (upward force) exerted by the water is more than the weight of the object, it will float. If the object weighs more than the buoyancy, it will sink. Buoyancy makes things feel lighter under water than they are in the air.

For example, a 32-pound piece of metal made into the shape of an empty box displaces one cubic foot of water. Salt water weighs 64 pounds per cubic foot, so the weight of water displaced is 64 pounds. Since the weight of the water displaced is more than the weight of the box, the box will float.

If a 32-pound piece of the same metal is made into a solid block, it occupies only $\frac{1}{8}$ cubic foot. Therefore, it will displace only $\frac{1}{8}$ cubic foot of water, or 8 pounds of water. Since the

weight of the water displaced, 8 pounds, is less than the weight of the object, 32 pounds, the object will sink.

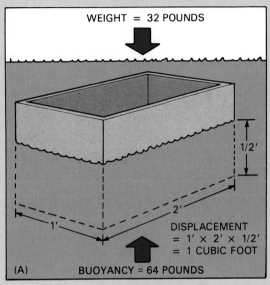

WEIGHT = 32 POUNDS

1/2'

2'

1'

DISPLACEMENT
= 1' × 2' × 1/2'
= 1 CUBIC FOOT

(A)   BUOYANCY = 64 POUNDS

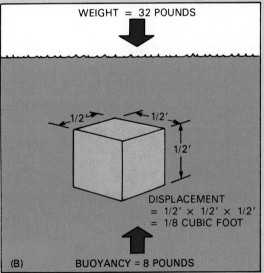

WEIGHT = 32 POUNDS

1/2'   1/2'

1/2'

DISPLACEMENT
= 1/2' × 1/2' × 1/2'
= 1/8 CUBIC FOOT

(B)   BUOYANCY = 8 POUNDS

Buoyancy equals the weight of the water that is displaced by an object. (A) A box floats in water. (B) A solid piece of metal weighing the same as the box sinks in water.

Steam-Powered Ships

In 1807, Robert Fulton built a ship powered by steam. He put it into service carrying people and cargo between New York City and Albany, N.Y. It was not the first steam-powered boat, but it was the first to be used successfully on a regular basis. The steam engine turned a paddle wheel that pushed against the water, moving the boat ahead. The *Clermont* was the first in a long history of the development of steam-powered boats.

Many steamship designs followed, using paddle wheels on the sides of ships, paddle wheels on the back, and screw propellers instead of paddle wheels. Some ships used both sails and steam. They were able to take advantage of a good wind when it was blowing and to keep under way when the wind died down. As both ships and the steam engines they used grew larger, sails started disappearing from commercial ocean-going ships. Very large passenger ships were built and used on the most heavily traveled routes.

Modern Ships

Better hull design and more efficient and powerful engines decreased the time needed for ocean travel. In 1952, the S.S. United States, with four turbines capable of producing 240,000 horsepower, made the Atlantic crossing in three days, ten hours, and forty minutes. Air travel, however, taking a matter of hours for the same trip, was attracting many of the passengers who had previously traveled by ship. Almost all of the great ocean liners have disappeared, but smaller cruise ships are still popular with vacationers.

Ships, however, still carry most of the intercontinental freight. Crude oil is carried from the producing sites to the using sites by large tankers. Cars exported from one country to another are carried by freighters. Bananas are shipped from tropical countries to world markets by ship. Ships can carry large, heavy cargoes much more economically than airplanes.

While other passenger ship travel has declined, cruise ships remain popular.
(Photo by Jeremy Plant)

Large tankers like this one carry oil from the producing sites to user countries. (Courtesy of Exxon Corp.)

Submersibles

Most ships are intended to operate on the surface of the water, and their buoyancy is designed to ensure that they do. Some ships, however, operate below the surface of the water. Ships that operate below the surface are called **submersibles** or **submarines**. They are able to operate either on the surface or at some depth below the surface because they can change their weight without changing their buoyancy. Weight is changed by filling special tanks with water (to increase weight, allowing them to sink) or with air (decreasing weight, allowing them to rise). By adjusting the amount of air in the tanks, they can "float" at any depth below or on the surface.

Objects float if their buoyancy is more than their weight.

A special consideration in designing a ship for underwater operation is the tremendous pressure exerted by water at depth. We live with pressure all around us. It is exerted by the weight of the air above us. Because water is heavier than air, water will exert a much greater pressure than the same "depth" (height) of air. At a depth of 33 feet, water exerts 14.7 pounds per square inch, the same pressure as the entire height of the atmosphere (more than 100,000 feet). At a depth of 66 feet, water pressure is twice that of the air above us. At 99 feet, water pressure is three times that of the air above. Because spheres and cylinders are very strong shapes for hollow containers, most submersibles are made of one or a combination of these shapes.

Nuclear-powered submarines can stay submerged for months at a time.
(Courtesy of the U.S. Navy)

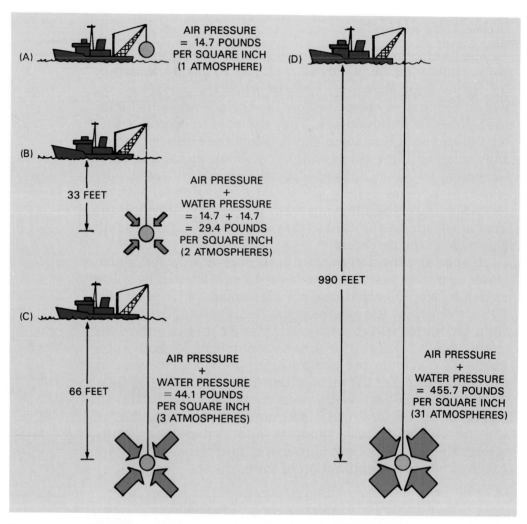

(A) AIR PRESSURE = 14.7 POUNDS PER SQUARE INCH (1 ATMOSPHERE)

(B) 33 FEET

AIR PRESSURE + WATER PRESSURE = 14.7 + 14.7 = 29.4 POUNDS PER SQUARE INCH (2 ATMOSPHERES)

(C) 66 FEET

AIR PRESSURE + WATER PRESSURE = 44.1 POUNDS PER SQUARE INCH (3 ATMOSPHERES)

(D) 990 FEET

AIR PRESSURE + WATER PRESSURE = 455.7 POUNDS PER SQUARE INCH (31 ATMOSPHERES)

Pressure under water is a challenge to designers of underwater vehicles.

Hydrofoils and Air Cushion Vehicles

Ships that operate above the surface of the water include hydrofoils and air cushion vehicles (ACVs). Hydrofoils take advantage of the fact that a boat with a relatively flat bottom will rise up in the water as it goes faster. When small hydrofoils, or flat surfaces, are attached to the bottom of the boat, the boat will ride on them once it is going fast enough. Since they offer little water resistance, hydrofoils can go very fast. ACVs use large fans to push air down under the boat, lifting the boat on a cushion of air. The boat sits slightly above the surface of the water and does not roll with waves. Passengers get a smooth, fast ride.

The hull of a hydrofoil comes completely out of the water, reducing water resistance. (Boeing photo)

INTERMODAL TRANSPORTATION

Goods transported over long distances often travel on several different kinds of transportation systems. The best mode of transportation is used for each part of the trip. It is more and more common for freight to be loaded into containers at the sender's factory. The entire container is then shipped to the end destination, without being unloaded along the route. The container is usually a trailer-truck-size box that can be put on a truck or a railroad flat car. A truck can transport the container to a train terminal, where a train carries it to a seaport. There, it is loaded on a ship for delivery to another port, where it can be reloaded on a train, and then delivered to its end destination by truck.

This kind of integrated, containerized system is called an **intermodal** transportation system. The ships used to carry the containers are called **container ships**. The carrying of truck trailers on railroad trains is called **piggyback**. Intermodal systems involve less handling of goods in-transit, resulting in less damage and loss. A properly functioning intermodal system is so reliable that it can be used to supply parts to a "just-in-time" manufacturing plant. (See Chapter 8.)

Intermodal transportation systems make optimum use of each type of transportation used in the system.

Trucks are loaded piggyback onto a train
(Courtesy of Santa Fe Railway)

This container ship is used in an intermodal transportation system. (Courtesy of U.S. Department of the Navy)

AIR TRANSPORTATION

In ancient Greek legend, Icarus was able to fly by flapping wings made for him by his father. He flew so high that the heat from the sun melted the wax that held his wings together, and he fell into the sea. People have always dreamed of flying, but some of the earliest fliers were not willing participants. Marco Polo, an Italian who traveled in and wrote about China in the 1200s, reported that Chinese sailors would tie an unwilling person to a large kite and launch it into the wind. If the kite and the person flew well, the sailors believed that the weather and their trip would be good. If the kite crashed to the ground, indicating bad luck, the ship would remain in port for the rest of the year.

Lighter-Than-Air (LTA) Vehicles

Flying began for willing passengers in 1783 when two French inventors built a lighter-than-air balloon that was large enough to carry people. The first balloon was filled with hot air, but lightweight hydrogen gas was also used by balloonists. Just as objects in water have buoyancy, objects in the air have **lift**, or an upward pressure equal to the weight of the air displaced by the object. In order for an object to float in the air, it must weigh less than the air that it has displaced (pushed aside). This can be achieved by filling a lightweight container with a gas that is **lighter than air (LTA)**, so that the two taken together weigh less than the air that used to occupy their space.

In the early 1900s, very large lighter-than-air ships called **dirigibles** were built to carry passengers and cargo around the world. These airships had rigid frames made of metal. They used hydrogen as the lightweight gas. The largest of these was the Hindenburg. It measured more than 800 feet in length, and it could carry 100 people on long trips. Unfortunately, hydrogen burns rapidly if it is ignited, and many of the large dirigibles met with spectacular but tragic ends. The most famous of these was the Hindenburg itself. It exploded and burned just as it was docking in New Jersey after crossing the Atlantic from Germany. Thirty-six of its ninety-seven passengers and crew died.

Today, LTA ships called **blimps** use helium gas. Helium gas is not as light as hydrogen, but it is much safer because it doesn't burn. Blimps do not have the rigid structure that the early dirigibles had. They are used to provide high platforms for observation cameras, for advertising, and for some cargo-lifting jobs.

Balloons float because they are filled with a gas that is lighter than air. (Courtesy of Albuquerque Convention and Visitors Bureau)

The hydrogen gas in the Hindenberg caught fire and burned as the dirigible was docking in New Jersey after a trans-Atlantic crossing.
(Courtesy of New York Daily News)

LTA vehicles use **passive lift**; that is, they float in the air due to their volume and weight. Vehicles that create lift by their movement through the air are called **active-lift** vehicles. Active-lift vehicles are said to be in **powered flight** because they must supply power to fly.

Active-Lift Aircraft

The first sustained powered flight was made by Orville Wright on December 17, 1903. Orville and his brother Wilbur had been carefully experimenting with gliders (unpowered planes) for some time. They had added a 12 horsepower engine driving two propellers to a glider. The first flight was only 120 feet, but other, longer flights followed.

Early planes were made of wood and cloth, and had two or three wings to increase lift. As engines became more powerful, heavier but stronger materials were used for construction of the plane, and the number of wings was reduced to one. Passenger airline service started in the United States in 1914. The U.S. Post Office started delivering airmail in 1919.

World War II brought about many advances in airplane design and manufacture. Each side tried to produce superior planes to help with the war effort. Many new developments in airframe design, electronics, and mass production resulted. One of the most significant developments of the war years was the jet engine. Jet engines were used on military planes right after World War II, but were not used on commercial passenger planes until 1952.

How Planes Fly

The four forces that act on an airplane are weight, lift, drag, and thrust. **Weight** is caused by gravity. It is the force pulling the plane down to the earth. **Drag** is the wind resistance that tends to hold back the plane when it moves forward. **Lift** is the upward force that must be created to get the airplane to fly. **Thrust** is the forward force produced by the engine(s) that moves the airplane.

Propeller, turbo-prop, jet, or other engines move the plane ahead at speeds of 100 to over 1000 miles per hour. Because of the shape of the wing, air rushing over it produces a higher pressure on its bottom side than on its top side. This is due to an effect described by Bernoulli's Principle. It says that as air flows over a surface, its pressure decreases in places where the speed of the air increases. The curvature of the top of the wing makes the air rush over the top faster than the bottom, creating a lower pressure on the top. The higher pressure on the bottom of the wing and the lower pressure on the top of the wing result in upward force on the wing, lifting it.

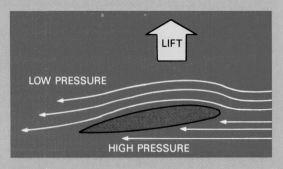

Airflow over an airplane wing

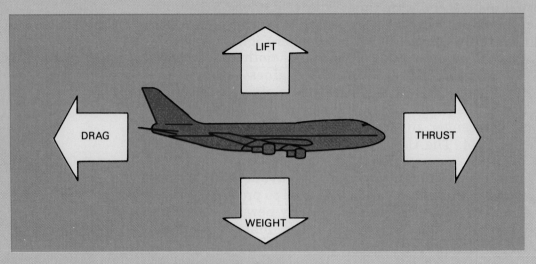

The four forces acting on an airplane

Aircraft Engines

Beginning with the earliest airplanes, many planes have used internal combustion piston engines turning **propellers** to provide thrust. Airplane propellers work in a similar way to the propeller on a ship. As they turn, they cut into the air, and move the air from front to back through the propeller. On an airplane, the propeller can be used as a pusher (mounted on the back of the wing) or as a puller (mounted on the front of the wing). Most propeller planes today use pullers.

The **jet** engine takes advantage of Newton's third law of motion. Newton's third law states that **to every action there is an equal and opposite reaction**. If, for example, you are in a small boat, and you push on the dock, the boat will move in the opposite direction, away from the dock. A balloon filled with air, suddenly let go to fly around the room, is another example of Newton's third law in action. The force of air rushing out of the balloon in one direction moves the balloon in the opposite direction.

In a jet engine, a **compressor** forces air through the engine to a combustion chamber, where the air is mixed with a fine spray of fuel and ignited. The burning air/fuel mixture expands rapidly, and rushes out through the exhaust. The force of the burning gases rushing out the exhaust produces an opposite force on the engine, moving the plane forward. The burning gases also turn a turbine as they exit the engine. The turbine is connected to the intake compressor and to other devices such as generators. In a **turbo-prop** engine, the turbine drives a propeller.

The jet engine uses the air it is flying through to supply the oxygen needed to burn the jet fuel. A **rocket** carries its own oxygen, so it can fly in places where there is little or no oxygen, such as at very high altitudes or in space. The rocket does not need a turbine, as there is no air compressor to drive.

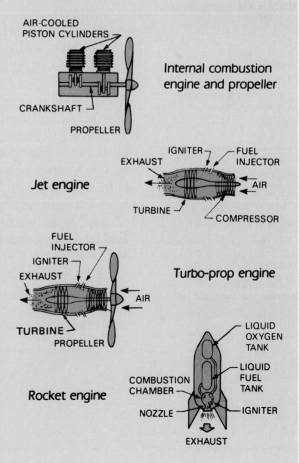

AIR-COOLED PISTON CYLINDERS

CRANKSHAFT

PROPELLER

Internal combustion engine and propeller

IGNITER — FUEL INJECTOR

EXHAUST

Jet engine

AIR

TURBINE

COMPRESSOR

FUEL INJECTOR

IGNITER

EXHAUST

Turbo-prop engine

AIR

TURBINE

PROPELLER

Rocket engine

LIQUID OXYGEN TANK

LIQUID FUEL TANK

COMBUSTION CHAMBER

NOZZLE

IGNITER

EXHAUST

The first person to fly at more than the speed of sound was Chuck Yeager. He piloted an X-1 experimental rocket plane to more than Mach 1 (about 700 miles per hour) in 1947. **Mach 1** means the speed of sound. **Mach 2** is twice the speed of sound. When an airplane approaches the speed of sound, the air that it is pushing ahead of it forms a shock wave called the sound barrier. Just before breaking through the sound barrier, the plane becomes difficult to handle due to buffeting of the shock wave. After the plane is going more than the speed of sound, the ride becomes smoother and the plane much easier to control.

The first passenger jets, the British Comets, suffered many accidents during their early lives. The problems were traced to metal parts that failed because of the unexpectedly high stress put on them as the planes climbed and accelerated. The price of this new knowledge was very high, both in lives lost and in monetary terms. Based on the knowledge gained from the crash investigations, not only was the Comet rebuilt as a much safer plane, but all other planes that came after it were built better.

Since the Comet, there has been a long list of successful passenger jets. Almost all long-distance flying is now done by jets that have built an impressive safety record. This is due to many things, including the low maintenance required for jet engines, extensive training of pilots and other flight crew members, and an increasingly sophisticated air traffic control system that keeps planes apart in the sky.

Many different kinds of planes are used today. Each is built for a specific job. Small private planes can carry two to six people economically over short distances. Airliners that are designed to take off and land on short runways are used for commuter traffic. Large **jumbo jets** carry hundreds of people at a time over long distances. The jointly built British and French **Concorde** carries passengers at nearly twice the speed of sound. Military planes can carry large cargoes and refuel in-flight. They can deliver their loads anywhere in the world without having to land to take on fuel. Small jets can travel at more than the speed of sound in order to protect our borders.

Planes like this one carry hundreds of people over several thousand miles economically. (Boeing photo)

The Concorde Supersonic Transport (SST) carries people from New York to Europe in three and one-half hours. (Courtesy of British Airways)

The Wright brothers' first flight could have taken place in the cargo compartment of this C-5A. (Courtesy of Lockheed Corporation)

Military planes like this travel at more than the speed of sound and are very maneuverable. (Courtesy of U.S. Department of the Navy)

Transportation Safety

Most transportation systems that are designed to carry people treat the safety of the passengers as the first priority. Punctuality, comfort, and other issues are secondary. New ways of making transportation safer are constantly being designed and tested.

Sometimes, new approaches to safety come about because of terrible accidents. In 1912, the largest steamship in the world at that time, the *Titanic*, struck an iceberg on her first voyage and sank in the North Atlantic. Some 1,513 of the 2,207 passengers and crew aboard died in the tragedy. The accident resulted in the setting of new regulations regarding lifeboats, lifejackets, and radios. These regulations have made travel by ship much safer.

Many safety features are the result of advance planning by designers who try to anticipate what could happen. Most vehicles must now meet safety standards before being sold to or used by the public.

Often, however, the most important safety precautions must be taken by the passengers themselves. The National Highway Traffic Safety Administration estimates that about 17,000 less people would be killed each year in traffic accidents if people used their seat belts. Only about 14% of all adults, however, use their seat belts.

An airbag being tested in an automobile (Courtesy of Ford Motor Company)

Always buckle up. (Courtesy of Ford Motor Company)

Planes are dropped from this test stand to test for crash worthiness. (Courtesy of NASA)

This plane has just been dropped from the test stand. (Courtesy of NASA)

SPACE TRANSPORTATION

Rockets are used to attain the high speeds needed to achieve orbit. (*Courtesy of Lockheed Corp.*)

With the successful launch of Sputnik by the U.S.S.R. in 1957, the world entered the space age. Sputnik was put into low earth orbit, circling the earth once every 90 minutes, sending a simple message. Yuri Gagarin of the U.S.S.R. made the first manned space flight in 1961. Two American astronauts followed him later that year. Throughout the 1960s, both the U.S.S.R. and the United States made increasingly more ambitious space flights. The culmination of the space flights was the first exploration of the moon by the American astronauts Neil Armstrong and Edwin Aldrin in July 1969.

Vehicles that travel in space need to carry a supply of oxygen to burn their fuel because there is no air in space. To date, all space vehicles have used rocket engines. The fuels used have been liquid fuel, solid fuel, and hydrogen. Other engines, based on very different principles, have been proposed, but they have not yet been used.

In order to escape the Earth's gravity, a speed of more than 17,000 miles per hour must be reached. This is called **escape velocity**. Escape velocity puts a space vehicle into orbit. Once in orbit around the earth, a small amount of additional thrust will move the vehicle out of orbit and away from the Earth.

Rapid developments were made in such areas as metallurgy, computers, integrated circuits, life support, navigation, and launch safety during the program to land people on the moon (Apollo). Many of these developments have spawned new products or techniques used in everyday life by many people. These include the intensive care unit used in hospitals, new fire fighting equipment, and reliable integrated circuits used in many of our electronic appliances.

Space travel poses many problems not faced by people before. In space, there is no air so space travelers must carry their atmosphere with them. Once away from the Earth's gravity, space travelers experience weightlessness. This can be fun and exciting at first, but practical problems, such as how to take a shower and how to drink a liquid, rapidly become apparent.

As we think about traveling to other planets, or further, the enormous distances to be traveled will pose a problem because of the long time needed for the trips. It took Voyager, an exploratory space vehicle, nine years to travel from the Earth to Uranus, where it took photographs and sent them back to Earth.

The space shuttle can travel in space, glide back to Earth like a plane, and be used again for another trip.
(Courtesy of NASA)

The space shuttle uses three main engines and two solid fuel rocket boosters to attain escape velocity. During its climb to orbit, it drops the boosters and an auxiliary fuel tank into the ocean. When it returns, it uses its small wings to glide in the atmosphere. The space shuttle is the first re-usable space vehicle. It is used as a space truck, carrying things and people back and forth from the Earth to low Earth orbit.

NONVEHICLE TRANSPORTATION SYSTEMS

All of the transportation systems described so far use vehicles to carry people or cargo from one destination to another. The vehicles can be guided vehicles, such as trains, that move over a fixed path, or free vehicles, such as airplanes, that can go in any direction.

Most transportation systems use vehicles to carry people and/or cargo.

Other forms of transportation systems that move oil, coal, electronic parts, sheet metal assemblies, or other goods without any vehicles also exist. These transportation systems include pipelines, conveyor belts, and other machines for guiding material and moving it.

Systems without vehicles are sometimes used to move people and materials from one place to another.

Pipelines

Pipelines are used for moving crude oil or natural gas from the field where it was pumped to a location where it is refined and/or loaded on tanker ships or trucks for further shipment. The Trans-Alaska pipeline moves crude oil from the northern part of Alaska to waiting tankers at Valdeez, a port on Alaska's southern coast. The pipeline must be heated and insulated to keep the crude oil liquid enough so that the pumps can move it, even in cold weather.

Materials are automatically moved along an assembly line by a conveyor system. (Courtesy of Cincinnati Milacron Inc.)

Conveyors

In automated or semiautomated assembly lines, parts that are being worked on are moved from workstation to workstation by conveyers. Most cars are assembled this way. Large metal parts for various industries are also cut, drilled, and machined while being moved from station to station on a material transport system.

Small pieces can be moved along a guide path, with their motion driven by vibration of the path itself. This technique is used for electronic parts, pieces of coal, and other relatively small objects.

PEOPLE MOVERS

Transportation systems for moving small numbers of people at one time over relatively short distances have evolved during the last century. These **people movers** include **escalators, elevators,** and **personal rapid transit systems (PRTs).** Not only do people movers make it easier for people to get from one place to another, but they have helped shape our major cities.

Freight elevators had been in use for some time before 1852 when Elisha Otis invented a safety mechanism to keep

Escalators provide quick, easy movement from one level to another. (Courtesy of Otis Elevator Company)

them from falling. The introduction of this safety device and other advances in elevator technology helped to make high-rise buildings feasible and popular. Without elevators, some cities would be more spread out, rather than densely populated by skyscrapers (very tall buildings).

While elevators move people vertically, personal rapid transit systems move people horizontally from one place to another. PRTs use small cars that people generally stand in. They move along tracks from one part of an airport to another or from one part of a city to another. They are usually automatically controlled and do not require an operator.

Elevators and PRTs are people movers that use vehicles. Escalators and moving sidewalks are people movers that do not use vehicles. They are used to move people a little more quickly than walking or climbing stairs. They are often used where people are likely to be carrying packages or luggage, such as in department stores or airports.

SUMMARY

Using human muscle power, animal power, and the wind, people have been building transportation systems since ancient times. Transportation has enabled people to explore their world. Exploration continues in space and under the sea.

Transportation and communication technologies have promoted the exchange of goods and ideas among peoples who had little or no previous contact. This exchange has brought about the economic interdependence of countries. Transportation technology has also changed our way of living by making regular travel to work and other activities an accepted way of life.

Most transportation systems use vehicles to carry people and/or cargo. An engine or motor in the vehicle converts energy into motion. The vehicle is built to carry the expected load, and the engine is selected to be appropriate for the job and the environment.

People are often part of the system, as well as users of it. People drive the vehicles, collect the tickets, maintain and manage the system. Many transportation systems use roadways or ports built with public money, while supplying vehicles bought with private (company) money. Accurate scheduling is often an important feature of a transportation system.

Transportation systems may be classified as land, sea, air, space, and nonvehicle transportation systems. These systems all have similarities, even though their uses are very different. The way in which the different subsystems are put together is determined by the needs of the user. The user's needs include size of load, distance to be traveled, speed required, and whether the load is to be moved across water or land.

Early land vehicles that were mechanically powered were powered by steam. Railroads were a great advance in transportation, and changed the way many people lived. The railroad provided a transportation bridge across America. Today, railroads carry a higher percentage of freight and a lower percentage of passengers as other types of transportation technologies have matured.

Automobiles have become a necessity to many Americans. While other technologies have been tried, the most common power plant in cars is the internal combustion engine.

As with land vehicles, the first mechanically powered ships were powered by steam. Other power sources have included steam turbines, diesel engines, and nuclear reactors. Passenger travel by ship has declined due to airline competition, but cruise ships and freight shipping are still in high demand.

Intermodal transportation combines ships, railroads, trucks, and sometimes airplanes. It takes advantage of the best features of each over different segments of a trip.

Freight is packed into containers at the shipping point, and the containers are carried intact on each of the carriers. They arrive at their destination without having been opened.

The Wright brothers' first powered flight in 1903 opened the way for a new technology to develop. Air transportation vehicles, first powered by internal combustion engines, are now also powered by jet engines, turbo-prop engines, and rocket engines. Jet engines and rocket engines enable planes to fly at more than the speed of sound.

Safety is of primary importance in transportation systems. Improved safety measures come about as a result of accidents or as a result of careful advance planning. Many times, however, the best safety precautions are the ones that you, the passenger, take.

Space transportation systems must deal with special problems of life support in a harsh environment, weightlessness, and resource conservation on long journeys. Vehicles are able to achieve an orbit around the Earth if they can reach the escape velocity of more than 17,000 miles per hour.

Transportation systems such as pipelines and conveyors do not use vehicles to move material from one location to another. Such systems are used in diverse applications ranging from moving small components on an assembly line to moving large volumes of crude oil over hundreds of miles of rugged terrain.

People movers carry people over relatively short distances. People movers such as elevators and PRTs use vehicles, while escalators and moving sidewalks do not. People movers have not only provided convenience to travelers, but have also helped to shape our environment.

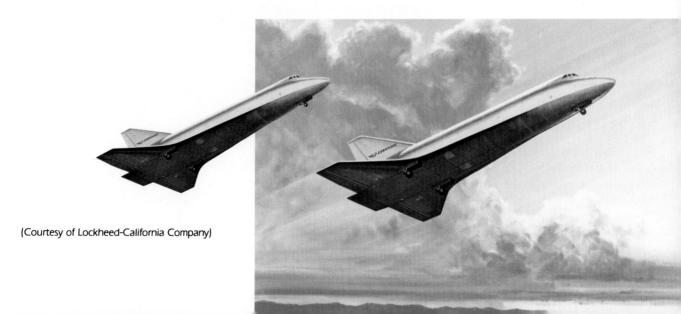

(Courtesy of Lockheed-California Company)

AIR FLIGHT

Setting the Stage

A lightweight spacecraft capable of traveling at speeds of over 4000 miles per hour is being developed. Once in orbit, this national aerospace plane would be capable of speeds in excess of 17,000 miles per hour. It would be like an airplane and a space shuttle; however, it would not have the cumbersome rocket boosters like the shuttle. Unlike an airplane, its capabilities would include travel in orbit over 300 miles above the earth and its atmosphere.

The space shuttle was designed to carry payloads of satellites, experimental equipment, and a space station into orbit. The aerospace plane is being developed as a more fuel efficient and faster means of air travel. Airplanes, the space shuttle, and the national aerospace plane all have led to the exploration and development of the new frontier—space.

Our pioneers in space, such as John Glenn, Robert Goddard, and Christa McAuliffe, once were students like you in school. They all contributed to some aspect of our bridging earth to outer space. You are the future astronauts, engineers, research scientists, and aircraft designers.

Your Challenge

Design and construct a model of an aircraft that could travel within or above the earth's atmosphere.

(Photo by Dennis Moller)

Suggested Resources

Styrofoam trays
Sketching paper
Carbon paper
Glue
Scissors
Utility or X-Acto® knife
Weight (dime or penny)

(Courtesy of NASA)

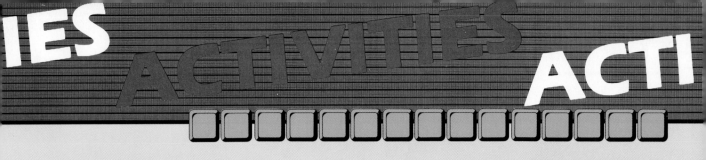
Procedure

1. After receiving proper instruction on lab safety and studying available literature on aircraft, sketch five designs for an aircraft.
2. Choose one of these and draw the parts of the aircraft to be used as the pattern.
3. Place the carbon paper onto the styrofoam tray. Place your sketch on top of the carbon paper. Trace the pattern onto the styrofoam.
4. Cut all parts and slots of the aircraft. Score areas requiring bending.
5. Assemble the aircraft. Attach weight as needed for flight.
6. With permission from the instructor, fly the aircraft.

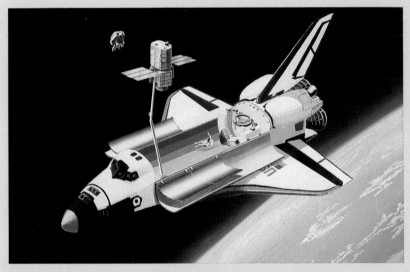

(Artist's concept by Robert W. Womack for NASA)

Technology Connections

1. What are vehicles that travel above the land and water surfaces of the earth called?
2. What is a vehicle with rocket boosters capable of traveling above the earth's atmosphere carrying large payloads called?
3. Describe what the projected national aerospace plane will be like.
4. Explain the difference among the following aircraft: airplane, space shuttle, national aerospace plane.

YOU'RE ON THE RIGHT TRACK!

Setting the Stage

Traffic jams have become more and more of a problem. Therefore, you've been asked to design a public transit system that can replace the automobile.

Your Challenge

Design and build a monorail car that can ride above a 1" wide flat rail track. The car must have a motor and be battery powered. If your teacher approves, teams of 2–3 students may work together to make one vehicle.

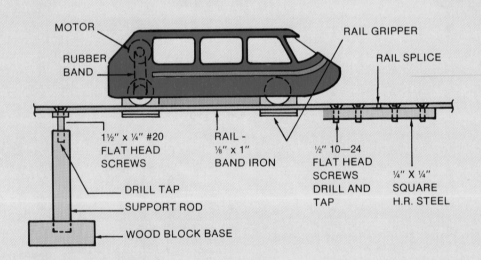

Suggested Resources

Safety glasses and lab aprons
9 volt DC electric motor with pulley
9 volt alkaline battery
9 volt battery snaps
9 volt battery holder
SPST miniature switch (optional)
Soldering pencils and solder
Band iron—⅛" × 1"
Hot rolled steel rods—½" diameter
Flat head machine screws with hex nuts—1½" × 1¼" #20
10-24 flat head machine screws—½" long
Assorted materials such as metal, wood, plastics, cardboard, fasteners, etc.
Assorted shop tools and machinery

Procedure

NOTE—If you have not been told how to use any tool or machine that you need to use in this activity, check with your teacher BEFORE going any further.

The amount of material needed for the track depends on its length. Plan accordingly.

1. Be sure to wear safety glasses and a lab coat.
2. The track can be designed and built by the entire class. It should be made from ⅛" × 1" band iron and supported by ½" diameter, 12" long steel rods. Support rods should be spaced approximately 36" apart.

3. The track can be made any length desired by joining the rail on the bottom with ¼" × ¼" square stock drilled and tapped with a 10-24 NC tap. Use ½", 10-24 flat head machine screws for this assembly.

4. Drill and tap one end of the ½" diameter support rods with a ¼" × 20 NC tap.

5. Cut blocks of wood 1½" × 6" × 9". Drill a ½" hole halfway through the center of a larger side of these base blocks. The untapped ends of the ½" support rods are inserted into these holes.

6. Assemble the track using screws and nuts. Be careful not to bend or kink the rail.

7. The first step in constructing the monorail car is to design the *drive wheel and undercarriage*. The electric motor comes with a pulley attached. You must determine how the motor will be mounted and how it will turn the drive wheel/axle. (Rubber bands are helpful. So are wing nuts for fast assembly/disassembly.)

8. Also, plan the mechanism that will keep the vehicle from falling off the track (small L-shaped pieces usually work fine.)

9. Make sketches and plans of your monorail car *before* beginning construction.

10. Consider the *weight* and *traction* of the monorail car during the design stages. If your monorail weighs too much, or if there is too much friction in the drive system, it will not run properly. Use washers and lubrication when possible.

11. The body of your monorail should be added last. It can be made from mat board, cardboard, etc. and painted as desired.

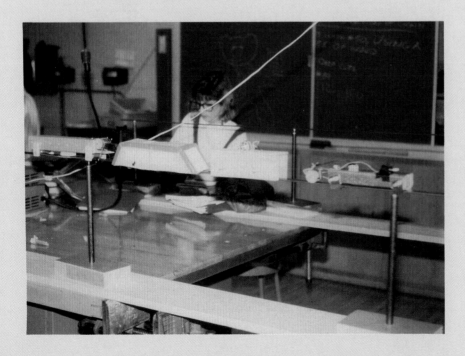

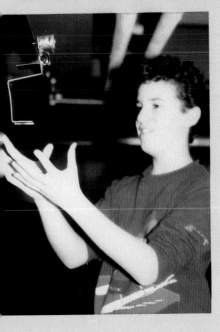

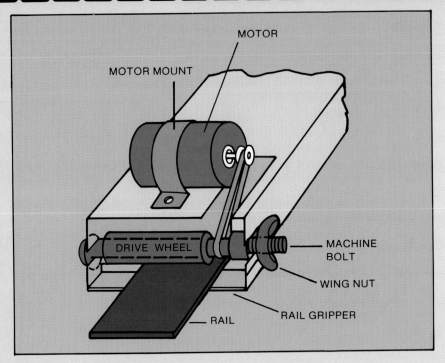

MOTOR

MOTOR MOUNT

MACHINE BOLT

WING NUT

RAIL GRIPPER

DRIVE WHEEL

RAIL

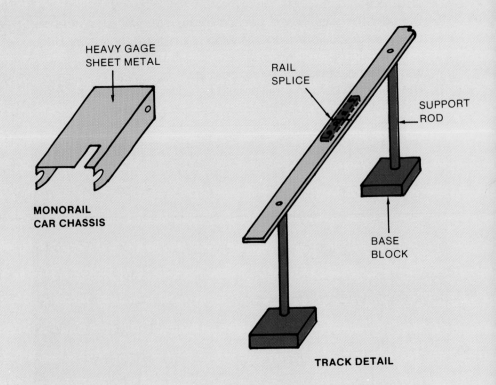

HEAVY GAGE SHEET METAL

MONORAIL CAR CHASSIS

RAIL SPLICE

SUPPORT ROD

BASE BLOCK

TRACK DETAIL

472

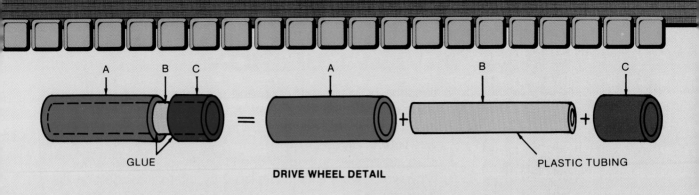

A   B   C   =   A   +   B   +   C

GLUE   PLASTIC TUBING

DRIVE WHEEL DETAIL

Technology Connections

1. Existing technological systems act together to produce new, more powerful technologies. What technologies did you use to make your monorail car?
2. High-speed monorail systems could replace the fairly slow rail system used today. What undesirable outcomes could result?
3. *Magnetic levitation (maglev)* can be used to lift a train off its track. Why does this allow the train to go faster?
4. The transportation systems of the future will make more and more use of computers for design and control. These systems will be much faster, safer, and more efficient than present ones. Super-conductors will mean even greater advances.

Science and Math Concepts

▶ *Magnetic levitation* uses the principle that similar magnetic poles repel each other.
▶ *Friction* is a force that opposes the motion of a body.
▶ An electric motor converts *electrical energy* from the battery into *mechanical energy*.
▶ *Superconductors* conduct electricity with little or no loss due to resistance.

473

REVIEW QUESTIONS

1. Describe how transportation and communication systems have made countries interdependent.
2. In what way(s) have modern transportation systems affected the way your family lives?
3. Is an automobile a necessity or a luxury in your family? Why?
4. Why were steam engines replaced by internal combustion engines in cars?
5. Why were steam engines replaced by diesel and electric engines in trains?
6. What role did trains play in the settling of the American west?
7. Why are electric engines used on railroads that travel into major cities?
8. Describe why a boat made out of steel and cement can float.
9. Describe how intermodal transportation works and what its advantages are.
10. What shapes do submarines and submersibles use? Why?
11. Describe how a wing enables a plane to fly.
12. Give at least two reasons why jet engines have replaced internal combustion engines on commercial passenger planes.
13. How is a rocket different from a jet engine?
14. Why are rocket engines used for space vehicles?
15. Name two kinds of transportation systems that don't use vehicles. Describe how they work.
16. Name a human-powered vehicle that is widely used for sport or leisure traveling in this country, but that is used as basic transportation in other countries.

KEY WORDS

Buoyancy	**Engine**	**Lift**	**Thrust**
Commute	**Intermodal**	**Piggyback**	**Transmission**
Container ship	**Internal**	**Pipeline**	**Turbo-prop**
Conveyor	**combustion**	**Propeller**	**Vehicle**
Diesel	**engine**	**Rocket**	**Weight**
Drag	**Jet**	**Steam engine**	

SEE YOUR TEACHER FOR THE CROSSTECH PUZZLE

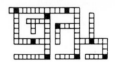

CAREERS IN ENERGY, POWER, AND TRANSPORTATION

CAREER PAGE

OCCUPATION	FORMAL EDUCATION OR TRAINING	SKILLS NEEDED	EMPLOYMENT OPPORTUNITIES
AIRCRAFT PILOT—Transports passengers, cargo, and mail. Some pilots dust crops, spread seed, test aircraft, and take photographs.	A commercial pilot's license issued by the Federal Aviation Administration (FAA) requires at least 250 hours of flight experience. Flying can be learned in military or civilian flying schools.	An understanding of flight theory and the ability to interpret data provided by instruments.	Above average. The growth in cargo and passenger traffic will create a need for more pilots, and more flight instructors.
PETROLEUM ENGINEER— Explores and drills for oil and gas. Determines the most efficient methods to recover oil and gas from petroleum deposits.	Four-year college engineering degree. Courses in energy technology, mathematics, physics, chemistry, mechanical drawing, and computers are useful.	Excellent mathematical and computer skills. Good background in chemistry.	Average. Oil and gas are becoming harder to find. More people will be needed to explore new sources, like the oceans and the polar regions.
TRUCKDRIVER—Transports goods from producer to consumer. Long-distance truckdrivers spend most of their time behind the wheel. Local truckdrivers spend much time loading and unloading.	In most states, a chauffeur's license is required. Employers prefer applicants with a good driving record. New drivers often start on small panel trucks and advance to larger trucks.	Ability to drive large vehicles in crowded areas and in highway traffic; knowledge of federal, state, and local regulations. Ability to inspect trucks and freight.	Average. However, this occupation is among the largest. The number of job openings each year will be very high.
VEHICLE MECHANIC—Repairs and maintains motor vehicles and construction equipment. Types of mechanics include aircraft, automotive and motorcycle, diesel, farm equipment, and heavy equipment.	Mechanics training is provided by the military, by private trade schools, or public vocational schools. Aircraft mechanics must be licensed by the FAA. Auto mechanics still can learn the trade by working with experienced mechanics, but formal training is becoming more important due to the complexity of new cars.	Knowledge of electronics and engine technology; knowledge of mechanical, hydraulic, pneumatic, and electrical systems. Good manual dexterity.	Above average. Rising incomes and a growing population will stimulate the demand for airline transportation. Expansion of the driving population, and more complex automotive systems will require more skilled mechanics.

Data from *Occupational Outlook Handbook, 1986-87,* U.S. Department of Labor.

SCRAMBLER

Objectives

When you have finished this activity, you should be able to :
- Identify the subsystems of a land transportation vehicle.
- Explain the role of management in an engineering project.

Concepts and Information

In your reading you have learned that there are many types of energy converters. Though some of these may be more efficient as a power source for transportation, none have proven as reliable and as easy to operate as the internal combustion engine. Because of this, it is still used to power the majority of transportation systems.

The internal combustion engine is referred to as a heat engine. It is so called this because as it burns fuel it produces hot gases. These hot gases expand and are converted into mechanical power. In most internal combustion engines, the fuel is ignited by a spark plug, but in this activity you will use an engine called a glow plug engine. Its name is derived from the glowing hot "plug" it uses to ignite the fuel. The glow plug is initially heated with a battery connected to it. Once the engine is operating, the burning gases from the combustion cycle continuously heat the glow plug.

Equipment and Supplies

NOTE: The following equipment and supplies list contains just some of the materials your group may use to build your vehicle. Except for the chassis, you may use whatever is available, provided it is safe. Do not limit yourself only to the materials suggested here.

Chassis
Plywood — ¼" thick
2" × 4" scraps
Glue
Wood screws

Wheels
Hardwood — ¾" × 2¼"
Purchased rubber wheels

Body
Wood
Fiberglass supplies
Vacuum forming sheets
Paper and wire

Miscellaneous
Balsa wood
Band iron
Sheet metal
Acrylic
Miscellaneous bolts and nuts
Screws
Pop rivets and riveter
Copper tubing
Clay
Brass welding rod
Plaster
Springs
Swivels and pulleys

Activity

The internal combustion engine is only one of several subsystems in a vehicle. Steering, lighting, suspension, brakes, and heating are examples of other subsystems in a transportation vehicle. As a group, your task is to form a design team which will design and construct a vehicle powered by a .049 two-cycle engine. This vehicle will contain steering and suspension, guidance, body, and chassis/engine mount subsystems. The body may be tested in a wind tunnel before final construction. A management team will oversee the design and construction of the vehicle. The vehicle will then be raced on a 12' radius track in competition with vehicles made by the other technology classes. A well-designed vehicle will achieve speeds of 35 mph or more.

Procedure

1. Your instructor will divide each group into the following four teams.
   1. **Management Team**
   2. **Steering and Suspension Subsystem Team**
   3. **Chassis and Engine Mount Subsystem Team**
   4. **Body Subsystem Team**
2. Read the following descriptions of each team's responsibilities so that you have a clear understanding of each team's responsibilities:

   Management team — Your team will be responsible for overseeing the design and construction of the vehicle. You will coordinate the activities of the other three teams and monitor their progress. For effective monitoring, you may want to assign a consultant/manager to each of the other three teams. You must be able to foresee and prevent any interference one subsystem may have with another subsystem. To do this, you must continuously take careful measurements to create a master drawing of the vehicle.

   You will also be the track manager. You are responsible for making sure the guidance subsystem is constructed and assembled. The guidance subsystem must guide the glow plug vehicle in a circular path 24 feet in diameter. This must be done by using a tether which will connect the vehicle to a central pivot. You will also assign the track personnel.

   Above all, you must be aware of all safety rules and help the teacher enforce them among the other teams.

Equipment and Supplies

Power System
.049 COX Black Widow glow plug engine
Glow plug fuel
1.5-volt glow plug power supply

Safety Equipment for Operating the Engine and Vehicle
Safety glasses
Leather gloves (2 pairs)
Large soft cloths for stopping the engine
Nylon utility line (165 lb. test) for Tether block — 12' long
Fishing swivel
Blacktop or concrete running surface — 30' in diameter

477

Steering and Suspension Subsystem Team — Your team will have the following major responsibilities. First, you will design a steering system for the front axle and wheels. Second, a stationary mounting system for the rear wheels of the vehicle must be designed. Third, you must design a method for attaching the tether line to, and detaching the tether line from, the chassis of the vehicle. Be sure to inform the body subsystem team how you will do this since it may affect their design.

Your team is also responsible for designing and constructing a suspension system for the glow plug vehicle. The smoother the travel of the vehicle, the faster it will go. Therefore, the suspension system you design must be able to absorb the shocks caused by any bumps or dips in the track.

Chassis and Engine Mount Subsystem Team — Your team will be responsible for constructing the chassis and engine mount. Refer to the chassis drawing for specific guidance as to how to build the chassis.

Body Subsystem Team — Your team is responsible for designing and constructing a body for the glow plug vehicle. One of the most important factors in determining the speed of a car is drag. **Drag** is the friction caused by the air as it flows over the body of a vehicle. You need to design the body so that it is smooth and sleek, and you may want to make clay prototypes to test in a wind tunnel. Be sure to engineer a method of attaching the body to, and detaching the body from, the chassis of the vehicle. You may paint or decorate the body of the vehicle in an attractive way.

3. Discuss with the rest of your class the design limitations listed below. You may want to add to these.
   a. **Chassis** — To construct the chassis, follow the design outlined in the chassis drawing. The chassis may be drilled and/or cut, but not in such a way as to alter its outside dimensions.

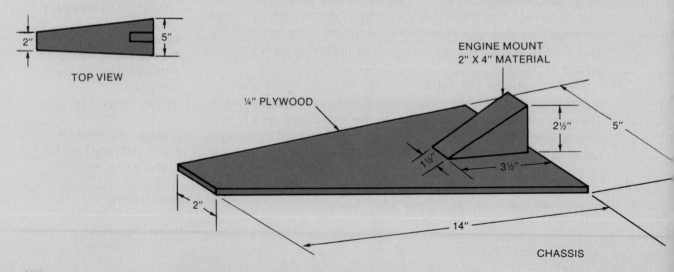

TOP VIEW

ENGINE MOUNT
2" X 4" MATERIAL

¼" PLYWOOD

2½"   5"

1½"

3½"

2"

14"

CHASSIS

b. **Power Plant** — .049 COX Black Widow glow plug engine. Must be mounted so that the propeller maintains one inch minimum ground clearance. Also, the engine should be mounted at a 5° angle. This will help the vehicle steer to the outside of the circle until centrifugal force takes over.

c. **Wheels** — Wood (turned on a machine lathe), rubber, or hard plastic (purchased).

d. **Body** — Material such as:
   - Fiberglass, drape-form molded over clay, paper-mache, or plastic. Use paraffin wax for a release agent.
   - Vacuum-formed plastic
   - Newspaper soaked in wallpaper paste applied to a frame of wood and/or wire
   - Carve the body out of balsa wood

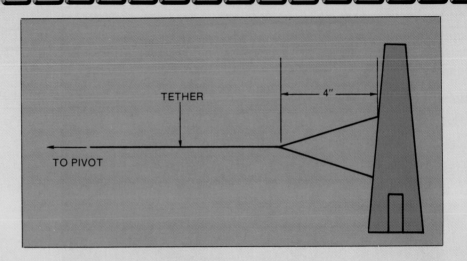

TETHER

4"

TO PIVOT

e. **Tether** — Must end in a "V" shape four inches in length, so as to attach to the vehicle chassis in two places. (See tether drawing.)

f. **Central Pivot** — Must allow the tether to turn freely without wrapping. The height of the tether where it attaches to the pivot must be no more than 8 inches.

4. After all teams have completed their design work, construct the vehicle. Take it outside to test and race it.

5. Once outside, test the steering subsystem. To do this, hold a piece of chalk at one end of the tether line. Draw a half circle on the pavement by keeping the tether line taut. Push the car to see if it will track the chalkline. Make any necessary steering adjustments.

6. Test run the car. For safety, reverse the propeller for any test run. This will reduce the maximum speed the vehicle will travel. The following personnel are needed for the test run: head starter and fuel adjuster; holder of car; safety inspector (each person should wear safety glasses and leather gloves).

7. Race the car. The following personnel are needed for the race: head starter and fuel adjuster; person who holds and releases the car; safety inspector; three track supervisors (each carries a large cloth to stop cars not tracking properly); timer track recorder; spectator supervisor.

Review Questions

1. Use the following formula to calculate how fast your vehicle traveled.

   Circumference of a circle $= 2\pi r$ ($r =$ radius of a circle; $\pi = 3.14$)

   Velocity $= \dfrac{\text{Distance traveled in feet}}{\text{Time (in seconds)}}$

2. How do local speedways compensate for centrifugal force that is developed on ¼-mile and ½-mile oval tracks.
3. Change the weight of your vehicle and time it again. How much does weight affect the speed?
4. Test the speed of the vehicle with the body removed. Does the speed change? (Remember, you are also reducing the weight of the vehicle when you remove the body.)
5. Tape a white piece of paper to the wall. Start the glow plug engine and let it run with the exhaust aimed at the paper. What is the result? Why does this happen?

SECTION 5

CONCLUSION: LOOKING INTO THE FUTURE

CHAPTER 16

IMPACTS FOR
TODAY AND
TOMORROW

MAJOR CONCEPTS

After reading this chapter, you will know that:

- Outputs of a technological system can be desired, undesired, expected, or unexpected.
- People determine whether technology is good or bad by the way they use it.
- Technology produces many positive outputs and solves many problems. Sometimes, however, negative outputs create new problems.
- Technology must be fitted to human needs.
- Technology must be adapted to the environment.
- Existing technological systems will act together to produce new, more powerful technologies.
- Using futuring techniques, people can anticipate the consequences of a new technology.

INTRODUCTION

In the old days, travel across country took weeks or even months. As recently as the mid-nineteenth century, the pony express was the quickest way of delivering mail. People had to make many products by hand. Today, modern production technology provides store-bought goods. Supersonic jets take us across the ocean in hours. Communication technology moves information almost at the speed of light.

There is, however, another side to technology. In December 1984, poison gas leaked from a Union Carbide chemical plant in Bhopal, India. The gas killed more than 2,000 people. In January 1986, the space shuttle Challenger exploded seconds after launch. All seven astronauts were killed, among them schoolteacher Christa McAuliffe. In April 1986, an accident occurred at a nuclear power plant in Chernobyl, near Kiev. People died from the high levels of radiation emitted by radioactive iodine and cesium.

Some people argue that we have lost control over our technology. They blame technology for tragedies like the explosion of the Challenger. Some say that we should stop all nuclear research because it brings on the hazards of radiation, the problems of nuclear waste disposal, and the threat of nuclear war. Others blame technology for pollution, since cars and industrial-processing plants pollute the air and water. Technology is also blamed for noise and for crowded cities and highways.

Is technology to blame for tragedies like Bhopal, Challenger, and Chernobyl? Is technology good or evil? Would we want to go back to a life style with less dependence on technology? Indeed, could we? What effects will technology have on our world in the future? These questions are the subject of this chapter.

The Challenger crew included (clockwise from top left) Ellison S. Onizuka, Sharon Christa McAuliffe, Gregory Jarvis, Judith Resnick, Ronald McNair, Francis R. Scobee, and Michael J. Smith. (Courtesy of NASA)

IMPACTS OF TECHNOLOGY

Outputs of a technological system can be desired, undesired, expected, or unexpected.

There are four possible combinations of outputs from technological systems: expected and desirable, expected and undesirable, unexpected and desirable, and unexpected and undesirable.

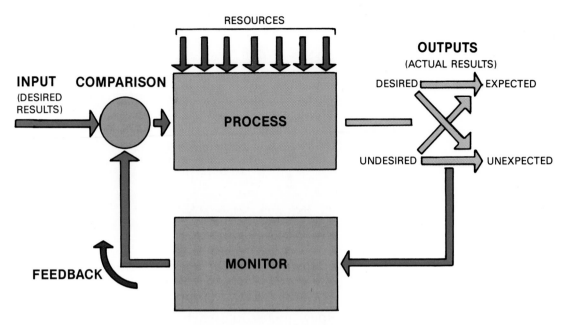

Technological systems create four possible kinds of outputs.

It is the unexpected and undesirable output that is of greatest concern to humans. If we cannot plan for an event, we may not have the necessary resources to deal with it.

The nuclear accident at Chernobyl was unexpected and undesirable. It took the Soviets several days to evacuate people near the plant. They were unprepared for such a disaster. We may not know all the effects of the accident for a long time, since doses of radiation can cause cancer deaths over many years. Likewise, the unexpected, undesirable poison gas leak in Bhopal resulted in injury and death. Firms must anticipate the negative effects of technology. They must provide healthy and safe work environments.

Technology is neither good nor evil. It can be applied to our benefit, or to our destruction. For example, nuclear power can help provide the solution to our energy needs. It can also cause radiation poisoning. Nuclear technology, like all technology, must be used wisely.

People determine whether technology is good or bad by the way they use it.

People can control technology. In the United States, the Nuclear Regulatory Commission (NRC) sets standards for construction and operation of nuclear power plants. To be approved by the NRC, a nuclear power plant must have a 9-inch-thick steel shell and a 3-foot-thick concrete structure around the nuclear core. The Chernobyl reactor was built without any protection around the radioactive core. Of course, a nuclear accident could occur even in the best-

This automated control room in a nuclear power plant is watched over by human workers. (*Courtesy of Rochester Gas & Electric Co.*)

designed plants. However, proper use of technology lowers the risks.

Today, **pollution** is a serious worldwide issue. Millions of motor vehicles spew gases like carbon monoxide into the atmosphere. These gases pollute the air we breathe. Some industries have been carelessly dumping their industrial wastes into rivers and oceans. These wastes have polluted our water supply.

It is not the technology that is harmful. It is misuse of the technology that causes problems. Just as technology can create a problem, it can provide the solution. Automobile pollution has been lessened by the development of pollution control devices for cars. Some polluted rivers have been cleaned up. Smoke from industrial plants can be cleaned with scrubbers.

Technology produces many positive outputs and solves many problems. Sometimes, however, negative outputs create new problems.

Should We Go Back to Nature?

If we went back to a simpler life, would problems like pollution disappear? Probably not. Pollution has been a problem since Roman times. The rivers around Rome were so full of human and animal waste products that people were forbidden to bathe in the waters. One hundred years ago, before there were automobiles, 150,000 horses in New York City produced over 500,000 tons of manure each year.

Our technological society faces problems other than pollution. For one thing, our environment is noisy. Those who live near airports hear jet planes taking off and landing. We ride on subways that screech. Car horns and tires generate noises that not only disturb us, but also create stress. Factory workers are bothered by the noise of machines. Even loud music is a threat to our hearing.

In earlier days, however, life was much more difficult. The factories were noisy, and children often worked in them

Sanitation man sweeping manure in New York City (*Courtesy of the Museum of the City of New York*)

for fourteen or more hours a day. Medical care was crude. Many people lived with poor sanitary conditions. There were no noisy airports, but the trains that ran through the cities made so much noise that buildings trembled. Although our society has technological problems, earlier societies did also.

We have become very dependent upon technology. Our lives and our routines are influenced by technological products and services. We could not go back to a life without automobiles, telephones, flush toilets, and modern medical care. Technology is here to stay. But it is up to us to determine how technology will be applied.

People can influence the development of technology. They can pass laws that promote or restrict a particular technology. They can use a technology for peaceful purposes or for destructive purposes. The use of technology is a human decision.

Matching Technology to the Individual

Technology can be used to make our lives more comfortable. Products, for example, can be engineered with human comfort in mind. The field of **ergonomics** or **human factors engineering** applies technology to the physical needs of human beings. Chairs fit the shape of the back. Drinking glasses are easy to hold. Tables are just the right height for us to eat without stretching or bending over. Auto seats provide a good driving position.

Computer keyboards are good examples of ergonomic design. Before one company produced its keyboards, it hired hundreds of secretaries to try out various keyboard designs. The secretaries typed on the keys and provided feedback about how comfortable the typing felt. The shape of the keys, the pressure needed to push them down, and the way they sprang back were things that determined how "user friendly" the keyboard was.

Products must be matched to human needs. Making products safe is another example of good ergonomic design. When kitchen appliances like food processors are designed, the cutters and blades must be guarded properly. When automobiles are designed, the safety of the driver and the passengers must be considered.

People's special needs can be satisfied by ergonomic design. Vehicles and sporting equipment can be made for people with disabilities. There are bicycles and skis designed for people who do not have the use of their legs.

Technology must be fitted to human needs.

Handbikes and Sunbursts

Bicycling is like walking: people must move continuously or they will begin to tip over. Bicycling, therefore, can help people learn good balance. It is also an excellent form of exercise. Some people, however, cannot ride a standard, foot-powered bicycle. They are left behind while others enjoy bicycle riding.

An arm-powered bicycle has been designed at the Veterans Administration Rehabilitation Research and Development Center in Palo Alto, California. The Handbike, as it is called, comes in adult, child, and racing versions. The bike combines the latest ideas from the design of standard bicycles, human-powered vehicles, and sport wheelchairs.

The Sunburst Tandem is a natural spin-off from the Handbike. The front rider pedals with any combination of arms or legs. The back rider steers, sitting high enough to see over the front rider. The Handbike and Sunburst are examples of how technology can help the physically challenged to do activities that would otherwise not be possible.

Macy Tackitt on her Handbike. Macy was partly paralyzed as a result of an auto accident. (Courtesy of VA Rehabilitation R & D Center, Palo Alto, California)

Designer Douglas Schwandt, a biomedical engineer, and friend Dianna Gubber riding the Sunburst (Courtesy of VA Rehabilitation R & D Center, Palo Alto, California)

Matching Technology to the Environment

Throughout history, people have used technology in ways that could harm the environment. Entire forests were cut down for fuel in the 1600s. Garbage was burned in open fires. Water was polluted by waste materials. In the past, however, these technological events occurred on a small scale. In modern times we use technology on a much larger scale. If not carefully managed, our modern technology could harm the world environment.

For example, we burn over 700 million tons of coal each year to provide energy. Coal-fired plants produce pollutants like **sulphur dioxide.** In addition, automobile exhaust contains chemicals like carbon monoxide. In the United States alone, nearly 170 million automobiles, buses, and trucks burn about 125 billion gallons of fuel each year. In the process, the vehicles produce vast quantities of pollutants.

Chemicals given off by automobiles and industrial plants can rise into the atmosphere and return to the earth as acid rain. **Acid rain** has been responsible for killing trees and forests in some parts of the United States, Canada, and Scandinavia. It has also killed fish in freshwater lakes. In New York State, an estimated 200 lakes in the Adirondack region can no longer support fish life. Hundreds more are in danger of losing their fish populations.

Over the last thirty years, new chemicals have been developed for agricultural use. These chemicals control pests and plant disease. Such chemicals, however, may remain in the soil and air. In some cases these chemicals can get into the food we eat and the water we drink.

Some industrial plants manufacture chemicals that may be harmful to humans. Concerned companies do **technology assessment studies.** For example, laboratory animals are sometimes exposed to the chemicals to see if negative effects occur. Such studies determine the effect the chemicals might have on humans. When negative effects are not considered, unexpected and undesired outputs may result. In Niagara Falls, New York, chemicals dumped by local industries caused a high rate of illness and death. About 800 families living in an area called Love Canal had to leave their homes.

Our environment must be treated with care. Resources are available to us only in limited supply. We must use them wisely and seek alternatives for scarce resources.

We can also develop technologies that harmonize with the environment. Solar energy can satisfy energy demand without posing a danger to the environment. Cars with electric motors are nonpolluting. Proper methods of farming can prevent land from eroding. As inhabitants of the earth, we can support an appropriate use of technology.

Today, federal laws like the Clean Air Act ban industries from polluting the environment. (Courtesy of Carnegie Library of Pittsburgh—Pittsburgh Photographic Library)

Technology must be adapted to the environment.

FUTURE OUTLOOKS IN TECHNOLOGY

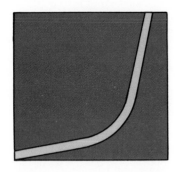

Technology is growing at an exponential (ever-increasing) rate.

What is in store for us in the future? How will technology affect the way we live? What will our cities look like? What kind of food will we be eating? Will we find cures for cancer and heart disease?

Stop and think about recent technological changes. Twenty-five years ago there were no video games, pocket calculators, home computers, cassette tapes, or industrial robots. Artificial hearts had never been used. Genetic engineering was undeveloped. Only within the last decade have we had space shuttles, test-tube babies, the Walkman®, pocket TVs, fully automated factories, worldwide satellite communication, and digital watches.

Remember that technology is growing exponentially (at an ever-increasing rate). Tremendous changes have occurred in the last few years. Can we even dream what kinds of changes might take place in the next twenty-five years? You have studied various technological systems while reading this book. Let's go to the future and see what these systems may be like.

COMMUNICATION IN THE FUTURE

As we move into the future, we will be able to communicate not only with each other, but with machines that can understand our voice and respond. Already, computers can speak like humans. The technology that enables them to make sounds like the human voice is called **speech synthesis.**

You may have come across a similar term related to music. Synthesized music is music produced by electrical circuits rather than by acoustic instruments. Telephone companies are already using speech synthesis. When you call directory assistance and ask for someone's telephone number, a human operator answers the phone. After you request the number, the operator switches you to a computerized device. The device provides the telephone number. The voice that you hear is a computer-generated voice. A voice synthesizer produces the telephone number electronically.

Soon **voice recognition technology** will be widespread. Computers will be able to recognize human voices. At that point, the telephone operator may not be necessary at all.

In the future, secretaries will probably not need typing skills. Typewriters will become intelligent and respond to voice commands. You will be able to speak to a typewriter, and it will print your words directly. Voice recognition technology will aid the disabled. Computers, robots, and wheelchairs will be controlled by voice commands. Already, a machine for the blind has been developed that can scan a printed page and read the words aloud.

This typewriter can recognize 1,000 spoken words. (Courtesy of Kurzweil Applied Intelligence, Inc.)

Communication at Home

Communicating with machines will help us do many things better. Pet robots will fetch a glass of water or shine shoes upon verbal command. Intelligent robots will do our housework and watch children or the elderly. Robots will continue to assume human qualities. Who will be the first person to marry a robot?

WABOT II, The Keyboard-Playing Robot

In the future, robots will not only be factory workers, but will provide personal services for people. Service robots will look and act more like people.

WABOT II is an intelligent robot musician. It can converse. It can read sheet music and perform on an electronic organ, using both hands and legs together. The robot was constructed by a research group at Waseda University in Tokyo, Japan, under the direction of Professor Ichiro Kato.

The fingers of both hands strike the keyboards of the organ. The feet work the bass keyboards and the pedals. A video camera recognizes the printed musical score. WABOT II can also accompany a singer. It plays in tune with the singer's pitch.

(Courtesy of Ichiro Kato)

We will be able to communicate with appliances in the house. We will be able to tell our lamps to turn on and our televisions to turn off. Our telephones will talk to us. When we come home, the telephone will tell us who called, at what time, and what the message was.

Today, we have Walkman® radios and Watchman® TVs. In the future, we may also have the Talkman. The Talkman will be a telephone you can carry in your pocket. Devices like this exist now. They are called cellular telephones and can be used from the car. Such devices will become more widespread in the future as the technology becomes less expensive.

Telephones will also become much more powerful. Not only will we be able to dial up a friend, but we will be able to see the friend in full color and in three dimensions. Holographic telephones will use laser beams to create 3-D images. As space travel becomes more common, we will use holographic telephones to visit with friends in space colonies.

Communication technology will affect the way our cameras work. Instead of film, cameras will use videotape cassettes no larger than audio cassette tapes. Cameras will become miniature videocameras. You will be able to see the pictures on a home TV monitor. A specially designed printer will also be able to print the picture in color.

Our television sets will be bigger and better in the future. With a new technology called HDTV (high-definition television), we will be able to see clearer, sharper pictures. If you look very closely at a TV picture, you will notice very fine horizontal lines across the screen. These lines are called **scanning lines.** Presently, 525 scanning lines make up our TV picture. HDTV will use 1,125 scanning lines. It will provide us with much more detail.

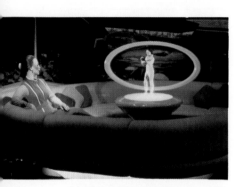

Holographic telephones will create 3-D images. (Courtesy of General Electric Company)

Communication at Work

We will continue to depend upon satellites for worldwide communication in the future. Satellites are already used to transmit telephone conversations, sporting events, and military information. In the future, they will be used to help people commute by television. Some people are already holding face-to-face meetings by satellite with people in other parts of the world.

This technology (**videoconferencing**) will become more widespread. Using television signals beamed from one part of the earth to another via satellite, we will be able to hold videoconferences with our friends. Videoconferencing will allow people to transmit voice and pictures. An electronic blackboard will permit a participant to draw diagrams. These diagrams will be transmitted electronically to people in another country. The drawings will appear on a screen in the conference room there.

It may be possible to implant a microchip into the brain. We may then be able to connect humans directly up to huge computer data bases. Perhaps we may be able to record our thought patterns and preserve them for generations to come.

Some people are afraid that computers will be used to monitor job performance. With chips implanted in the brain, it might be possible to monitor an employee's thoughts. Employers might also be able to control the emotions of their workers by sending signals to the device.

Would microchip implants allow employers to produce worker-slaves? Would the location of these worker-slaves always be known? Would they be totally dependent upon their employers? Already some unions like the Newspaper Guild, the United Autoworkers, and the Communication Workers of America have insisted on antimonitoring clauses in their contracts with employers.

Only uses of technology that enhance the quality of life should be encouraged. Computers, for example, have made telephone operators more productive. However, the operators report that they are now expected to answer more calls per minute than ever before. This has made their jobs more stressful.

Communication and the Consumer

Computers connected to telephone lines will change the way we shop. From your own home computer, you will be able to search through store catalogs and order merchandise without leaving home. You will be able to pay bills and transfer money from a checking account to a savings account. You will be able to reserve a flight and buy airline tickets. You will even be able to take a tour of another city through videodisk images. You will decide where you'd like to visit, in which hotel you'd like to stay, and what sights you'd like to see.

Computer and television technology will change even the way we get our hair cut and buy clothes. Suppose you think you'd like a certain hair style, but you're not sure it will look

Personal computers can communicate with mainframe computers over telephone lines. (Courtesy of AT&T, Bell Laboratories)

This woman is using a computerized mirror to "try on" a new outfit. (Courtesy of L. S. Ayres)

One of the new "smart cards" (Courtesy of Motorola, Inc.)

good on you. Your hair stylist will be able to generate a computer image of the hairstyle on a TV screen. The image will show you exactly how you would look. In the same manner, you will be able to try on different outfits by looking into a computerized mirror.

Computers and credit cards could turn our country into a cashless society. Credit cards will have miniature computers built into them. The cards will be used to purchase groceries, to buy household items, and to withdraw money from automated banking machines.

These **smart cards** have an integrated circuit memory chip within the card. The chip keeps track of the consumer's bank account balance. Electrical contacts in a card reader connect to the contacts on the card. After the card is read, the chip's memory is changed to reflect the new bank balance. Smart cards will be used to pay for public telephone calls and pay TV. The cards will also store vital medical information such as drug allergies, blood type, and medical insurance data.

MANUFACTURING IN THE FUTURE

Robots will continue to take the place of people in industry. Because of competition from developing nations, companies in advanced industrial societies like the United States will not be able to afford to pay human workers when robots can do the job instead. More of our factories will be virtually humanless.

At present, assembling products requires human skill. If parts don't fit together, skilled human workers can try to put them together another way. Robots are used to weld parts, paint them, machine them, and transfer them from one location to another.

In the future, robots will become more intelligent. They will be able to see and feel. They will therefore be able to assemble products as well as make the parts. When this happens, more people will lose their factory jobs. These people will need to be retrained. Factories using human labor will move to less-developed nations, where labor costs are low.

Workers and employers will have to work together to plan retraining programs. New kinds of arrangements may permit people to work fewer hours. One hundred years ago, textile workers worked 75 hours per week. Today, the average worker works 42 hours per week. Futurists forecast that fifty years from now, the work week will be reduced to 20 hours. Many more jobs will be available at "mother's hours" (9:00 a.m.—3:00 p.m.), or "student's hours" (3:00 p.m.—6:00 p.m.).

As robots become more intelligent workers, will they have their own unions? Will groups organize to protect the rights of robots? Will turning off the power to a robot be considered murder?

Manufacturing in Space

In the future, manufacturing will be done in space as well as on the earth. Manufacturing in space has some advantages over manufacturing on earth. Some materials, for example, can be processed more easily in a low-gravity environment.

Gravity tends to pull the molecules of materials toward the earth, making it difficult to separate and purify the materials. Suppose you want to separate certain biological cells to develop a new cure for a disease. On earth, because of gravity, only a tiny bit of the cells could be extracted at one time.

There is almost no gravity in space. Therefore, materials in space can be given an electrical charge and be "pulled apart" by an electric field. The pure materials can then be collected in separation trays. In space, it is possible to purify 700 times as much of certain materials as on earth.

To make electronic chips for some integrated circuits, highly purified forms of a material called galium arsenide are needed. Presently, a high percentage of earth-made galium arsenide chips are rejected because of impurities. In space, very pure crystals of galium arsenide can be grown.

Optical glass used for microscope lenses must be very, very pure. However, to make glass on earth, materials like sand and limestone must be heated to very high temperatures (about 3100°F). At these high temperatures, the molten glass is very corrosive. It will chemically attack whatever container is being used to melt it. Little bits of the container dissolve and flow into the glass. The glass, therefore, is no longer pure.

In space, glass can be melted without containers. Since there is no gravity, the molten glass can be contained by sound waves. These sound waves hold the molten glass together in a clump. Because there is no container, extremely pure optical glass can be produced.

This symbolic photograph shows a group of pure, needle-shaped urea crystals The crystals were grown aboard the orbiter "Discovery" in a gravity-free environment. (Courtesy of NASA)

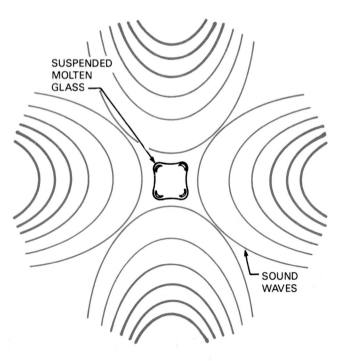

SUSPENDED MOLTEN GLASS

SOUND WAVES

Acoustic waves (sound) hold molten glass during gravity-free processing in space.

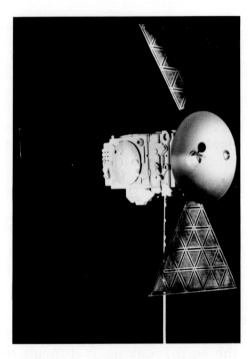

Model of a space factory (Courtesy of NASA)

Resources are often chosen on the basis of their cost. It costs a great deal to produce products in space. We make cost-benefit trade-offs when we decide to produce very pure optical glass in space at very high cost or less pure glass on earth at much lower cost.

The first Industrial Space Facility (ISF) will be launched by a space shuttle in the 1990s. It will serve as a free-flying platform to develop new kinds of metal alloys. It will also be used to grow pure crystals for use in computers and other electronic devices. As space-related businesses increase, new jobs will be created. NASA already has agreements with over twenty companies for commercial activities in space.

To take advantage of zero-gravity material processing, factories will be established in space. These factories will use minerals available on the moon and nearby planets. They will use solar energy for power. The factories will be visited by humans. Most of the time, however, they will operate automatically. A factory could even be built right on the moon. Since the moon's soil is rich in minerals like aluminum, silicon, and titanium, the factory could use natural moon materials to build other factories.

CONSTRUCTION IN THE FUTURE

To support new space-based industries, space construction will become a big business. Orbital space stations will serve as bases for humans, who will come into space to service satellites and space factories.

As space industry grows, space stations will get larger. They will need to provide services for more people. Soon space cities will appear. People will live in space for months or even years at a time.

One design for a space city shows a large cylinder that spins slowly about its axis. The spinning creates artificial gravity. As you look up from where you are living in the city, people and structures on the other side of the cylinder would appear to be upside down. The views of the earth, however, will be spectacular.

On earth, more of our homes will be produced in factories and transported to the construction site for assembly. It now takes five people two days to frame a house built "stick by stick." A factory manufactured house can be assembled on site by two people in two days.

As energy costs continue to rise, communities of energy-efficient homes will be built. One such community is Arcosanti in the Arizona desert. Arcosanti is a town for 5,000 people. The town is concentrated in a small area and contains high-rise apartment buildings.

Arcosanti is designed to be a complete community. It includes residential, business, and office spaces. Spaces have also been built for parks, recreation, performances, and small meetings. The apartment buildings surround a shopping mall. From the apartments, residents can look

A futuristic space station (Courtesy of Boeing Aerospace Company)

inside at the shops and people, or outside toward a beautiful wilderness. Inside the community, there are no roads and no cars. People are expected to walk and interact with each other. Greenhouses allow residents to grow their own food year round. Garbage is converted into an energy source. Water is recycled and purified. The architect, Paolo Soleri, said, "The need to plan for tomorrow is a part of all life."

Our homes and buildings will benefit from modern technology. Built-in computers will control many household systems. We will live in homes with bathrooms where you can select a shower, a spring rain, or a warm breeze. Bedrooms will come equipped with voice-activated audio and video systems. Kitchens will include computer-controlled appliances. Soon we will live in **smart houses** and work in **intelligent buildings.** Future houses will be designed so that residents can both live and work at home.

Intelligent office buildings will be wired with data communications lines as well as electrical wiring. Each building will be connected to a central computer that will tie together all the offices. People in one office will be able to communicate with people in other offices via a desk-top terminal. The buildings will include computerized energy management systems, safety and security systems, word-processing centers, long-distance data communication systems, and plain old telephone service systems (POTS).

A city of the future, complete with parking for your personal hovercraft. (Courtesy of General Electric Company)

The community of Arcosanti, located in the Arizona desert (Courtesy of Ivan Pintar)

TRANSPORTATION IN THE FUTURE

The speed of transportation has increased exponentially over the years. In 1889 Nellie Bly traveled around the world. It took 72 days, 6 hours and 11 minutes to do so by coach, ship, train, and camel. In 1977, a Boeing 747 made the round trip in just over 54 hours. It now takes 3½ hours to travel from New York City to Paris in the Concorde. The Concorde is a supersonic aircraft that flies at a speed of over 1,000 miles per hour.

Air Transportation

New means of travel and transport will be developed to meet the needs of people in the twenty-first century. One such new vehicle is the LTA (lighter-than-air) airship. This vehicle is designed to take off and land vertically. It will also be able to carry very heavy loads. LTAs have been made possible by earlier technological developments, such as balloons and dirigibles.

The Magnus LTA will be able to carry over 50 tons, more than the largest helicopters, and do it for much less cost. The airship will be made of a sphere of nylon filled with helium. A stingray-shaped cabin will carry two engines, which will be mounted on the wing tips. The engines will provide thrust for takeoff. They will also cause the huge sphere to spin. When a sphere spins in air, forces cause it to rise. This is known as the Magnus effect, discovered by German physicist H. G. Magnus in 1853. The Magnus effect causes a spinning baseball to rise as it nears the plate. The Magnus LTA will be about 200 feet in diameter. It will travel at a maximum speed of 70 miles per hour. It will be able to travel 500 miles while carrying its payload.

Before long, a transatmospheric aircraft will be built for military use. It will cross the atmosphere and then travel in a low earth orbit. The aircraft will blast off like a rocket but travel like an airplane. In less than two hours, it will come down on the other side of the world.

The lighter-than-air Magnus
(Courtesy of Magnus Aerospace Corp.)

A transatmospheric aircraft
(Courtesy of Lockheed-California Company)

Space Transportation

Travel has always fascinated people. From the early days of sailing ships, travel has offered us the means to explore new places. We continue to dream about far-off places, but now those places are in space.

Future travel will take us far from earth. Already travel agencies are selling tickets for rides on the space shuttle. The tickets cost $50,000. The first flight was originally planned for 1992 but most likely will not occur that soon.

Not long after this space shuttle flight, tourists will be going on space safaris. They will visit the moon and nearby planets. They will be coming home with the most wonderful photographs and home videos.

Personal space vehicles will carry people to and from space stations to visit friends and pick up supplies. Space service stations will be constructed to repair all kinds of interplanetary spacecraft.

Ground Transportation

On the ground, vehicles will become faster and more energy efficient. Future railroads will make use of **magnetic levitation. Levitation** means floating above the ground. Magnetic levitation (maglev) trains will use very strong magnets to repel the train from its track. This will create an air space that will greatly reduce the friction of the train on the track. The train will ride just high enough above the track to avoid touching it.

Automobiles of the future will continue to be used for personal transportation. They will be sleeker, safer, and more efficient vehicles. They will get excellent gas mileage because of body shapes that reduce wind resistance. Cars will also be much more electronic. Many of the systems will be computer controlled. The systems will talk to us. The voice of the computer will tell us that the lights are on or ask who we are before opening the door. Your voice-print will be sensed by the computer. If you are an authorized driver, the door will unlock.

In addition, new electronic systems will be added. A sonar system will detect obstacles behind the car as the car backs up. A new four-wheel steering system will let the driver turn all four wheels toward the curb to make parking easier. Doze alerts on the steering wheel will sound an alarm if you get sleepy and relax your grip on the wheel. Electronic windshield wipers will start automatically as soon as rain falls on the windshield. Electronic navigating systems will show a car's present location on a video screen map display. It will also indicate the car's final destination and the best way to get there.

Cars of the future will also run on energy sources other than gasoline. Electric cars will become popular. These will run on large banks of batteries. Each battery bank will

A maglev train (Courtesy of General Motors Corporation)

This satellite navigational system is combined with a laser disk map that indicates the car's position. (Courtesy of Chrysler Motors)

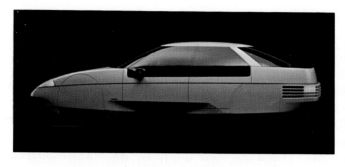

The Aero 2000 has a surface with very low wind resistance. It can travel about 71 miles on a gallon of gas and has an experimental turbo-charged, 68-horsepower engine. (Courtesy of Ford Motor Company)

A PRT system (Courtesy of West Virginia University)

provide several hundred miles of driving. When the batteries run down, the driver will pull into a service station, the batteries will then be exchanged for a fully charged set.

Future highways will have computerized systems that will send signals to automatic cars. The driver will simply shift the car into "autopilot" position. The car will then cruise safely down the road until the exit for which it was programmed. Once back on local roads, the driver will again take control.

Personal rapid transport vehicles **(PRTs)** will become quite popular as we move toward the twenty-first century. PRTs are automated vehicles that run on tracks or guideways. PRTs are built either above or below the ground. People will be able to use PRTs for transportation and not have to worry about traffic on the roads.

BIOTECHNICAL SYSTEMS IN THE FUTURE

Many of the greatest developments of the twenty-first century will come from biotechnology. The diseases that plague us in the twentieth century may be curable within the next few decades. People will also take better care of themselves in the future. Presently, many people willingly take in poisons like tobacco smoke. Smokers have more than five times the risk of dying from lung cancer and twice the risk of dying from heart disease than do nonsmokers. Because of better public awareness and new drugs like interferon, heart disease and cancer will become very rare.

As we move into the future, we may find the cause of old age. Hormones that cause the aging process could be identified. Antidotes to these hormones would be developed. It may be possible to slow the process of aging. Already, scientists have had success with rats. The hearts and lungs of older rats have become as strong as those of young rats.

As we have been able to clone plants, we may be able to clone human beings. We could splice the nucleus of a human cell from a person we wish to duplicate into a human egg cell. In this way, people with identical characteristics could be genetically engineered. It also may be possible to engineer humans to live in hostile environments. For example, a fish-like gill could be implanted in humans. We could then breathe under water.

Ethics

Ethics are the standards of right and wrong that people use to govern their behavior. Although new technologies may be possible in the future, questions of ethics will cause a public debate. For example, suppose people do research on laboratory-grown human embryos. Some that are produced might be badly deformed. Would we have the right to discard them? Is the research justified if our goal is to improve techniques of genetic engineering? Would dictators be able to use these techniques to engineer superhumans that could serve in their armies? With our fear that nuclear war might wipe out the world's population, would we be justified in producing a whole society of frozen embryos? These could be kept for years and grown in artificial wombs if a disaster occurs.

Biotechnology can improve the quality of the human species. But who should make the decisions to develop improved varieties of humans? Should the decisions be made by individual governments? By the United Nations? By scientists? By religious leaders? Would we ever be able to reach worldwide agreement? Should scientists and technologists have the right to tamper with human life?

Food Production

How would you like a delicious piece of newsprint? Doesn't sound terrific? OK. How about a nice maple leaf? What? You'd rather have a pizza? Quite understandable. But foods of the future may very well come from presently inedible materials. Chemists are attempting to break down common materials into their basic elements, such as carbon, hydrogen, and oxygen. The elements will then be put together in new combinations to form artificial foods. These foods will be just as healthful as the foods we eat today. Probably, we would still prefer the good old-fashioned hamburger. In parts of the world where people do not have enough food to eat, however, these new artificial foods could prevent starvation.

In the next 30 years, we will need to produce as much food as humans have produced in their entire history. Farmers will use **growth hormones** to produce larger, less fatty animals. Cows will be able to produce more milk. It may be possible to produce animals that have only the parts we need. For example, chickens without necks or feathers could be developed by using only the genes for the meaty parts of the bird. Our vegetables will be improved as well. Soon, we will be able to transfer beet genes to tomatoes. The beet genes will make the tomato skins hard for shipping. Tomatoes will be juicy on the inside, but hard on the outside. We may even be able to implant genes from cows into fish and raise sea steaks!

Feeding people in underdeveloped nations will become a new priority in the future. We will try to find nutritional foods that are easy and inexpensive to grow. Most of our planet is covered by oceans and deserts. Within the next several decades, edible plants will grow in the desert. They will be irrigated with salt water. These salt-water plants, called **halophytes,** will be used first to feed animals and as a source of plant oil. Later, these crops will be used to make a variety of processed foods.

From the control panel on a desert farm, people will harvest crops with robotic harvestors. (Courtesy of General Electric Company)

Salt-tolerant plants, called halophytes, will be grown with seawater. (Courtesy of Environmental Research Laboratory, University of Arizona)

ENERGY IN THE FUTURE

An artist's idea of an algae farm (Courtesy of Solar Energy Research Institute)

As nations become more technological, the global energy supply will have to increase. Substitutes for fossil fuels will have to be found. In the future, renewable energy sources will fulfill most of our energy needs. Space-based power stations will beam concentrated microwave energy down to the earth.

Green algae plants may provide a source of energy. Algae contain large amounts of oils and fats. These oils and fats are easily separated from the algae and can be converted into fuels. One acre of water or ground covered with algae can produce as much as 100 barrels of oil per year. These oils are a rich source of chemicals for making gasoline.

One of the best energy hopes for the future is **fusion power.** Fusion is a technology that joins atoms together. This process creates a new form of matter. However, the new matter has less mass than the original two parts did together. The fusion reaction converts some of the matter into energy. Fusion reactions require huge amounts of energy. Researchers are trying to find ways to obtain at least as much energy from the reaction as is needed to start the reaction. This balance is called **breakeven.** Once the reaction produces more energy than is needed to start it, the excess energy will be used to generate electricity. Some researchers think that commercial fusion reactors will be built by 2020. The great advantage of nuclear fusion over nuclear fission is that fusion produces clean energy. There will be little or no nuclear waste or radiation.

FUTURING—FORECASTING NEW TECHNOLOGIES

Existing technological systems will act together to produce new, more powerful technologies.

As you have read this book, you have probably noticed that the newest technologies are also the most complex. Satellites, space stations, robotic factories, maglev trains, and lunar habitats are possible only because many people and many different kinds of technology work together.

The newest technologies, and certainly the technologies of the future, will depend on contributions from many of our present technological systems. Future technologies will also rely on the development of new and different resources. However, the resources will come from the seven basic categories.

Technological systems will flow together and produce new, more powerful systems. The flowing together (**con-**

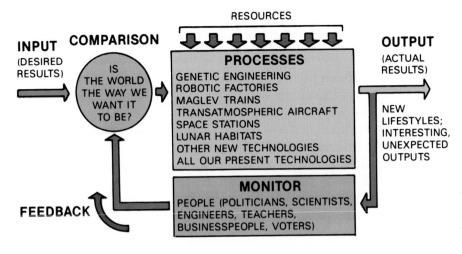

This system diagram shows how future technologies will affect our lives.

fluence) of systems will mean that new technologies will not be separate and distinct. No longer will there be pure transportation or communication systems. A transportation system like the space shuttle, for example, needs construction and manufacturing technologies to build it. The system also needs communication technology to guide it.

Modern biotechnology is also a confluence of systems. Genetic engineering is really a manufacturing process. New kinds of organisms are produced. When doing genetic engineering, technologists and scientists make use of construction, manufacturing, and communication technologies. They construct structures (like fermentation vats) to house processing operations. They use manufactured tools and instruments. They use communication and information technology when they do computer modeling.

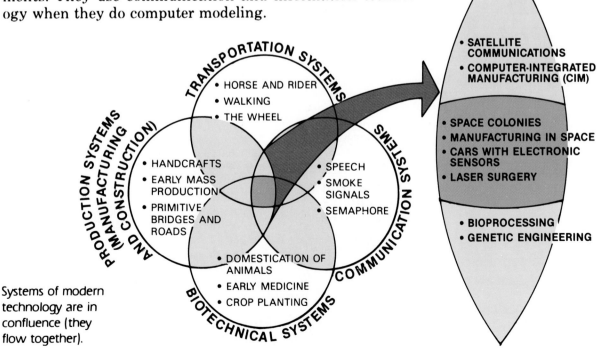

Systems of modern technology are in confluence (they flow together).

Futuring Techniques

Combined technologies act in a more powerful way together than each of them acts individually. **Futurists** predict what technology will be like in the future. To do this, they use several interesting techniques.

One technique of forecasting the future uses a **futures wheel.** A futures wheel starts out with an idea at the center. Spokes radiate outward from the central idea. The central idea leads to other ideas. Each idea has several possible outcomes of its own. From each outcome, there may be additional outcomes.

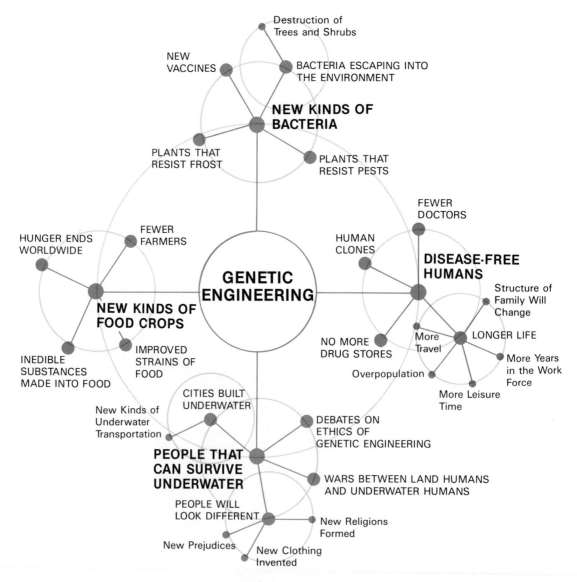

A futures wheel helps us determine possible outcomes of a new idea or technology. In this case, possible outcomes of genetic engineering are forecast.

	GENETICALLY ENGINEERED HUMANS	CITIES ON THE MOON	LIFE ON OTHER PLANETS	CITIES UNDER THE SEA
AIR POLLUTION	• Humans will be able to breathe sulphur dioxide	• Moon societies will safeguard against the kind of pollution that occurred on earth	• People will vacation on nonpolluted planets	• Polluted air will be used as fuel by sea cities
RISING ENERGY COSTS	• Humans will be able to survive on fewer calories	• Energy will be beamed to earth from a moon base	• Energy sources will be exported from other planets to earth	• There will be new uses of sea water to provide energy
OVERPOPULATION	• Child-bearing will be forbidden • Only genetically engineered humans needed for specific purposes will be authorized • People will debate whether we should alter human life	• Societies will move to the moon	• Marriages will occur between earthlings and space beings • Attempts to transfer populations to other planets will meet with resistance; interspace wars will result	• People will work in new industries like mining sea beds
LONGER LIFE SPANS	• People will be altered several times during their lifetimes as conditions change on earth	• People will commute to "summer homes" on the moon • New travel agencies will specialize in moon-earth travel	• People will spend more time traveling to distant planets	• People will live part of their lives on land and part in sea cities to learn more about all forms of life

Another forecasting method is a **cross-impact analysis.** Here, we identify several possible futures. We then try to think about what impact each of the futures would have if the futures occurred in combination.

People also forecast the future by **Delphi surveys** and **trend analysis.** In a Delphi survey, a panel of experts is asked to list the ten most likely futures. The list is then circulated among all the experts. Each ranks the ideas in the order he or she thinks they are most likely to occur. The results are tabulated. A new list representing the thinking of the entire panel is generated. The new list is again circulated. The experts again rank the ideas. The results are tabulated once more. The final tabulation represents what the experts think will be the most likely futures.

A trend analysis forecasts the future based upon past performance. For example, the cost of computers has dropped at an increasing rate over the last ten years. Forecasters might suggest that the price will continue to drop at an increasing rate in the future. In this way, computer prices in the year 2000 might be predicted to be very low. Trend analysis does have its drawbacks. For example, just because things happened one way in the past does not mean they will continue along the same path.

A cross-impact analysis

Using futuring techniques, people can anticipate the consequences of a new technology.

SUMMARY

Some of the impacts of technology are positive, and some are negative. Whether technology is good or evil depends upon the way humans use it. Proper planning can reduce the undesired outputs of a technological system.

We have become very dependent upon technology. It would not be easy to go back to a simpler life-style. It is up to us to determine how technology will be applied in the decades ahead. Technology must be fitted to the needs of human beings. Likewise, we must treat nature with care. Our technologies must be adapted to the environment.

The outlook for technology in the future is very promising. Technological changes will continue to increase exponentially. Computers, telephones, satellites, and televisions will work in harmony to provide new and exciting communication technologies. Voice recognition technologies will enable us to speak to our typewriters. We will hold videoconferences with people in distant lands. Our shopping and our studying will increasingly depend upon the computer.

In the United States and other advanced nations, manufacturing will be done more by robots and less by people. Only in less developed nations will a large percentage of the workforce work on factory production lines. Intelligent robots will provide more and more services. They will assume more and more human characteristics. New manufacturing facilities will be built in space. New materials will come from the sea, the moon, and other planets.

Space will provide a home for new industries and new structures. Space stations will beam energy back to earth. They will also serve as resting stations for people on the way to new cities in space.

On earth, new construction technology will be more energy efficient. New housing developments will blend the conveniences of city life with the advantages of rural living. We will live in "smart houses" and work in "intelligent buildings" with computer-controlled appliances and advanced communication systems.

Transportation in the future will get faster and will use less energy. Maglev trains will reach speeds of over 300 miles per hour. Personal space vehicles will carry us from space stations to colonies in space. On earth, personal vehicle systems like PRTs will become quite popular.

Biotechnical systems will make great contributions in the twenty-first century. In addition to providing cures for diseases, biotechnology will extend our life spans. We may be able to genetically engineer human beings with desired traits. We will be able to grow new kinds of foods and produce artificial foods to feed the world's hungry.

By early in the twenty-first century, we will have developed fusion power. Fusion power could provide us with a never-ending supply of clean energy.

Futuring techniques like futures wheels, cross-impact analysis, Delphi surveys, and trend analysis can help us forecast the consequences of future technologies. We can try to promote only those technologies that will enhance the quality of human life.

CONCLUSION

One thing is certain. Technology will continue to grow at an ever-increasing rate. Because of this, humans must plan and direct technological change. We cannot afford to leave the applications of technology to chance.

Life in the future will depend upon how well we plan the use of new technologies. As space travel, space cities, and satellites become more common, we must think more in global terms. We must think more about the entire human race and less about individuals, or even individual nations. We all inhabit one very small planet in a huge solar system. The quality of life on our planet should be the main concern of all humans. We must work together, as a human race, to ensure the survival of our species.

Technology is not magic. As you study it more in detail and as you understand more mathematics and science, you will see that technology is within our control. It is up to us to see that technology is used to enhance the quality of life for future generations.

(Courtesy of General Electric Company)

DUST OR BUST!

Setting the Stage

When you were sent from Earth to work on the desert planet Nala, you were told about the dust. But you hadn't thought it would be as bad as this. Without the air conditioning, you could choke to death in a matter of days.

Suddenly the air conditioning and emergency backup system fail. You have only a short time to find a way to remove the dust being carried in by the air ducts. You try filter paper and cloth, but the dust is so fine it goes right through.

Your Challenge

Find a way to remove the dust particles before your classroom becomes a death trap.

Procedure

CAUTION: As we are dealing with a high-voltage static electricity charge, the chances of getting a shock are high. Normally this is not dangerous, but those with health problems should not try this activity.

1. Be sure to wear safety glasses and a lab coat.
2. Cut the Plexiglas tube to length (24–26" long).
3. Drill two ⅜" holes in the wall of the tube 3–4" from the ends.
4. Cut the brass rod 3" longer than the Plexiglas tube.
5. Push the one-hole stoppers firmly into both ends of the tube.
6. Insert the brass rod through the stopper holes so that the extra length is all at one end.
7. Use hot glue to fasten the brass rod at the FLUSH end.
8. Using the copper wire, wrap a coil around the length of the tube. Tape both ends of the coil in place leaving 4–5' of extra wire at one end.
9. Fasten the rubber bulb or plastic squeeze bottle to one of the holes drilled in the Plexiglas.
10. Fasten the loose end of the copper wire to the base of the Van de Graaff generator.
11. Using the rubber bulb or plastic squeeze bottle, suck in some smoke or chalk dust (the more, the better).

Suggested Resources

Safety glasses and lab apron
Clear Plexiglas tubing (24–26" long, approximately 1½" diameter)
2 large one-hole rubber stoppers that fit tightly into the ends of the Plexiglas tubing
Brass rod (at least 3" longer than the Plexiglas tubing)—⅛" diameter
Copper wire (single strand, light gauge, 15–20')
Chalk dust (or other source of smoke or dust)
Electrical tape
Van de Graaff generator
Large rubber squeeze bulb or plastic squeeze bottle
Wire cutters
Drill and drill bits
Hot glue gun and glue stick

12. Turn on the Van de Graaff generator.
13. Slowly bring the protruding end of the brass rod near the top of the generator.
14. Be sure to keep your eyes on the dust or smoke to see what happens.

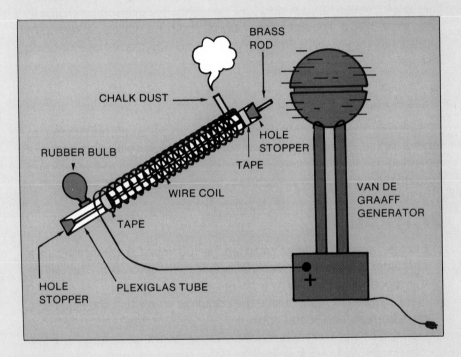

BRASS ROD

CHALK DUST

HOLE STOPPER

TAPE

RUBBER BULB

WIRE COIL

VAN DE GRAAFF GENERATOR

TAPE

HOLE STOPPER

PLEXIGLAS TUBE

Technology Connections

1. The device you have made is called an *electrostatic precipitator.* It filters tiny particles from the air by using static electricity.
2. What happened to the dust or smoke particles inside the tube when the brass rod was brought near the Van de Graaff generator? Why?
3. Would the electrostatic precipitator remove toxic (poisonous) *gases* from the air?
4. Many industries send smoke and other particles into the air as a result of production. Often they can add an electrostatic precipitator to their system to help filter out these pollutants. What industries do you think would make the most use of this device?
5. A good solution often requires a compromise. What is the trade-off (advantages *and* disadvantages) of adding an electrostatic precipitator to a power plant?

Science and Math Concepts

▶ Opposite electrical charges (+ and −) attract each other. Similar electrical charges (+ and +) or (− and −) repel each other.

▶ Plexiglas and rubber are *insulators.* Insulators are poor conductors of electricity.

▶ Metals such as copper and brass are good *conductors* of electricity.

▶ A negative (−) charge results from an excess of electrons. A positive (+) charge results from too few electrons.

511

SATEL-LOON

Setting the Stage

The storm has been building steadily. The waves are becoming gigantic. There seems to be no way to save the boat from the rocks. Hurriedly, you throw your gear into waterproof containers and toss them into the sea. At the signal, you and your fellow travelers jump into the water and make your way as best you can to shore.

By the light of day you spot some of the gear thrown up on shore. Luck is with you, and you find your Emergency Position Indicating Radio Beacons (EPIRBs). The emergency signal you send will be picked up by satellites as they pass overhead. At least you now have hope of being rescued.

Your Challenge

Simulate satellite communication by using a small transmitter to send a signal up to a satel-loon. The satel-loon will be capable of repeating the signal on another band or frequency and re-transmitting the signal back to another location on the ground.

Procedure

1. Enclose a small AM-FM radio and a 100 mw CB transceiver in a container with approximately 8" between the two units. This will be the repeater.
2. Place cushioning foam inside the container to protect the electronic gear from an accidental crash.
3. Use an FM transmitter and send a signal to the repeater. Tune the transmitter and receiver until you hear a clear signal.
4. Lock the CB located in the satel-loon in the transmit position. Use another CB about 30 feet away to receive the re-transmitted signal from the repeater.
5. Check all units for clarity.
6. Pick a clear day with no wind if you are using helium-filled balloons to fly the satel-loon. Choose a day with a steady wind if you are using a parafoil kite. Use anchor stakes for each system and 165 lb. test line.
7. After checking all units on the ground, send the satel-loon aloft.
8. Transmit a message using your satel-loon.
9. Invite the local newspaper to cover this activity. You are simulating the latest technology in communications.

Suggested Resources

AM or FM transmitter— 100 mw
2 CB transceivers
AM-FM radio
Helium-filled balloons or parafoil kite
165 lb. test nylon line
Anchor stake
Cushioning foam

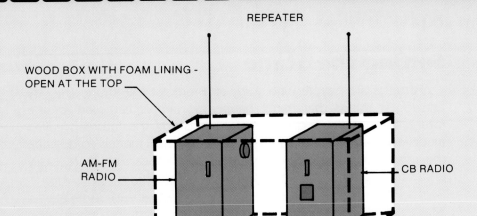

REPEATER

WOOD BOX WITH FOAM LINING - OPEN AT THE TOP

AM-FM RADIO

CB RADIO

8"

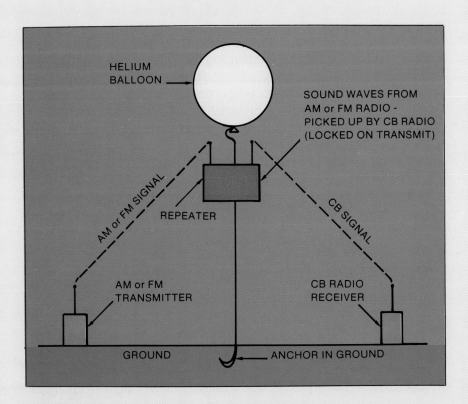

HELIUM BALLOON

SOUND WAVES FROM AM or FM RADIO - PICKED UP BY CB RADIO (LOCKED ON TRANSMIT)

AM or FM SIGNAL

CB SIGNAL

REPEATER

AM or FM TRANSMITTER

CB RADIO RECEIVER

GROUND

ANCHOR IN GROUND

Technology Connections

1. Who uses EPIRBs? Who else will receive the EPIRB's signals?
2. Compare the transmission of EPIRB signals via satellite with satellite communication using telephones or TVs.
3. What type of programs are transmitted via satellite?
4. What is videoconferencing? How are satellites used in videoconferencing?

Science and Math Concepts

▶ Radio signals travel at the speed of light.

MAPPING THE FUTURE

Setting the Stage

Futurists are people who study and forecast the future. They try to suggest things that might happen so people can plan for the future. One of the techniques used to study the future is called modeling. Futurists often use graphic or physical models to describe the relationships between their ideas so they can be more easily understood. One of the simplest models is called a futures wheel.

Your Challenge

Working in teams, develop a futures wheel that illustrates the effects a new technology might have on society in the future.

Procedure

1. Select one of the following emerging technologies for your futures wheel: plastic automobiles, electric automobiles, computer aided design, picture telephones, teleconferencing, robotics, computer integrated manufacturing, personal robots, super trains, lighter-than-air vehicles, wind generators, smart houses, industrial space facilities, and video conferencing.
2. Write the name of the topic on an 8 ½" × 11" sheet of paper. Post the name of your topic in the center of your bulletin board.
3. Identify the immediate consequences (effects) of using the new technology. For example, widespread use of solar energy might conserve fossil fuels, create new architectural styles, and lower home heating costs. Try to identify both positive and negative consequences.
4. Write the consequences on 3" × 5" cards. Post the cards around the central topic. Use string to create lines between the central topic and its immediate consequences.
5. Identify a second generation of consequences for each immediate consequence. For example, conserving fossil fuels might reduce air pollution, eliminate jobs, and cause fossil fuel prices to rise.
6. Write the secondary consequences on 3" × 5" cards. Post them around their immediate consequences. Using string, connect each immediate consequence to its secondary consequence. (Note: Can you use different-colored markers, different-shaped cards, or some other method to make the difference between the immediate and secondary consequences clear?)
7. Identify a third generation of consequences. For example, eliminating jobs might increase unemployment as well as create a need for retraining programs.

Suggested Resources

Composition board (sheathing)—½" × 4' × 4'
3" × 5" cards
String
Straight pins, staples, or thumb tacks
Magic markers

514

8. Write your third generation consequences on 3" × 5" cards. Post them near their secondary consequences. Using string, connect the secondary consequences to the next generation of consequences.

9. Share your futures wheel with the class. Assign each member of your team a section of the futures wheel that they can describe to the class.

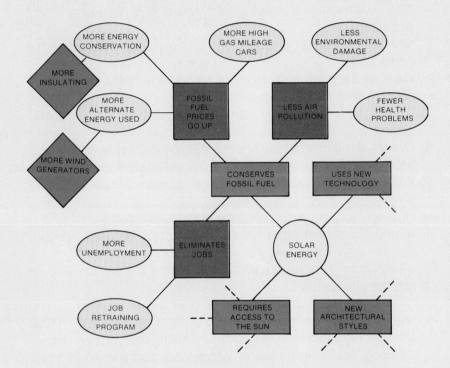

Technology Connections

1. *Futurists* are people who think about what might happen in the future. Are you a futurist?

2. Futures wheels help people discover and describe what might happen in the future. Could a futures wheel help you plan your future? What would you put in the center of your futures wheel?

3. Ever since humankind invented the wheel, technology has played an important role in the way we live and work. However, technological advancements are being made more rapidly than ever before. Will people continue to allow technology to shape the way they live and work in the future? Or, do you think people will reject technology and try to live simpler lives?

Science and Math Concepts

▶ Cause-and-effect relationships are events that result in changes. Cause is often a human activity and effect is the change caused by the activity.

▶ Modeling describes the relationships among a collection of ideas in a graphic, physical, or other form so it can be more easily understood.

REVIEW QUESTIONS

1. What four possible combinations of outputs can result from technological systems?
2. Do you believe that technology is good, evil, or neutral? Explain why.
3. How can people control the development of a technology they feel may be harmful?
4. Give two examples showing that pollution is not just a recent problem.
5. Give an example of a technology that is poorly matched to the human user.
6. Give an example of a problem technology has caused and a solution technology has provided.
7. Explain how technology can make life easier for a disabled person.
8. Draw a design for a futuristic communication system that would allow you to communicate with a class of students in Europe.
9. What would be two good products to manufacture in space? Why?
10. How will future travel differ from present-day travel?
11. Would a personal rapid transit (PRT) system be useful in your town or city? Explain why or why not.
12. What kinds of moral questions will arise if people are able to engineer human life?
13. Give an example of how a modern or future technology will require inputs from several existing technological systems. (**Hint:** Think of how biotechnical, communication, construction, manufacturing, and transportation systems act together to produce new technologies, such as manufacturing in space.)
14. Draw a futures wheel. At the center, place a skin cream that brings back youth. Predict the possible outcomes.

KEY WORDS

Acid rain	Futuring	Magnetic levita-	Pollution
Confluence	Growth hormone	tion (Maglev)	Smart houses
Environment	Halophyte	Manufacturing in	Speech synthesis
Ergonomics	Impact	space	Sulfur dioxide
Fusion	Intelligent	Personal rapid	Videoconference
Futures wheel	buildings	transit (PRT)	Voice recognition

**SEE YOUR TEACHER FOR
THE CROSSTECH PUZZLE**

516

TECHNOLOGICAL TIME LINE

STONE AGE

1,000,000 B.C.	Fire and the development of early stone tools
17,000 B.C.	Domestication of animals
12,000 B.C.	Early oil lamps that burned fish and animal oil
10,000 B.C.	Yeast for leavening bread

AGRICULTURAL ERA

8000 B.C.	Agriculture in the Fertile Crescent
6500 B.C.	Houses built from mud-bricks; dugout canoes
6000 B.C.	Pottery
3500 B.C.	Writing (cuneiform and hieroglyphics)

BRONZE AGE

3000 B.C.	Tools made from metal; development of the wheel; wine made from pressed grapes
1900 B.C.	Plowing with animals
1500 B.C.	Grinding of grain; glassmaking

IRON AGE

1200 B.C.	Smelting of iron; development of the Greek alphabet
700 B.C.	Irrigation and sewage in Rome
221 B.C.	Construction of the Great Wall of China
144 B.C.	First high-level aqueduct in Rome; use of cement
100 B.C.	Water wheels
410 A.D.	Sack of Rome by Visigoths
500 A.D.	Sailing ships that could travel with and against the wind
550 A.D.	Babylonian Talmud written
650 A.D.	Writing of the Koran
750 A.D.	Arabs learn papermaking from Chinese prisoners
765 A.D.	School of medicine founded in Baghdad
1024 A.D.	Chinese use first paper currency
1200 A.D.	Windmills
1281 A.D.	Chinese use gunpowder in war against Mongols
1400-1600 A.D.	Age of exploration and discovery; Balboa, Cabot, Columbus, de Gama, de Leon, Hudson, Magellan
1450 A.D.	Gutenburg invents movable type
1560 A.D.	Water-pumping machine
1600 A.D.	Magnetism discovered by William Gilbert
1610 A.D.	Galileo's telescope
1619 A.D.	First slaves brought to North America by British
1628 A.D.	Circulation of the blood discovered by William Harvey
1700 A.D.	Newton's Laws
1712 A.D.	Newcomen's steam engine
1721 A.D.	Bach completes Brandenburg concertos
1724 A.D.	Farenheit thermometer invented
1752 A.D.	Benjamin Franklin proves lightning is electricity

INDUSTRIAL ERA

1760–1840 A.D.	Industrial Revolution in Britain
1769 A.D.	James Watt's steam engine
1770 A.D.	Hargreaves's spinning jenny patented
1776 A.D.	U.S. Declaration of Independence
1780 A.D.	The lathe
1784 A.D.	Wrought-iron process by Henry Cort
1785 A.D.	Cartwright's loom
1789 A.D.	French Revolution
1789 A.D.	LaVoisier's theory of chemical combustion published
1793 A.D.	Eli Whitney's cotton gin
1798 A.D.	Eli Whitney's factory mass-produces firearms
1804 A.D.	Jacquard's loom
1807 A.D.	Britain abolishes the slave trade; Robert Fulton's steamboat
1808 A.D.	Slave trade ends in the United States
1811 A.D.	Luddites smash knitting machines in Britain
1812 A.D.	Napoleon invades Russia
1816 A.D.	Regular transatlantic sailing service begins between New York and Liverpool, England
1822 A.D.	First textile mill in United States, in Massachusetts
1829 A.D.	The steam locomotive
1835 A.D.	Bessemer process for making steel
1839 A.D.	Photography invented
1843 A.D.	Morse begins telegraph line between Baltimore, Md. and Washington, D.C.
1861–1865 A.D.	Civil War in the United States
1863 A.D.	Emancipation Proclamation
1876 A.D.	Alexander Graham Bell invents the telephone
1885 A.D.	Carl Benz, in Germany; first successful gasoline-driven motorcar
1887 A.D.	Daimler's internal combustion engine automobile
1888 A.D.	First Kodak hand-held camera
1893 A.D.	Diesel engine invented by Dr. Rudolf Diesel
1900 A.D.	EverReady flashlight invented; paper clip patented
1901 A.D.	Marconi sends first transatlantic radio signals; spinal anesthesia developed in France
1902 A.D.	Rayon patented
1903 A.D.	Wright brothers' airplane
1905 A.D.	Albert Einstein proposes his theory of relativity
1906 A.D.	Freeze-drying invented in France; hot dog gets its name from a cartoon of a dachshund in a bun
1908 A.D.	Model T Ford built by mass production
1909 A.D.	Beginning of the age of plastics; Bakelite patented
1910 A.D.	Cellophane invented
1911 A.D.	Rutherford proposes his theory of atomic structure
1912 A.D.	Sinking of the Titanic—1,513 people die; first heart attack diagnosed in a living patient
1914 A.D.	Panama Canal opens; bacteria used to treat sewage; outbreak of World War I
1915 A.D.	SONAR invented
1918 A.D.	Red, green, and yellow traffic lights in New York
1919 A.D.	First transatlantic flight—Newfoundland to Ireland

1920 A.D.   First U.S. commercial broadcasting station—KDKA from Pittsburgh goes on the air

1921 A.D.   Hybrid corn greatly improves crop yields; Band-Aids invented; first lie detector; first cultured pearls

1924 A.D.   First round-the-world flight; Kleenex tissues

1925 A.D.   First successful experiments with hydroponics

1926 A.D.   First television demonstration; pop-up toaster

1927 A.D.   The theory of negative feedback developed

1928 A.D.   Penicillin invented by Alexander Fleming; Rice Krispies marketed by Kellogg cereal company; Scotch Tape produced by 3M company

1929 A.D.   Electroencephalograph (EEG); foam rubber

1930 A.D.   Packaged frozen foods go on sale in Massachusetts

1931 A.D.   First regular TV broadcasts (station W6XAO in California); Empire State Building in New York City

1934 A.D.   The electron microscope

1935 A.D.   Radar developed; Nylon patented; B-17 bomber produced by Boeing

1936 A.D.   Vitamin pills; Polaroid sunglasses

1938 A.D.   The ball-point pen is patented

1939 A.D.   Outbreak of World War II; DDT produced; first jet aircraft flies in Germany; Pan Am begins first regular commercial transatlantic flights

1940 A.D.   Freeze-drying used to preserve foods in United States

1942 A.D.   LORAN developed; first color snapshots

1944 A.D.   First automatic digital computer, the Mark I; DNA isolated at the Rockefeller Institute

1945 A.D.   Atom bomb used on Hiroshima and Nagasaki; ENIAC computer built using 18,000 vacuum tubes

1946 A.D.   Timex watches; bikini bathing suits

1947 A.D.   Microwave oven; first Honda motorcycle; development of the transistor at Bell Laboratories

1948 A.D.   Polaroid camera; LP records; Teflon

1950 A.D.   Power steering; start of regular color TV broadcasts; first credit card (Diner's Club)

1952 A.D.   Salk polio vaccine; first sex-change operation—Christine Jorgensen; Sony pocket transistor radio; 3-D movies

1953 A.D.   First successful open heart surgery; supersonic jet aircraft built by Pratt and Whitney; Dacron developed in Britain; breeder reactor in Idaho; double helix model for structure of DNA proposed

1954 A.D.   Nuclear-powered submarine "The Nautilus"; numerical control machining; photocell developed

1955 A.D.   Velcro fasteners patented; birth-control pills; nuclear-powered electricity generation; fiber optics; domestic freezers; multitrack recording

1956 A.D.   FORTRAN computer language; hydrogen bomb exploded; first videotape recorder; first desktop computer (Burroughs E-101)

SPACE AGE

1957 A.D.   Sputnik; Sabin oral polio vaccine; high-speed dental drill; mercury batteries

INFORMATION AGE

1958 A.D.   First integrated circuit made by Texas Instruments; bifocal contact lenses; first U.S. satellite

1959 A.D.   Pilkington float glass process perfected; Sony produces first transistorized TV; Russia sends unmanned spacecraft to the moon (Luna 2); Xerox copier

1960 A.D.   Lasers; light-emitting diodes (LEDs); weather satellites; felt-tip pens

1962 A.D.   TV signals sent across the Atlantic via satellite; world's first industrial robot produced by Unimation Corporation; Tang orange juice

1963 A.D.   John F. Kennedy assassinated; cassette tapes; Instamatic cameras; the tranquilizer Valium produced; Artificial heart used during surgery

1964 A.D.   China tests nuclear bomb; IBM word processors; photochromic eyeglasses that change in response to the amount of sunlight

1965 A.D.   First space walk; Super 8 cameras; reports link cigarette smoking to cancer

1966 A.D.   Flashcubes; electronic fuel injection; Russians make successful soft landing on the moon (Luna 9)

1967 A.D.   Christian Barnard performs first heart transplant; energy-absorbing bumpers; Frisbees

1969 A.D.   American manned moon landing; first Concorde SST flight

1970 A.D.   Russians land moon robot (Lunokhod); floppy disks; jumbo jets

1971 A.D.   Microprocessor developed by Intel; Mariner 9 orbits Mars

1972 A.D.   CAT scan; photography from satellites (Landsat); videodiscs

1973 A.D.   Genetic engineering; Skylab orbiting space station launched; supermarket optical price scanning; Selectric self-correcting typewriter; push-through tabs on cans

1974 A.D.   Toronto Communications Tower (tallest building in the world) opens

1975 A.D.   Liquid crystal displays; video games; disposable razors; cloning of a rabbit; Americans and Russians dock in space

1976 A.D.   Birth of the Apple computer; electronic cameras

1977 A.D.   Space shuttle; trans-Alaska pipeline system completed; neutron bomb developed

1978 A.D.   First test-tube baby born; programmable washing machines; computerized chess; auto-focus cameras

1979 A.D.   Skylab, the orbiting space station, falls back to earth

1980 and beyond A.D.   Compact discs; voice synthesis; electronic offices; genetic engineering for humans; Sony walkman; electronic mail; NMR (nuclear magnetic resonance); automated factories; robots with sight; Rubik's cube; pocket-sized TV; personal computers; laser weapons; supercomputers; disc cameras; Sony Mavica fully electronic camera with disc storage; NASA space telescope; birth control pill for men

TECHNOLOGY EDUCATION STUDENT ASSOCIATIONS

Arvid Van Dyke

INTRODUCTION

Technology education student associations help you work with others in your technology class, your school, and your community. The technology student organization is the group in your school that uses technology in a variety of interesting activities, projects, and contests.

Think of the other groups in your school. Some, called "clubs," serve a special interest or offer after-school activities. Associations such as the Student Council serve all students in the school. The technology education student association, now called the American Industrial Arts Student Association (AIASA), serves all students taking Technology Education or Industrial Technology in their school.

Technology education student associations should meet the standards for technology education that have been established by an international association of teachers. These standards are helpful in making student associations serve the needs of students who must be technologically literate to live and work in our technological world.

Technology education student associations involve communication, construction, manufacturing, and transportation activities. Opportunities for creative thinking, problem solving, and decision making are also provided. Learning how technology works can be greatly enhanced by the student association's activities.

LEARNING TO BE LEADERS

Every group or organization has a set of leaders, called officers. Most officers are elected by group members. In a corporation, for example, a board of directors may appoint (or employ) a president to run the company. Vice presidents, treasurers, and other managers or leaders are the top people who make things happen in the organization.

Student organizations help build leadership and public speaking abilities. (Courtesy of Kramer Photos, Jerry Kramer, Box 87, 118 S. Main, Melvern, KS 66510)

Every student should serve at some time as an officer in the student association. You can volunteer to be a candidate or be nominated by someone else. Experience as an elected officer will improve your ability to lead and to work with others. Leadership skill is especially valuable when you are attending college or working. Employers look for people who are willing to learn and are able to get along well with others.

ASSOCIATIONS START IN CLASS

Many teachers recognize the benefit of student associations and involve all of their students in the associations. The students learn to lead and to use technology to make learning activities more significant. The officers and committees of an in-class student organization can help lead and manage activities during the class period. Each class may then want to take part in the school's technology student association activities.

THE SCHOOL ORGANIZATION OR CHAPTER

Most student groups in your school are organized to allow students to work together in activities related to a subject they take. A student association for technology education plans activities that relate to technology. By working together, the group accomplishes more than each member could accomplish individually.

The association forms committees to allow more of its members to lead activities. Each committee plans and leads at least one activity relating to the school's technology education classes. The activity adds to students' understanding of technology and its impact on our world.

The student association in your school **affiliates** (joins together) with the state association so that all school chapters throughout your state are stronger and can share information. In the same way, your state office affiliates with all other state offices to form a strong national association. The national association offers many services to its member states, schools, and students.

ADVANTAGES TO STUDENTS AND THEIR SCHOOL

The technology education student association can help you continue your exploration in the field of technology. You learn about career options and opportunities. By working in groups, you learn leadership skills and work cooperatively with other students and adults. Your school technology education program becomes better known because the student association attains recognition through the success of its members. You prepare for contests, follow the Achievement Programs, and travel to conferences in your state. Attending a national conference is an honor and a very educational experience.

LEARNING ABOUT TECHNOLOGY WITH THE TECHNOLOGY EDUCATION STUDENT ASSOCIATION

Your class and chapter officers can select several activities that will help students learn more about technology. Each activity allows students an opportunity to make technology work. Begin a brainstorming session, and find activities that are closely related to the technology area you are currently studying. Try to use technology in your student association activities. Refer to the "Suggested Technology Activities" section of this book for activity ideas.

CONTESTS TO MOTIVATE AND TEACH

Contests start in the classroom or laboratory, like many other activities of the technology education student association. The contest or project can be one of your class assignments. For example, all students might be required to make a safety poster during the laboratory safety unit of instruction. All posters are then graded by the teacher before a student committee judges the posters for in-class awards. The top posters are entered (registered) in the AIASA Safety Poster Contest for recognition at the state or national conferences.

Competitive events motivate students to learn and to use technology for problem solving. Contests should offer a challenge to students' minds, hands, and attitudes. Some types of contest test a student's understanding of technology and its impact. Other AIASA contests recognize leadership skills. Still others allow students to use communication skills in speaking or writing about technology and its significance.

Working on technology activities is an important part of your student organization membership.
(Courtesy of Kramer Photos, Jerry Kramer, Box 87,
118 S. Main, Melvern, KS 66510)

GLOSSARY

Absolute Zero The temperature at which all molecular motion stops, approximately -459° F or -273° C.

Acid Rain Rain containing chemicals given off by automobiles and factories, often hundreds of miles away, that damages trees and other plants.

Active Solar The use of solar energy to perform a specific task, such as heating water or generating electricity.

Agriculture The production of plants and animals and the processing of them into food, clothing, and other products.

Amplifier A device that makes a small mechanical force or electrical voltage larger.

Analog In electronics, a smoothly varying voltage or current that can be set to any desired value; as opposed to digital, which can be set only to a limited number of values (e.g., 1s and 0s in binary form).

Architect A person who designs buildings.

ASCII American Standard Code for Information Interchange. A digital code in which seven binary bits represent letters, numbers, and punctuation.

Asphalt A tarlike substance used to pave roads and produce roof shingles.

Assembly Line A manufacturing method in which each worker or machine does only a small part of the whole job, and the products being built move from worker to worker or machine to machine.

Automation The process of controlling machines automatically.

Bandwidth The measure of the ability of a communications channel to carry information. The larger (wider) the bandwidth, the more information the channel can carry.

Battery A device that stores energy chemically and converts it to electricity.

Biomass Vegetable and animal waste matter used for energy generation.

Biotechnology The use of living organisms to make commercial products; includes antibody production, bioprocessing, and genetic engineering.

Bit The smallest unit of digital information; a "1" or "0" in digital coding.

Bottom Plate A single 2 × 4 used to construct the bottom of a wall section.

Broadcast The sending of a message to many receivers at the same time.

British Thermal Unit The amount of energy needed to raise the temperature of one pound of water one degree Fahrenheit.

Brittle A material that breaks more easily than it bends.

Bronze Age The period (3000 B.C. to 1200 B.C.) in which people discovered how to combine metals to make alloys. The most common alloy was bronze, made from copper and tin.

Building Permit Permission received from a local authority to begin construction.

Buoyancy The upward force on an object in a liquid.

Business Plan A written plan that states the goals and objectives of a business, the strategy and methods to be used to achieve them, and the financial requirements of the business.

Byte A group of bits, usually eight, that is used to represent information in digital form.

CAD *Computer-Aided Design:* The use of a computer to assist in the process of designing a part, circuit, building, etc. *Computer-Aided Drafting:* The use of a computer to assist in the process of creating, storing, retrieving, modifying, and plotting of a technical drawing.

CAM   Computer-Aided Manufacturing. The use of computers to control a manufacturing process.

Capital   One of the seven resources used in technological systems. Capital is the money or other form of wealth used to provide the machines, materials, and other resources needed in the system.

Cash Flow Analysis   A projection of how much money will have to be spent each week or month of operation and how much income is expected in the same period.

Casting   Forming a product by pouring a liquid material into a mold, letting it harden into a solid, and removing it from the mold.

Cement   A construction material made of limestone and clay.

Ceramics   Materials that are made from clay or similar inorganic materials.

Channel   The path that information takes from the transmitter to the receiver.

CIM   Computer-Integrated Manufacturing. The use of computers to control both the business and production aspects of a manufacturing facility.

Circuit   A group of components connected to perform a function.

Circuit Breaker   An electromagnetic device that acts like a fuse and interrupts an electrical circuit when too much current flows (as when a short circuit occurs).

Closed-Loop System   A system that uses feedback to affect the process, based on a comparison of the system's output(s) to its command input(s).

Coal   A hard substance that comes from decayed plant and animal matter under great pressure for millions of years. It is burned to obtain energy.

Coating   A combining process used to beautify or protect the surface of a material.

Code   A set of signals or symbols that has some specific meaning to both the sender of the message and the receiver of the message.

Combining   The joining of two or more materials in one of several ways, including fastening, coating, and making composites.

Communication   Successfully sending a message or idea from one person, animal, or machine (the origin) to a second person, animal, or machine (the destination).

Compact Disk   A thin round disk on which information is stored digitally in the form of pits that reflect or absorb light from a laser.

Comparator   The part of a closed-loop system that compares the output (actual results) to the command input (desired results).

Component   A part that performs a specific function.

Composite   A synthetic material made of other materials.

Compression   A force that squeezes a material.

Concrete   A construction material made of stone, sand, water, and cement.

Conditioning   Changing the internal properties of a material.

Conductor   A material whose atoms easily give up their outer electrons, letting an electrical current flow easily through it.

Confluence   The flowing together of systems and technologies.

Construction   The building of a structure on a site.

Container Ship   A ship that carries freight prepacked in containers that are carried to and from the ship on railroad flat cars and trucks.

Contract   A formal agreement between any two people. In construction, the agreement that describes in detail what will be built, when and how it will be paid for, and who assumes risks if something goes wrong.

Coordination   In construction, the cooperation between two or more groups of people to avoid interference when installing different systems in a single building.

Current The flow of electrons through a material.

Daisy Wheel Printer A daisy wheel printer uses a print head shaped like a daisy. It prints with letter quality. A daisy wheel is made from plastic and includes an entire set of characters (numbers, letters, and punctuation).

Data Raw facts and figures. Data may be processed into information. (See Information.)

Design Elements Graphic designs like lines or bars that improve the appearance of a printed page.

Desired Results A system's command input; a statement of the expected output(s) of a system.

Desktop Publishing The linkup of a personal computer, special software, a mouse, and a high-quality laser printer to compose entire pages of text and pictures.

Diesel A type of internal combustion engine that does not use spark plugs.

Digital In electronics, the use of a limited number of values of voltage or current to represent information (e.g., 1s and 0s in a binary system).

Distributed Computing The use of two or more computers connected by data communications to perform information processing.

Dot Matrix Printer A computer printer that uses a group of pins, arranged in a rectangular format called a matrix, to press into a ribbon and print characters on paper.

Down Link The portion of a satellite communication link from the satellite to the ground station(s).

Drag The wind resistance that holds a plane back when it is moving forward.

Drilling A separating process that cuts holes in materials.

Ductile A material that bends more easily than it breaks is called ductile.

Educate To provide instruction or training. One of the reasons communication systems are used.

Elastic A ductile material that bends and then returns to its original shape is called elastic.

Electric Motor A device that converts electric energy into rotary motion.

Electron The negatively charged part of an atom that orbits the nucleus. The movement of electrons from one atom to another creates an electric current.

Electronic Communication Those methods of communication where the channel uses electrical energy.

Energy One of the seven resources used by technological systems. Energy is the capacity for doing work. It takes many forms (thermal, electrical, mechanical, etc.) and comes from many sources (solar, muscle, chemical, nuclear, etc.)

Energy Converter A device that changes one form of energy into another.

Engine A device that converts energy into motion and force.

Engineer A person with an engineering degree and/or a state license who designs buildings, roads, electronic circuits, airplanes, computers, and other technological systems.

Entertain To amuse or hold the attention of someone. One of the reasons communication systems are used.

Entrepreneur A person who forms and runs a business that is often based on new ideas and/or inventions.

Ergonomics Fitting technology to human needs.

Exponential Increasing or decreasing at a changing rate (as opposed to *linear,* increasing or decreasing at a constant rate).

Extruding A method of forming parts in which a softened material is squeezed through an opening, giving the part the shape of the opening.

Fastening The process of attaching one part to another.

Feedback The use of information about the output(s) of a system to modify the process.

Fermentation   A process in which living organisms digest starch and sugar, producing by-products like alcohol and carbon dioxide gas.

Ferrous Metals   Metals made from more than 50 percent iron.

Fiber Optic   The passing of light through a flexible fiberglass or plastic light guide.

Finite   Having limits.

Fission   The splitting of an atomic nucleus into two smaller nuclei, neutrons, and energy.

Flexible Manufacturing   A manufacturing process in which the tools and machines can be easily reprogrammed to produce different parts, making it economically feasible to make small quantities of any given part.

Floor Joists   The long boards that support the floor in a house.

Footing   The base of the foundation; generally made from concrete.

Forging   Heating a metal part and hammering it into shape.

Forming   Changing a material's shape without cutting it.

Forms   In construction, molds that give concrete its shape while it is hardening.

Fossil Fuel   A fuel formed from the partially decomposed remains of plants and animals buried in the earth over extremely long periods of time. Examples of fossil fuels include coal, oil, and natural gas.

Foundation   In construction, the portion of a structure that supports its weight; the substructure. The foundation includes the footing and the foundation wall.

Foundation Wall   The part of the foundation above the footing.

Framing   The process of constructing the skeleton of the house.

Frequency   The number of times an electromagnetic wave changes polarities in a specified period of time, usually cycles per second (hertz).

Frost Line   The depth to which the ground freezes in a particular location.

Fusion   The combining of two atomic nuclei into a single nucleus plus a large amount of energy.

Futuring   Projecting possible future outcomes of a given system or situation and analyzing the effects of each.

Gantt Chart   A type of bar chart used to plan and track a job's schedule.

Gas   *Natural gas:* A mixture of methane and other gasses that can be obtained from the earth and burned to provide energy. *Gasoline:* A liquid, obtained by refining crude oil, that is used as a fuel in internal combustion engines.

Gasahol   A fuel used in internal combustion engines made from combining gasoline with alcohol.

Gasification   The process of deriving methane gas from biomass or coal.

General Contractor   A person or company who accepts the total and complete responsibility for building a construction project.

Generator   A device that converts rotary motion into electric energy.

Geothermal   An energy source derived from the heat of the earth.

Girder   A strong horizontal structural support made from wood or steel.

Gluing   A method of fastening materials using glue, which creates chemical bonds between itself and the materials being glued.

Goal   In problem solving, the desired result of the proposed solution.

Graphic Communication   Those methods of communication where the channel carries images or printed words.

Gravure Printing   Printing from a recessed surface; intaglio printing.

Grinding   The process in which small amounts of a material are removed by rubbing it with an abrasive.

Halophyte   A plant that can grow using salt water irrigation.

Hardness   The ability of a material to resist being dented or scratched.

Header   A wooden support used on top of windows and doors to distribute the load from above to the studs.

Heat Treating   A conditioning process in which a material's internal properties are changed through the use of heat.

Horsepower   A measure of power equal to 550 foot-pounds per second, 746 watts, or 33,000 foot-pounds per minute.

Hydraulic   Activated by fluid pressure.

Hydroelectricity   Electricity produced by turbines driven by falling water.

Implementation   In problem solving, trying out the proposed solution.

Inclined Plane   A surface placed at an angle to a flat surface. It is used to enable a small force to lift a heavy object from one level to another.

Industrial Materials   The intermediate step between primary raw materials and finished products. Primary raw materials are made into industrial materials, which are made into end products.

Industrial Revolution   A period of inventive activity, beginning around 1750 in Britain. During this time machines mechanized what had previously been manual work. The Industrial Revolution was responsible for many social changes, as well as changes in the way things were manufactured.

Inform   To impart knowledge of facts or circumstances. One of the reasons communication systems are used.

Information   One of the seven resources used by technological systems. Data is raw facts and figures; information is data that has been processed (recorded, classified, calculated, stored and/or retrieved). Knowledge is gained when different kinds of information are compared and conclusions are drawn.

Input   The command entered into a system; the desired results of the system.

Insight   A process of conceiving possible solutions to a problem in which solutions leap into the mind of the problem solver.

Insulation   Material that does not conduct heat or electricity very well.

Insulator   A material whose atoms hold their outer electrons tightly, resisting the flow of electrical current through it.

Integrated Circuit   A complete electronic circuit built on a single piece of semiconductor material. Integrated circuits contain from dozens to over 500,000 transistors and other circuit components. They can perform extremely complex functions.

Intelligent Building   A building with advanced communications capabilities built in to be shared by different tenants.

Intermodal   A transportation system that uses more than one type (mode) of transportation (e.g., ships, railroads, trucks).

Internal Combustion Engine   An engine that burns its fuel within a totally enclosed chamber.

Iron Age   The period during which people learned to extract iron from iron ore and to make tools with it. In the Middle East, this happened around 1200 B.C.

Isometric   A drawing within an isometric axis. This axis includes a framework of three lines; one is vertical, the other two are drawn at angles of 30 degrees to the horizontal.

Jet   An aircraft that uses a jet engine. A jet engine is one that provides thrust from the burning of fuel and air, and the rapid expansion of the burning gases.

Just in Time Manufacturing   A manufacturing process in which the raw materials and other required parts arrive at the factory just before they are needed in the assembly process.

Kinetic Energy   The energy of an object due to its motion.

Laser   Light Amplification by Stimulated Emission of Radiation. A laser is a source of very pure (single color) light that is focused into a very narrow beam.

Laser Printer   A laser printer uses a laser beam to expose a photosensitive drum. It can be used to print letter quality text and graphics very rapidly.

Lift   The upward force on an object in the air, resulting from its weight and volume (passive lift) or its shape and movement through the air (active lift).

Machines   With tools, one of the seven resources used in technological systems. Machines change the amount, speed, or direction of a force.

Machine Communication   Communication between people and machines (such as the use of a joystick or a mouse with a computer); or between two or more machines (such as when computers control manufacturing machinery).

Manufactured Housing   Housing that is built in modules in a factory, transported to the construction site, and placed on a foundation, where the utilities are installed.

Manufacturing   The building of products in a workshop or factory.

Mason   A craftsperson who is skilled in the art of working with concrete.

Mass Media   Communication systems such as magazines, newspapers, radio, and television, that are used to reach large numbers of people.

Mass Production   The manufacture of many goods of the same type at one time, frequently involving interchangeable parts and the use of an assembly line.

Materials   One of the seven resources used in technological systems. The physical substances (e.g., wood, iron, oil, water, sand) that are used in a process.

Modeling   The testing of a problem solution or a system without building the solution or system itself. Modeling includes using small physical replicas of the solution (scale models) and intangible representations of the solution (mathematical models, computer models, etc.).

Modem   (MODulator-DEModulator) A device used to send data signals over analog communication channels (such as telephone circuits).

Momentum   The tendency of an object in motion to stay in motion or an object at rest to stay at rest.

Monitor   To observe the ouput of a system.

Mortar   A mushy cementlike substance that hardens and acts like glue to hold concrete blocks together.

Mortgage   A loan that is secured by a home, building, or other large physical asset.

National Electrical Code   A set of standards developed by the National Fire Protection Association which establishes safe methods of installing electrical wiring and equipment.

Negative   A reversed image on photographic film, used for printing pictures.

Network   Information transmitters and two or more receivers connected by channels. A data network connects computers and/or computer devices. A voice network connects telephones. A video network connects television cameras and receivers.

Noise   An imperfection in a communication channel or equipment. Noise makes the message more likely to be misunderstood.

Nuclear Energy   Energy derived from the splitting (fission) or combining (fusion) of atoms.

Numerical Control   The control of manufacturing machines by punched tape.

Oblique   A pictorial drawing where one surface is seen straight-on.

Office Automation   The use of computers and communications in an office setting to increase productivity.

Offset Printing   A printing process in which the image to be printed is photographically placed on a metal plate, transferred to a cylinder covered by a rubber blanket, and then transferred (offset) to paper.

Oil   Any of a number of liquids that can be burned to obtain energy, or that can be used as lubricants. Oils come from animals, vegetables, and minerals. Petroleum oil is found under the ground and may be refined to kerosene, jet fuel, and gasoline.

Open-Loop System   A system that does not use information about the output(s) to affect the process.

Operating System   Computer software that allows the user or other programs to access a computer's memories (disk, tape, semiconductor, etc.), printers, and other attached devices.

Optical   Having to do with light or sight.

Orthographic   A way of drawing an object using several straight-on views. Usually, the top, front, and side views are drawn.

Output   The actual result obtained from a system.

Parabolic Reflector   A curved reflector that focuses light, heat, or radio waves to a single point, called the focal point.

Passive Solar   Taking the heating effects of the sun into account when designing walls, doors, and window placement in buildings.

People   One of the seven resources in a technological system. People design systems, operate them, and benefit from them.

Perspective   A method of drawing that makes things look realistic. Parts of an object or scene that are futher away appear smaller.

Persuade   To convince someone to do something. One of the reasons communication systems are used.

Pert Chart   A flow chart used to plan and track tasks within a project.

Phototypesetter   A machine that uses a photographic process to convert computer output to high-quality printed text.

Pictorial   Pictorial drawings show an object in three dimensions. Three common types of pictorial drawings are isometric, oblique, and perspective.

Piggyback   The use of railroad flat cars to carry truck trailers.

Pitch   The amount a roof slopes. It is equal to the number of inches the roof rises vertically for each foot of horizontal span.

Plasticity   The property of a material that allows it to bend and stay bent.

Plumb   The condition that exists when a structural element is perfectly vertical.

Plumbing   The systems in a structure that have to do with carrying water and waste materials.

Pneumatic   Activated by air pressure.

Polymers   Materials, like plastics, which are made from long chains of molecules. (*Poly* means many, *Meros* means parts in Greek).

Potential Energy   The energy stored in an object due to its position, shape, or other feature.

Power   The amount of work done in a given period of time; the time rate of doing work.

Prefabricate   In construction, to build a building or a portion of it at a location other than the construction site.

Pressing   A forming process where a plunger forces material into a mold.

Pressure   The amount of force exerted on an object's surface, divided by the area over which it is exerted.

Process   The part of a system that combines resources to produce an output in response to a command input.

Processor   In computers, the part that controls the flow, storage, and manipulation of data.

Program   A list of instructions that directs a computer's activities.

Project Manager   A person or company whose job it is to oversee contracts, scheduling, material deliveries, and overall progress on a construction job.

Properties of Materials   The characteristics of materials (such as hardness and plasticity) that make them suitable or not suitable for certain applications.

Prototype   A model of a final product or structure that is built to help evaluate the soundness of a design and to discover unanticipated problems.

Quad   One quadrillion British Thermal Units of energy. (1,000,000,000,000,000)

Quality Control   The act of testing for faults in a product and correcting their causes.

Rafters   The sloping elements that make up a framed roof.

RAM   Random Access Memory. The largest portion of a computer's main memory, in which the program is stored while it is being worked on, and in which at least some of the data and results are temporarily stored.

Raw Materials   Those extracted from the earth and processed into basic industrial materials or finished products.

Receiver   The part of a communication system that accepts the message from the channel and presents it to the destination.

Recycle   To reuse all or portions of a substance.

Relief Printing   Printing from a raised surface; letterpress printing.

Resistance   The opposition to electrical current flow.

Resources   The things needed to get a job done. In a technological system, seven types of resources are processed to produce outputs. The seven types of resources are: people, information, materials, tools and machines, energy, capital, and time.

Robot   A multifunction, reprogrammable machine capable of movement.

Rocket Engine   An engine that provides thrust from the rapid expansion of burning fuel and oxygen, which it must carry with it.

Sawing   Sawing involves separating material with a blade that has teeth.

Science   The study and description of natural phenomena.

Screen Printing   A printing process in which ink is pressed onto the paper or object to be imprinted through holes in a master stencil (the screen).

Semiconductor   A material that is neither a good conductor nor a good insulator. Transistors, diodes, integrated circuits, and some other electronic components are made of semiconductor material.

Separating   A category of processes that divides or puts apart materials. Separating processes include shearing, sawing, drilling, grinding, shaping, turning, filtering, and chemical and magnetic separation.

Shaping   Processes used to change the shape or contour of materials.

Shear   A pair of forces that act on an object in opposite directions along the same line or plane, as the two blades on a pair of scissors.

Shearing   Using shears to separate materials.

Sheathing   The outer layer of material (often plywood, particleboard, or foamboard) that covers and protects the walls and the roof.

Sill Plate   A wooden board that is installed on top of the foundation before the joists are attached.

Smart House   A home with computer control of many routine functions.

Solar   Coming from the sun.

Solar Cell   A device that converts light energy into electrical energy.

Soldering   A method of joining two wires together by melting a metal called solder onto them.

Spackle   A plasterlike substance used to finish interior walls and ceilings.

Specification   Detailed statement of requirements. In problem solving, a goal is a broad or general statement of requirements, and a specification is a detailed and specific statement of requirements.

Speech Synthesis   The simulation of speech by a computer or electronic circuit.

Steam Engine An engine that delivers power generated from the expansion of steam created by boiling water.

Stone Age The period during which people used stones to make tools and weapons. In the Middle East, this was before 3000 B.C.

Studs The vertical members that make up a wall section.

Subcontractor A person or company whose job it is to build a part of a construction project.

Subfloor The rough flooring, usually plywood or particleboard, that is nailed to the floor joists before finish flooring is applied.

Subroutine A routine (sequence of actions) that is a part of a larger routine. A subroutine may be examined separately from the larger routine.

Subsystem A small system that, together with other subsystems, makes up a larger system.

Supercomputer A computer vastly superior in size and speed to the mainframe computers in use at any given time. Because of the rapid progress in computer technology, supercomputer performance of yesterday is commonplace today.

Superconductor A material whose electrical resistance suddenly drops to nearly zero at a certain low temperature.

Superstructure In construction, that part of the structure above the foundation; it is usually the part of the structure that is visible above the ground.

Synthetic Human-made; not occurring in nature.

System A means of achieving a desired result through the processing of resources in response to a command input. A system may be open-loop (no feedback) or closed-loop (using feedback).

Technical Drawing Drawing using instruments and tools to accurately communicate information about the size and shape of objects.

Technology The use of accumulated knowledge to process resources to satisfy human needs and wants.

Telecommute Working at home using computers, terminals, and data communications rather than physically traveling to work.

Teleconference A meeting conducted by people located at different sites, using communications to link them together. The attendees can use voice only (telephone conference call), voice with still pictures, or full motion pictures and voice (videoconference).

Tension A force that stretches an object.

Thermal Having to do with heat or the transfer of heat.

Thermoplastic Plastic material that softens when heated.

Thermoset Plastic Plastic material that will char and burn when heated. It does not soften.

Thrust The force developed to move an airplane forward through the air.

Time One of the seven resources used in technological systems. In modern systems, time ranges from less than one billionth of a second to human lifetimes.

Tools With machines, one of the seven resources used in technological systems. Tools extend the natural capabilities of people, and they are used to process or maintain other resources in systems.

Top Plate A double layer of 2 × 4s used to construct the top of a wall section.

Torsion A force that twists an object.

Toughness The ability of a material to absorb shocks without breaking.

Trade-off An exchange of the benefits in one solution for the disadvantages in another solution.

Transformer A device used in alternating current electric power systems to increase (step-up) or decrease (step-down) voltage.

Transistor A three terminal electronic component made of semiconductor material that enables the control of a large amount of current with a small amount of control current.

Transmitter The part of a communication system that accepts the message from the originator and places it on the channel.

Trial and Error A method of solving problems in which many solutions are tried until one is found to be acceptable.

Truss A carefully engineered, large prefabricated wooden triangle used as a section of a roof.

Turning A separating process where the material, not the cutting tool, moves. A lathe is an example of a turning tool.

Union A labor organization representing a group of workers that bargains with employers to set wages and work practices for its members.

Up Link The part of a satellite communication system from a ground station to the satellite.

Vapor Barrier A layer of plastic, paper, or foil that covers insulation. It prevents warm moist air inside the house from condensing on cooler studs, thereby preventing rot.

Venture Capitalist A person or company who invests money in a new or growing business in exchange for part ownership in the business.

Voice Recognition The technology of using human voice as an input to a computer or a machine.

Voltage The force necessary to move electrons from one atom to another in a material.

Watt A measure of power equal to one kilogram-meter per second. One watt also equals one ampere × one volt.

Wavelength The distance covered by one cycle of an electromagnetic or other wave.

Weight In aeronautics, the downward force on a plane, opposing lift.

Wind Load The effect of wind as it blows against a structure.

Word Processor A machine that combines typing and computer technologies to help an operator create, store, retrieve, modify, and print text.

Work The product of a force needed to move an object and the distance that it is moved in the direction of the force. If a force is applied against an object, but the object does not move, then no work is done.

ADDITIONAL
ACTIVITIES

HURRICANE ALLEY

Setting the Stage

Hurricane racing does not use motor cars, motorcycles, rockets or antigravity engines for power. It uses wind. Windmobiles, of all shapes and designs, mounted with high-density woven metallic fiber sails, are the only vehicles permitted on the track. A windmobile hitting the smallest bump on the track at such high speeds would probably be destroyed.

Your Challenge

Design and construct a wind-powered vehicle that will roll the longest distance when propelled by the exhaust of a heavy duty shop vacuum. You may only use the supplies listed below.

Suggested Resources

Safety glasses and lab apron

Three pieces of cherry or other hardwood—
1/4" × 1/4" × 36"

Four plastic wheels

One 9" piece of 1/8" diameter steel axle rod

One 12" x 12" piece of mirrored mylar for the sail

One soda straw—for the axle bearing

Hot glue gun and glue sticks

Model cement

Drill press, jig saw and/or band saw, belt sander, table saw, miter box, hack saw, files, and other miscellaneous hand tools and machines

Assorted fasteners

Graphite lubricant

Heavy duty shop vacuum

Procedure

1. Be sure to wear safety glasses when working with tools and machinery.
2. Design your wind-powered vehicle on graph paper. Draw a number of sketches before you settle on one particular design.
3. Try to develop as many alternative design solutions as you can. As your supplies are limited, you may have to make some compromises.
4. Do not use any tools or equipment until your teacher has shown you the safe and proper operating procedures. SAFETY FIRST!
5. Carefully cut the pieces of hardwood to size. Use a very fine-toothed saw.
6. Drill 3/16" diameter axle holes where necessary.
7. Assemble the vehicle using hot glue. Be careful . . . hot glue is HOT!
8. Use the straw as an axle bearing. This will cut down on *friction*.
9. As hot glue doesn't stick very well to the mylar sail material, model cement can be used as an alternative. Let the cement dry thoroughly.
10. Give your vehicle a test run. Evaluate the results and make any necessary design changes.

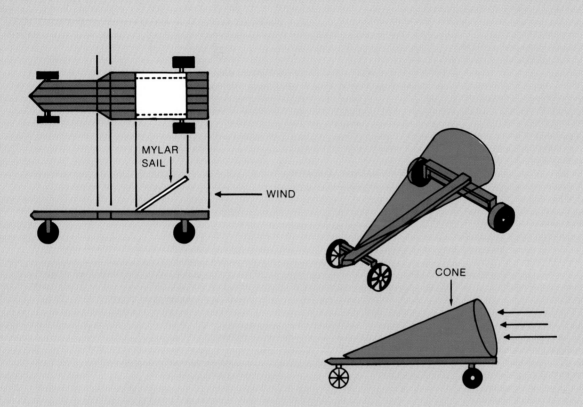

MYLAR SAIL

WIND

CONE

Technology Connections

1. Can you draw a systems diagram that reflects the problem-solving approach used with the wind-powered vehicle? Label the input, process, output, monitor, feedback, and comparison steps in the diagram.
2. The design, selection, and construction of the wind-powered vehicle fall into the *process* step of the systems diagram. What method is used to monitor the *output* results?
3. What do we mean by *feedback* in the problem-solving system? How is this used to optimize the performance of the wind-powered vehicle?
4. What is brainstorming? What is trial and error?
5. Why will the same vehicle go a greater distance if the wheels turn more freely?
6. Why does the wind-powered vehicle continue to roll after the 'wind' is removed?

Science and Math Concepts

▶ *Friction* is a force opposing the motion of a body.
▶ *Inertia* is the property of matter that resists a change in motion. An object at rest tends to stay at rest, and an object in motion tends to stay in motion unless acted upon by an outside force.

POLE POSITION

Setting the Stage

The Formula I Race takes place in a few days. You are one of the top drivers and are due at the course in one hour for the last qualifying session. At the end of the day yesterday, your car suffered engine damage. The head mechanic called last night to say it won't be ready in time for the race. Two friends have offered their extra cars for you to use. They have the same engines, but look very different. In order to have a chance to win, you must select the best car.

Your Challenge

Design and build the body for a race car. Test the results in the wind tunnel. Then compare the results with other students' results. The body design that disturbs the smoke the least will have the least amount of drag and will be the fastest car.

Procedure

1. Experiment with the drag form factors of various shapes you may wish to use for your car body design. First, shape clay into a teardrop approximately 1¼" diameter × 3½" long. Test this shape in the wind tunnel by lighting a piece of firecracker punk or stick incense to create smoke. Do the same test on other clay shapes such as spheres, cubes, and rectangles. Use shapes of different sizes. (Note: Do not change the air flow in the tunnel when you are doing comparison testing.)
2. Now do a test to determine the best angle of transition between two different shapes. Form the clay test shapes on a small ⅛" piece of acrylic. The closer the smoke trail is to the surface of the clay, the better the drag form factor. If the smoke curls behind the clay, this displays wind turbulence and increased drag.
3. Choose the type of car body and the scale you would like to use.
4. Sketch the front, side, and top views of the vehicle.

Suggested Resources

Modeling clay
Ceramic clay
Plaster of paris
Paper mache
Basswood
Wind tunnel
Architect's scale
Drafting equipment

5. Using the data from your wind tunnel tests, modify your drawing to decrease the drag.
6. Using modeling materials, form a scale model of the design you have chosen.
7. Test your model in the wind tunnel at various wind speeds. (Note: All models should be tested at the same wind speed before adjusting the air pressure for another wind speed.)

ANGLE OF TRANSITION

Technology Connections

1. Research the car companies that publish their drag form factors. Which have the best drag factors?
2. What is the purpose of each part of the wind tunnel?
3. Compare the volumes of the model cars, using a water displacement method. Does volume affect drag? How does this affect surface area?
4. Why are wings used on Formula I cars?
5. Sand an airfoil shape used for airplane wings. Test this shape in the wind tunnel. Can you determine from the smoke pattern why lift is created with this shape?
6. What other factors influence what cars we purchase?

Why do car companies use these same modeling techniques and similar wind tunnel tests?

What else could you use to simulate tests and mechanical operations?

Science and Math Concepts

▶ The amount of pressure drag and friction drag depends on the shape of the body.
▶ Turbulence—large fluctuations in speed and the mixing of the different layers of air.
▶ Laminar—different layers of air flow smoothly over each other.
▶ The product of the aerodynamic drag coefficient (C_D) and the frontal area of the object (A) is commonly referred to as the drag form factor (C_DA) of the object.

GLASSMAKING

Setting the Stage

Materials are an important resource for technology. Materials are often processed to be made more useful. Sand, for example, can be processed into glass. The glass can then be made into useful objects by means of forming processes.

Your Challenge

Mix (batch) materials and produce a small glass object.

Safety Considerations

1. Be very careful as the materials are at extremely high temperatures.
2. Wear eye protection at all times.
3. Wear insulating gloves.
4. Avoid taking crucibles in and out of the kiln. Rapid changes in temperature can cause cracking.
5. If a crucible develops cracks, discontinue using it immediately, and call the instructor to assist you.
6. Hot glass looks just like cold glass. Never leave newly formed pieces out in the open where an unsuspecting individual could touch them.
7. Always use tongs to handle hot crucibles. Never reach into the kiln with your hands to insert or remove a crucible, even while wearing insulated gloves.

Suggested Resources

Enameling kiln
Porcelain crucibles
Aluminum or graphite molds
Glassmaking chemicals
Scales
Mortar and pestle
Storage containers
Polariscope
Drilling and boring tools

8. Finished objects must be properly annealed, or they are liable to explode. Test for stress using polariscope.

9. For borosilicate glass: It should be noted that chlorine gas will be liberated during the melt. This gas is poisonous. However, in the quantities used here, this gas should not be dangerous. Therefore, measurement of the NaCl component should be strictly monitored. Any fumes should be vented. It is possible to produce the glass without adding NaCl if desired.

Procedure

1. Review the safety considerations.
2. Carefully weigh and mix the ingredients for the glass. This is called **batching.** Various metals can be introduced to the batch to provide color. (See batch ingredients table.)
3. With the kiln off, practice the procedure for melting and casting the glass. This will contribute to safe operation when you are actually doing the procedure.
4. **Calcine** the batch in a crucible by heating it to a temperature lower than its melting temperature. The batch will bubble vigorously as the gases escape. Calcining occurs at approximately 400°C (750°F).
5. Once all the gases have been liberated during calcining, the temperature of the batch may be increased. This melts the remaining chemicals. Melting temperature is approximately 815°C (1500°F).

MODIFIED BOROSILICATE GLASS BATCH INGREDIENTS*

Material	Oxide	# Grams in 200 gm Batch
Georgia Feldspar $(K_2O + Al_2O_3 + SiO_2 + Na_2O + CaO)$	SiO_2 Al_2O_3 K_2O Na_2O CaO	88.4
Sodium Tetraborate $(Na_2B_4O_7)$	B_2O_3 Na_2O	65
Sodium Chloride (NaCl)	Cl Na_2O	2.0
Sodium Carbonate (Na_2CO_3)	Na_2O	44.6
TOTAL		200 grams

For blue color, add 1% by weight of black copper oxide (2 grams) to batch. For green color, add 0.32% chromium oxide (0.65 grams) and 0.1% black copper oxide (0.2 grams).

*This glass was developed by Elton Harris, Corning Glass Works Experimental Melting Division.

6. When all the batch material is completely melted, **fine** the glass by heating it to a temperature slightly above the melting temperature. The high temperature materials in the batch need time to dissolve completely. Also, any remaining gas bubbles need to escape. Fining occurs slightly above the melting temperature and is necessary to achieve clear, uniform, bubble-free glass.

7. Form the glass. To form a coaster, use a **press mold** machined from a piece of 1/2" aluminum plate in two halves. You can also try a **dump mold** made by milling or drilling shapes into the top of an aluminum block. Using a pair of tongs and insulating gloves, remove the crucible containing the molten glass from the kiln. Pour the glass into the mold and then dump it once it's solidified.

8. The glass must be placed immediately into an annealing oven in order to relieve stress and avoid shattering. While you were forming the glass, the rapid cooling put stress on the material. Annealing relieves that stress. To anneal the glass, heat it at an intermediate temperature for half an hour, and then permit it to cool slowly. Annealing temperature is 464°C (870°F).

9. You can do rotational casting by machining an aluminum or graphite cylinder measuring two inches in outside diameter and four inches in length. Use a boring tool to machine the inside into the desired shape. Machine a shaft onto the mold. Pour the molten glass into the mold, cap the top, and spin with an electric drill or vertically mounted motor at 1,750 RPM. Centrifugal force conforms the glass to the inside shape.

10. Inspect the glass visually to determine if the stress has been relieved. To do this, construct a polariscope by placing two polarizing filters (or sheets of polarizing film) at 90° to each other, separated by a space into which the glass sample is inserted (see drawing). Inspect the sample by looking through the polariscope toward an incandescent light source. If you see bands of color or streaks of black, the glass is unsafe and may shatter. It needs to be annealed again.

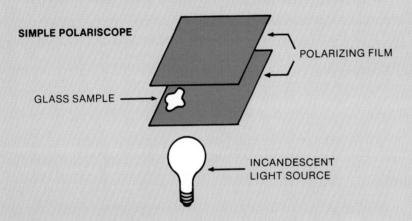

SIMPLE POLARISCOPE

POLARIZING FILM

GLASS SAMPLE

INCANDESCENT LIGHT SOURCE

Technology Connections

1. Bring in samples of glass. Identify the origin of the samples and what processing method was used.
2. Research and describe how window glass is produced. Contrast window glass with some glass produced for special purposes, such as Corning's Steuben glass.
3. Place the operations you used in this activity in categories of forming, separating, combining, and conditioning.
4. Processing materials involves changing their form to make them more useful to people.

Science and Math Concepts

▶ Changes occur in the molecular structure of ceramic materials when they are fired in a kiln. The material melts and fuses together. This forms a new molecular structure. The process cannot be reversed.

▶ Glass is a supercooled liquid. Antique window glass is thicker on the bottom of the pane than on the top because the glass molecules have slowly moved down the pane.

THE MEDUSA SYNDROME

Setting the Stage

Medusa was a Greek goddess who was so ugly that anyone who looked upon her for even a second instantly turned to stone. As everyone knows, Medusa was a myth, a fantasy of some ancient story teller. But was she?

In the year 1999, at the archaeological excavations on the Greek Isle of Gorgon, scientists discovered a sealed chamber. As they broke the seal of the door and opened it, everyone who looked directly into the dark abyss went instantly blind. Within days the word was out— Medusa was on the loose.

Your Challenge

Design and mass produce a portable device that can be used to look over walls and around corners to see if Medusa is coming. Since each student in the class must be protected, a sufficient number of these devices must be manufactured.

Suggested Resources

Safety glass and lab apron
1 piece of 1 1/2" diameter rigid plastic pipe—18" to 24" per student
1 plastic 35 mm film cartridge—(1 per student)
Mirrored plexiglass 2 1/2" x 3 1/2"—two per student
Bench rules—12" and 36"
Combination square
Fine point felt-tip pen
Miter boxes
Drill press
Belt and/or disk sander
1 3/8" diameter hole saw
1 1/4" diameter hole saw
Hammer
Center punch
Jig saws and/or band saws
Files
Hot glue guns and glue sticks
Assorted jigs and fixtures (determined and constructed by the teacher)

Procedure

Note: The assembly line stations, quality control devices, jigs and fixtures, exact specifications, etc., can be determined and constructed by the entire class.

1. Be sure to wear safety glasses and a lab coat.
2. All students must make sketches and plans of the periscope device before beginning the mass production manufacturing and assembly line process.
3. Computer-aided drafting (CAD) is recommended after an initial sketch is made.
4. If you have not been instructed in the proper and safe use of any tool or machine that you need to use with this activity, check with your teacher BEFORE going any further.
5. Using a miter box, carefully cut 45 degree angles on both ends of the 1 1/2" diameter pipe. These angles must parallel and 'lean' in the same direction. The pipes should be 18-24" long after cutting.
6. Cut or sand off the bottom of the 35 mm film cartridge. (This is the eyepiece!)
7. Using the drill press and the 1 3/8" hole saw, drill a hole opposite the 45 degree angle cut on the 1 1/2" pipe. (DO THIS ONLY ON ONE END!)
8. Repeat step 7 using the 1 1/4" drill on the other end of the pipe. (For the eyepiece!)
9. Using the files, sander, sandpaper, etc., remove any rough edges from the holes and pieces of pipe.

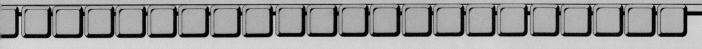

10. On the mirrored plexiglass, use the felt-tipped pen to duplicate the elliptical shape of the 45 degree angles cut on the end of the 1 1/2" pipe. (Make a template or pattern!)
11. Use the band or jigsaw to cut out the ellipses of the mirrored plexiglass. STAY OUTSIDE THE LINES!
12. Use the belt or disc sander to bring the mirrors to their final shape.
13. Use the hot glue gun to fasten the film cartidge eyepiece tube in place on the inside.
14. Temporarily hold or tape the mirrors in place. Look through the eyepiece to make sure the periscope is working properly. Make any needed adjustments.
15. Use the hot glue gun to fasten the mirrors in place. (Shiny side in!). Remember—Hot glue is HOT!

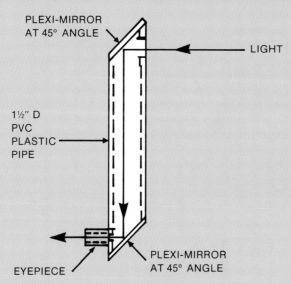

PLEXI-MIRROR AT 45° ANGLE

LIGHT

1½" D PVC PLASTIC PIPE

EYEPIECE

PLEXI-MIRROR AT 45° ANGLE

Technology Connections

1. How does *mass production* and the factory system help bring prices down?
2. What is an *assembly line?* What is a *conveyor belt?*
3. Why are *interchangeable parts* important to an assembly line?
4. What happens if you look through the periscope eyepiece with a small telescope?
5. What happens if you look through two periscopes at the same time? (One for each eye, of course!) Now you know how a newt views the world!

Science and Math Concepts

▶ Plane mirrors obey the *Law of Reflection*—The angle of reflection equals the angle of incidence. A ball bouncing on a flat surface also obeys the Law of Reflection.
▶ When light strikes a smooth surface, the light is reflected in an even pattern.
▶ Lenses *refract* (or bend) light. This is because light travels at different speeds through different substances.

545

SPACE COMMUNITY

Setting the Stage

The President's announcement at the press conference sent a shockwave throughout the room. "Ladies and gentlemen of the press . . . the scientists at the Advanced Systems Lab have received and delivered to me this evening a message from an extraterrestrial life source.

The extraterrestrials have requested an urgent meeting with representatives of our entire planet. This meeting will be held on the fourth planet in our solar system—the planet Mars. Our rocket scientists have assured me that we have the ability to reach that planet safely. The only problem that remains is to design and construct a structure that can be transported, assembled, and used by the first permanent settlers of the 'red' planet Mars."

Your Challenge

Design and construct a model of a structure that will be transported, assembled, and used by the first permanent settlers on the planet Mars. Your structure must provide systems to support human life, including a pressurized oxygen environment with temperature control; a place to grow food, perhaps hydroponically; a source for water; and recreational facilities. If desired, groups of students may combine their structures to develop a small Martian community.

Suggested Resources

Safety glasses and lab apron

Balloons—various shapes and sizes (Don't use the cheap, thin kind!)

Pariscraft—A gauze material coated with plaster of paris

Large bucket or bowl of water

Masking tape

Tempera paints—assorted colors

Paint brushes—assorted sizes

1/2" sheet of plywood

Modeling clay

Cardboard paper towel rolls

Wood, plastic, and metal supplies as needed

Assorted tools and machinery as needed

Hot glue, model cement, assorted fasteners

Procedure

1. Be sure to wear safety glasses and a lab coat.
2. Keeping in mind that balloons will be used as forms, sketch some possible designs for the Martian structure. Don't forget the air-locks, passageways, windows, supports, etc.
3. Inflate the appropriate shape balloons. Attach multiple balloon structures together with masking tape. Paper towel rolls can be used, if desired, as passageways.
4. Cut short strips of the Pariscraft material with a scissors.
5. Dip the strips into a bucket of water and carefully coat the entire balloon structure with two to three layers of Pariscraft. Carefully rub the Pariscraft to smooth out any wrinkles. Make sure there are no loose ends.
6. If possible, hang the structure to dry overnight.
7. Make any minor design changes at this time. Pariscraft can be lightly sanded to smooth surface areas. Be careful you don't sand completely through the shell.
8. Paint the completed structure as desired. Windows can be drilled and cut into the structure. Clear acetate can be used for glass. Use your imagination!

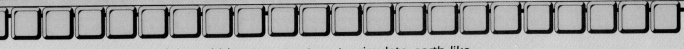

9. Construct modules within your structure to simulate earth-like conditions.

10. Using a plywood base, determine the location of your structure. Construct a support 'foundation' using pieces of wood and/or plastic.

11. Use hot glue to fasten the structure and foundation in place.

12. Go to the library and find a picture of the Martian surface. (NASA has published photographs of the Viking Lander.)

13. Your Martian landscape can be developed using clay, Pariscraft, sandpaper (for texture), paints, assorted scraps, etc. Again, use your imagination.

Technology Connections

1. Construction sites must be chosen to fit in with the needs of people and the environment.

2. How does the design of your structure meet the demands of the hostile Martian environment?

3. What special life-support systems would probably be needed on Mars? (i.e. temperature, oxygen, water, food, etc.)

4. Do you think the balloon-type construction technique used in this activity could be used on a full-scale structure? Why shouldn't plaster be used for exterior walls on Earth?

Science and Math Concepts

▶ Plaster is made from a rock called *gypsum*. Gypsum is the mineral *calcium sulfate* ($CaSO_4 \cdot 2H_2O$).

▶ When gypsum is heated, it loses part of its water and becomes plaster of paris (($CaSO_4)_2 \cdot H_2O$).

▶ When water is added to plaster of paris, the plaster hardens back into gypsum.

547

CYCLOID CURVE

Setting the Stage

The name and type of a new roller coaster for the amusement park has already been determined. The next stage is to design the best curves, turns, and elevations for a thrilling ride. The project on the drawing board is a curve that will allow the person to ride down an incline, gaining the greatest speed in order to be propelled up the next incline. This will take some modeling and trial runs to find the best curve.

Your Challenge

Design and construct a ramp that will allow an object to move down an incline and obtain the greatest speed. The ramp will be exactly 12" high and will end at the floor. Any length ramp may be used. The ramp that propels the object across the finish line first and is constructed using the best structural engineering techniques and craftsmanship wins!

Procedure

1. A common cycloid curve is developed when the path of a fixed point on the circumference of a circle is recorded as the circle is rolled along a straight line. To develop a cycloid curve, cut out a cardboard circle 12" in diameter. Tape a pencil perpendicular to the circle. Make sure the lead is flush with the outside edge of the circle. Holding the pencil and circle against a large piece of paper taped to the wall, trace the curve that develps as you roll the circle along a straight line. Turn this tracing upside down and you will have the profile or side view of the cycloid curve ramp.
2. Experiment with designs for the sides of the ramp. To do this, place popsicle sticks on your tracing in positions that will give the ramp structure its greatest strength. Try several different popsicle-stick configurations before deciding on your final design for the sides of the ramp.
3. Trace the final design for the sides. (Note: Allow in your drawing for the thickness of any support structures that will underlay the ramp. The final ramp surface must match perfectly the cycloid curve that you developed.)
4. Place wax paper over your drawing. Glue popsicle sticks together according to your design. (The popsicle sticks and glue will release easily from the wax paper.)
5. Lift the wax paper, along with the first side, off the drawing. Place a new sheet of wax paper on the drawing and construct the second side.

Suggested Resources

White glue
Popsicle sticks
Drafting equipment
Oak tag
Illustration board
Steel ball bearing or Hot
  Wheels® car
Tape

6. Fasten the two constructed sides together with glue and popsicle sticks. Install the ramp surface and side walls to hold objects in place as they roll down the ramp.

7. Conduct your contest for the fastest ramp. Remember, craftsmanship counts, too!

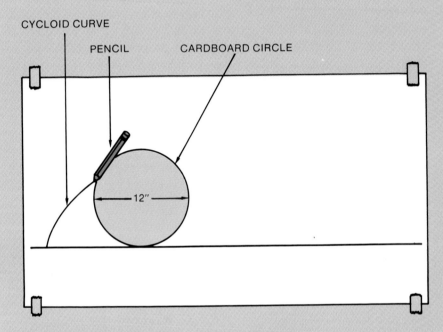

CYCLOID CURVE

PENCIL   CARDBOARD CIRCLE

12"

Technology Connections

1. Make several different types of ramps by cutting the profile curve out of ¾" thick wood. Using a shaper bit, your teacher will cut a groove in the ramp for a marble. Compare the performance of these samples with the performance of the cycloid curve by rolling marbles down the ramps and trying to hit a target approximately 4 feet away.

2. Could you hook a timing device to the computer and record the marble's speed or time?

3. Based on your research of structures, determine the materials that would be used to construct an amusement ride.

4. List other structures that would utilize the same construction techniques as the cycloid amusement ride.

5. Sketch the shape of the foundation that would be used for a roller coaster amusement ride.

Science and Math Concepts

▶ In about 1696, Jacques Bernoulli discovered that a cycloid curve is the curve along which a particle can move under gravity from a high point to a lower point in the shortest possible time.

▶ As a roller coaster and its occupants descend, their gravitational potential energy changes to kinetic energy.

HOVERCRAFT

Setting the Stage

Your engineering firm has been asked to assist the Navy in designing an innovative Hovercraft to deliver cargo in large amounts. The cargo would be able to move faster and be larger than that carried in a conventional freighter. One of the main problems the Navy has to overcome is leakage of air from the cushion that keeps the craft aloft. If your firm is successful in solving these challenges, the next step may be to replace many ships with fast transoceanic vessels.

Your Challenge

Design, find, and assemble the materials necessary to build a model Hovercraft that will be capable of traveling over land or water.

Procedure

1. Bring pictures of Hovercraft into class. You will use these pictures to help determine the location of the major parts.
2. On one of the styrofoam food trays, draw a center line in both directions, dividing it into four parts. You may want to add a few more reference lines. (One styrofoam food tray will be used for a second try.)
3. Cut a hole for the hair dryer motor one-third of the distance in from one side and on the center line. Tape the hair dryer motor and fan blades in place.
4. Use the top half of a styrofoam cup to house the fan, and fasten it to the body of the Hovercraft with glue or tape so no air will escape to the sides when the motor is running. (Adding this seal directs the air flow toward the underside of the Hovercraft.)
5. Tape a plastic-wrap skirt approximately 1" wide around the edge of the container.
6. Using tape, temporarily mount the battery case near the center of the container. (In the photograph, the battery case has been permanently affixed to the underside of the Hovercraft.)
7. Mount a small propeller on the AFX® or Sizzlers® engine shaft.
8. Design a supporting system to hold the AFX® or Sizzlers® engine and the propeller. Tape this subsystem near the rear of the container.
9. Add a styrofoam or balsa wood rudder for steering.
10. Add a switch and wire the motors to the batteries.
11. Now you must become a true designer, making adjustments just as professionals do, to make your Hovercraft fly.

Suggested Resources

Two styrofoam food trays, or a similarly shaped styrofoam substitute

Hair dryer fan and motor (3 to 6 volt, DC)

AFX®, Sizzlers®, or similar motor from a model car

Glue and tape

Cardboard

Plastic wrap

Styrofoam cups

Tape

Batteries (AA)

Balsa wood

Plastic battery case

Small propeller from a hobby store

12. Following are a few of the options you may want to try in order to properly balance and operate the Hovercraft:

- Keep sketches of the location of the subsystems and how the Hovercraft functions. You may have to refer back to your data to improve the design.
- Reposition some of the subsystems if the Hovercraft is unbalanced.
- Change the size of the batteries. (Use AA batteries first.)
- Change the number of batteries used.
- Change the skirt thickness.
- Use a different motor in the hair dryer fan.
- Use a motor for solar cells for the drive system. Be careful when you test the Hovercraft as its speed and inability to stop against a wall will surprise you.
- When all your subsystems are properly balanced and the motors are operating properly, carefully remount them permanently in a new container.

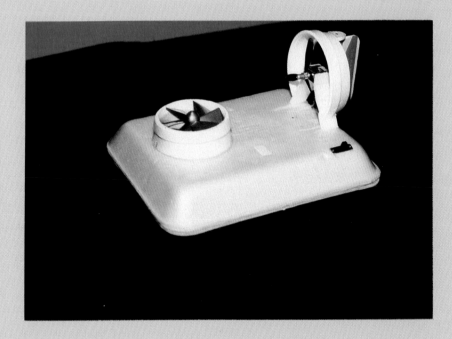

Technology Connections

1. How closely does your Hovercraft resemble full-size ones?
2. Hovercraft usually have one motor. How do they move ahead? How do they control direction? Can Hovercraft climb inclines? Why or why not?
3. Air-cushion vehicles (ACVs) or Hovercraft are used to cross the English Channel. Research this area and fill in a resource chart with information relating to Hovercraft. (Ideas can be found in this chapter.)

Science and Math Concepts

▶ The ACV operates on a long-known principle: the resistance of air to compression. Fans blow downward through the hull and establish a cushion of air between the ACV and the water.

▶ There is a direct ratio between the dimensions of the hull and the power needed to lift the ACV. If the beam and length are doubled, the power must be doubled to maintain the same height. When the beam and length are doubled, however, the load-carrying area is quadrupled. Therefore, the larger the craft, the more efficiently power is used.

PHOTO ACKNOWLEDGMENTS

Cover Photos (clockwise from upper left):
Bob Brooks/THE IMAGE BANK; Greg Heisler/THE IMAGE BANK;
Ford Motor Company; Tim Bieber/THE IMAGE BANK

Title Page Photos (clockwise from upper left):
Caldwell Banker, John Hancock Center; Tandem Computers; ASEA
Robotics Inc.; Greg Heisler

Section Opening Photos:
New York Convention and Visitor's Bureau;
Tandem Corporation, photo by Jay Freis; Cincinnati Milacron;
Amoco Corporation; NASA

Page 48 (clockwise from sailboat):
Pearson Yachts; Perini Corporation;
AT&T; NASA; and Cray Research, Inc.

Page 133: Michael Hacker

Page 140 (left to right):
Maxell Robots; Pepsi Cola, USA

Page 217-18:
All photos courtesy of U.S. Iron and Steel Institute

Page 277 (top to bottom):
Ford Motor Company; General Motors Design

Page 328:
Photos courtesy of Forest Products Lab

Page 535 (top left) and page 543:
Courtesy of Corning Glass Works

INDEX